W9-CPE-784

SPREADSHEET TOOLS
FOR ENGINEERS

EXCEL 97 VERSION

Byron S. Gottfried
University of Pittsburgh

WCB
McGraw-Hill

Boston, Massachusetts   Burr Ridge, Illinois   Dubuque, Iowa
Madison, Wisconsin   New York, New York   San Francisco, California   St. Louis, Missouri

WCB/McGraw-Hill

A Division of The **McGraw-Hill** Companies

In memory of my parents,
Faye and Sidney Gottfried

SPREADSHEET TOOLS FOR ENGINEERS: EXCEL 97 VERSION

Copyright © 1998 by The McGraw-Hill Companies, Inc. Printed in the United States of America. Except as permitted under the United States Copyright Act of 1976, no part of this publication may be reproduced or distributed in any form or by any means, or stored in a data base or retrieval system, without the prior written permission of the publisher.

This book is printed on acid-free paper.

1 2 3 4 5 6 7 8 9 0 DOC/DOC 9 0 9 8 7

ISBN 0-07-024654-8

Vice president and editorial director: *Kevin T. Kane*
Publisher: *Tom Casson*
Executive editor: *Eric Munson*
Marketing manager: *John T. Wannemacher*
Project manager: *Kari A. Geltemeyer*
Production supervisor: *Heather D. Burbridge*
Designer: *Felicia McGurren*
Printer: *R. R. Donnelley & Sons Company*

Library of Congress Cataloging-in-Publication Data

Gottfried, Byron S.
  Spreadsheet tools for engineers: EXCEL 97 version / Byron S.
Gottfried
    p. cm.
  Includes index
  ISBN (invalid) 0-07-246548-8
  1. Engineering--Data processing. 2. Microsoft Excel for Windows.
3. Electronic spreadsheets. I. Title.
TA345.G68 1998
519'.0285'5369--dc21

http://www.mhhe.com   97-29232

McGraw-Hill's *BEST*—Basic Engineering Series and Tools

Chapman, *Introduction to Fortran 90/95*
D'Orazio and Tan, *C Program Design for Engineers*
Eide et al., *Introduction to Engineering Problem Solving*
Eide et al., *Introduction to Engineering Design*
Eisenberg, *A Beginner's Guide to Technical Communication*
Gottfried, *Spreadsheet Tools for Engineers: Excel 97 Version*
Mathsoft's Student Edition of Mathcad 7.0
Palm, *Introduction to MATLAB® for Engineers*
Pritchard, *Mathcad: A Tool for Engineering Problem Solving*

TABLE OF CONTENTS

B.S. Gottfried

EXAMPLES

PREFACE

Everyone, it seems, uses spreadsheets these days. Spreadsheet use is one of the most popular personal computer activities, second only to word processing, in frequency of use and variety of applications. Budgeting, financial planning and record keeping are particularly common spreadsheet applications.

Yet relatively few people are aware of the mathematical problem-solving capabilities that are built into modern spreadsheet programs. Even engineering professors, who typically use spreadsheets to prepare budgets and maintain their class grades, are largely unaware that spreadsheet programs can conveniently be used to solve equations, fit curves through data sets, analyze data statistically, carry out studies in engineering economic analysis, and solve complicated optimization problems. Armed with these tools, a good spreadsheet program becomes the modern-day equivalent of the classical engineer's sliderule.

Excel, developed by the Microsoft Corporation, is one of the most widely used spreadsheet programs. It includes special features that offer all of the above problem-solving capabilities. Excel can also be used to solve other types of problems in engineering analysis, such as the evaluation of integrals or the solution of interpolation problems, even though it lacks special features that automate these tasks. And, like most other spreadsheet programs, it is especially well suited for displaying data in various graphical formats.

This book explains how to use Excel to solve simple problems that commonly arise in engineering analysis. It is intended as a supplementary textbook for use in introductory engineering courses, though it should also be of interest to practicing engineers as well as more advanced students. This second edition has been rewritten for Excel 97 (the version of Excel included in Microsoft's Office 97 suite). It also includes new material on getting help in Excel, and on creating and using macros. In addition, the earlier material on graphs, including semi-log and log-log graphs, has been expanded into a new, separate chapter. Finally, a brief chapter on sorting and filtering data has been added at the end of the book in response to reader suggestions.

We use much of this material with beginning engineering students in our Introduction to Engineering Analysis course at the University of Pittsburgh. This allows us to introduce various problem-solving techniques as they are needed in the course, without spending a great deal of time and effort explaining the mechanics of the underlying mathematical solution procedures. Thus, we are able to concentrate on engineering applications rather than detailed mathematical procedures. Student response to this approach has generally been very positive; in fact, we often have difficulty getting students to leave our computer-equipped classroom at the end of a two-hour period.

Chapter 1 sets the tone for engineering analysis in general by presenting a brief approach to problem-solving. Chapter 2 describes the basic rudiments of Excel. This chapter is by no means complete, but it contains enough background material so that the reader can understand and solve the problems in any of the subsequent chapters. A knowledge of the material in Chapter 2 is essential for reading the remainder of the book (though experienced Excel users need not necessarily go through this chapter in detail). Chapter 3 then discusses bar graphs (called Column charts in Excel) and *x-y* graphs (called XY charts or Scatter charts in Excel), including semi-log graphs and log-log graphs. The material in this chapter provides the graphical background required for some of the later chapters, particularly Chapters 4 and 5.

Chapters 4 through 12 address various analytical techniques that are commonly used by engineers. These chapters are essentially independent of one another and can be read in any order. Instructors using this book for a course can pick and choose among these chapters freely, in accordance with their own preferences. Each chapter includes several examples and lots of problems. Instructors can easily supplement these problems with other problem sets, reflecting their own disciplinary interests.

When I began this project I had hoped to include some topics in engineering analysis to motivate the spreadsheet applications. As the writing progressed, however, it became clear that I could not provide substantive material on engineering analysis, the use of Excel, and the ideas behind some of the mathematical procedures all in one brief, introductory-level book. Hence the book contains less material on engineering analysis than I would have liked, though many of the problems present simple engineering applications of the various spreadsheet-related topics.

The success of the earlier edition of *Spreadsheet Tools* was a factor in influencing McGraw-Hill to create its new BEST series (Basic Engineering Series and Tools) for beginning engineering students. The BEST series provides engineering educators with high-quality text material that is timely, affordable, and flexible in how it can be used. This second edition of *Spreadsheet Tools* is a member of that series, and it has been my privilege to serve as Consulting Editor.

In closing, I wish to thank the many readers of the first edition, particularly my faculty colleagues at Pitt and elsewhere who used the first edition in various introductory engineering courses, for their many helpful comments and suggestions. Finally, I wish to express my gratitude to Eric Munson and Holly Stark at McGraw-Hill for their close support and cooperation.

Byron S. Gottfried

CHAPTER 1

ENGINEERING ANALYSIS AND SPREADSHEETS

Engineering analysis is a systematic process for analyzing and understanding problems that arise in the various fields of engineering. To carry out this process successfully you must be familiar with general problem-solving techniques, you must have an overall understanding of your particular problem and its applicable engineering fundamentals, and you must have a working knowledge of the required mathematical solution procedures. It is also very helpful to have available a computer-based *spreadsheet program* that will solve your problem quickly and easily, once you have defined the problem and set it up properly.

This book discusses all of these items, with particular emphasis on spreadsheet solutions to a variety of introductory problems in engineering analysis. The associated mathematical solution procedures are also presented clearly, so that you have an understanding of what the spreadsheet does and how it goes about its business. Before launching into the details of spreadsheet solution procedures, however, let's consider what we mean by *general problem-solving techniques*, *applicable engineering fundamentals*, and *mathematical solution procedures*.

1.1 GENERAL PROBLEM-SOLVING TECHNIQUES

It is very important that you develop good problem-solving habits early in your career. Here are some general suggestions that will help you in the problem-solving process. These suggestions apply to all engineering problems, irrespective of any particular application or any special mathematical procedure.

1

1.   Your perspective of the problem will change over time. Therefore, you should *set aside some time to think about the problem before you attempt to solve it*. This will help you to understand the problem more clearly.

2.   Most engineering problems can be represented graphically or pictorially. Therefore, when solving a problem of this type, *draw a sketch of the problem* before you begin to solve it. This will assist you in visualizing the problem.

3.   Be sure that you understand the *overall purpose* of the problem and its *key points*. Don't allow yourself to become sidetracked by peripheral or irrelevant information.

4.   Ask yourself *what information is known* (input data) and *what information must be determined* (output data). List the input and output information in general terms. (Try to do this in terms of specific variables; do not simply write down numbers.)

5.   Ask yourself *what fundamental engineering principles* apply to the problem (see Sec. 1.2). Be certain that you understand how these principles apply to the particular problem you are trying to solve.

6.   *Think about how you will solve the problem before you begin the actual solution*. Many problems can be solved several different ways. What mathematical method will you use? Will a computer be required? How will you present the results? Some advance planning could save you a lot of time and grief.

7.   Take your time when actually solving the problem. *Develop your solution in an orderly and logical manner*. Be sure that the work is clearly labeled, particularly if you are solving the problem by hand.

8.   Once you have obtained a solution, *think about it. Does it make sense?* What assurance do you have that it is correct? Incorrect answers can often be detected in this manner.

9.   Be sure that your solution is *clear and complete*. Is the solution presented in an orderly manner? Are the results labeled? Are units included with the numerical answers? Is the logic used to obtain the solution clear? (Most professors are more interested in how you obtained your answers than the actual numerical results.) Is a table or graph required to present the results in a clear and concise manner?

Remember that *problem solving is a skill that takes time and practice to acquire*. You will become better at it as you acquire more experience with a greater variety of increasingly complex problems.

1.2 APPLICABLE ENGINEERING FUNDAMENTALS

Most engineering problems are based upon one of the following three underlying fundamental principles:

1. *Equilibrium.* Most *steady-state* problems (i.e., problems in which things remain constant with respect to time) are based upon some type of equilibrium. The following are some common forms of equilibrium that you are likely to see in elementary engineering problems:

    (*a*) Force equilibrium.

    (*b*) Flux equilibrium (see item 3 below).

    (*c*) Chemical equilibrium.

2. *Conservation laws.* The two common conservation laws are *conservation of mass* and *conservation of energy*. A great many problems in all fields of engineering are based upon one or both of these principles. (Certain problems are based upon a conservation of momentum principle, though you are unlikely to encounter these in a beginning-level course.)

3. *Rate phenomena.* There are many physically different rate phenomena, though all are represented in the same manner; i.e., a *potential* drives a *flux*. One common example is *Ohm's law* of electrical current flow ($i = \Delta V/R$), where i represents an electrical current (a flux) and ΔV represents a voltage difference (a potential). IIere the voltage difference (potential) drives the current flow (flux).

    Another common example of a rate phenomenon is *Fourier's law* of heat conduction ($q = k\Delta T/\Delta L$), where q represents a heat flux (expressed as heat flow per unit area per sec) and ΔT represents a temperature difference (a potential). Thus, the temperature difference (potential) drives the heat flux.

Example 1.1 Preparing to Solve a Problem

Suppose you wish to analyze the electrical circuit shown in Fig. 1.1.

(*a*) What is the overall purpose of the problem?

(*b*) What information is known?

(*c*) What information must be determined?

(*d*) What fundamental engineering principles apply to the problem?

(*e*) What will be the overall solution strategy?

(*a*) The purpose of the problem is to determine all of the unknown parameters that characterize the behavior of the circuit. These include the current flowing through each path and the voltage drop across each resistance.

(b)   The known information consists of the circuit configuration, the source voltage ($V = $ 12 volts), and the values of the individual resistances ($R_1 = 10\Omega$, $R_2 = 5\Omega$ and $R_3 = 20\Omega$).

(c)   The following items must be determined: The currents i, i_1, and i_2 and the voltage drops V_1, V_2, and V_3. (Note that V_1, V_2, and V_3 are the voltage drops across R_1, R_2, and R_3, respectively.)

(d)   The fundamental principles that apply are Ohm's law (rate phenomena) and Kirchoff's laws (equilibrium).

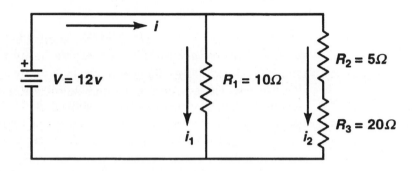

Figure 1.1

(e)   The order of the calculations must be determined carefully. At each step, an unknown quantity must be determined in terms of known information. Thus, the calculations can be carried out in the following manner:

(i)   Recognize that $V_1 = (V_2 + V_3) = V$.

(ii)   Determine the current i_1 flowing through the leftmost path (i.e., across R_1) as $i_1 = V/R_1$.

(iii)   Determine the overall resistance R_T of the rightmost path as $R_T = R_2 + R_3$.

(iv)   Determine the current i_2 flowing through the rightmost path as $i_2 = V/R_T$.

(v)   Determine the total current flow i as $i = i_1 + i_2$.

(vi)   Determine V_2 as $V_2 = i_2 \times R_2$.

(vii)   Determine V_3 as $V_3 = i_2 \times R_3$.

(viii) *Check*: Does $V_2 + V_3 = V$, as it should? (If not, something is wrong. Go back and find the error.)

(ix)   Determine R_{eq} as $1/R_{eq} = 1/R_1 + 1/R_T$.

(x)   *Check*: Does $i = V/R_{eq}$, as it should? (If not, go back and find the error.)

This problem is simple enough to solve by hand, with the aid of a calculator.

Example 1.2 Assessing the Solution

A group of students have obtained the following solution for the problem presented in Example 1.1:

i = 1.68 Amp i_1 = 0.48 Amp i_2 = 1.20 Amp

V_1 = 12.0 volts V_2 = 9.6 volts V_3 = 2.4 volts

(a) Is the solution clear and complete?

(b) Does the solution appear to be correct?

(a) The solution is clear and complete, though the detailed calculations are not shown. The unknown quantities are labeled with the proper units.

(b) When the students thought about the solution, they realized it is not correct. First, of the two loop currents, i_1 and i_2, the larger current (1.20 amps) is flowing through the rightmost path, which has the the higher resistance (25Ω). This does not make sense. Furthermore, within the rightmost path, the larger voltage drop (9.6 volts) occurs across the smaller resistance. This also does not make sense (because Ohm's law states that the *voltage drop is proportional to the resistance*).

Once the students examined their results more carefully, they realized that their calculations were correct but that their results had been transposed. The correct solution is

i = 1.68 Amp i_1 = 1.20 Amp i_2 = 0.48 Amp

V_1 = 12.0 volts V_2 = 2.4 volts V_3 = 9.6 volts

1.3 MATHEMATICAL SOLUTION PROCEDURES

Once a problem has been defined and properly formulated, it must, of course, be solved for the desired unknown quantities. This generally requires a particular mathematical procedure that is appropriate for the problem under consideration. Thus, you must know the commonly used mathematical procedures, and you must be able to identify which type of procedure to use with each specific problem.

The following types of mathematical procedures are used frequently in introductory-level (and more advanced) engineering analysis problems.

1. *Data analysis* techniques. These are simple statistical techniques that are used to analyze data (e.g., to calculate means, medians, modes, standard deviations, and histograms). Their use enables an engineer to draw meaningful conclusions about information that is hidden within a set of data.

2. *Curve-fitting* techniques. Here we are concerned with passing a curve through an *aggregate* of data rather than through individual data points. Think of this as a systematic way to "eyeball" a curve through a set of data. The resulting curve is generally of more value than the individual data points, particularly when the data points are subject to some *scatter* (as is usually the case when working with real data).

3. *Interpolation* techniques. These techniques allow an engineer to obtain accurate values for the dependent variable when the corresponding independent variable falls within a set of tabulated data points. Problems of this type arise frequently when working with measured data.

4. Techniques for solving *single algebraic equations*. Such equations appear with great frequency in virtually all areas of engineering. It is important to be able to solve them quickly and efficiently.

5. Techniques for solving *simultaneous linear algebraic equations*. Most of the applications that give rise to single algebraic equations also result in simultaneous algebraic equations once the problem conditions become somewhat more complicated. Moreover, many very complicated problems in engineering analysis can be represented as a set of simultaneous linear algebraic equations.

6. Techniques for *evaluating integrals*. In calculus, the classical interpretation of an integral is the area under a curve. (If you haven't learned this yet, you will very soon.) Many engineering applications require the evaluation of integrals in order to determine averages or to determine the cumulative effect of some process that varies with time or distance.

7. Techniques for carrying out *engineering economic analysis*. Most realistic engineering problems have more than one solution. (There are many ways to design a bridge.) The choice among them is often based upon economic considerations. Engineering economic analysis provides criteria and techniques for comparing one solution with another.

8. *Optimization* techniques. If a problem has multiple solutions, finding the best among them may involve lengthy and tedious search procedures. Optimization techniques provide efficient, systematic methods for carrying out these searches.

There are different approaches to each set of solution procedures. *Classical methods*, which are taught in beginning college math courses, are based upon the use of algebra and calculus. These methods tend to be very elegant but generally apply only to relatively simple problems.

Computers can solve more complicated problems, using either symbolic or numerical solution procedures. *Symbolic procedures* result in solutions that are expressed in algebraic terms and resemble the solutions obtained using classical methods. Such solutions can be difficult to obtain, however, and may be too complex to be of any practical value, particularly if the original problem is complicated. *Numerical procedures*, on the other hand, result in solutions expressed entirely in terms of numbers (or graphs based upon those numbers). Numerical solutions can usually be obtained quickly and easily, even for large and complicated problems. Therefore, the use of numerical procedures has become very common, both for engineering students and for practicing engineers.

This book describes a variety of numerical procedures that are commonly used to solve engineering problems. Our emphasis will be on the use of computerized spreadsheets to carry out these procedures. Examples are provided illustrating the use of these procedures for simple but representative engineering applications.

1.4 THE ROLE OF SPREADSHEETS

Before the arrival of personal computers, beginning engineering students were generally required to learn the mathematical details behind most of the commonly used numerical methods. They were then often required to program many of these methods for a large mainframe computer using a general-purpose programming language such as Fortran or Pascal. Or they might have written Fortran or Pascal programs that would access a pre-written library of routines for various numerical methods. In either case, the students were required to go through lengthy and tedious procedures. This provided a thorough indoctrination to the use of numerical methods in engineering, but many students would grow impatient and lose interest along the way.

During the 1980s, as personal computers became increasingly more common and dramatically more powerful, *spreadsheets* emerged as one of the principal types of personal computer applications. Though originally intended for carrying out financial calculations, the newer versions of most commercial spreadsheets include provisions for implementing many of the commonly used numerical methods. (If you don't know what a spreadsheet is, be patient — we'll get to that at the start of the next chapter.)

Most spreadsheets now have some numerical methods built directly into their command structure. In particular, modern spreadsheet programs allow you to do the following:

- Store and process data.

- Display data graphically.

- Analyze data statistically.

- Fit algebraic equations through data sets.

- Solve single and simultaneous algebraic equations.

- Solve optimization problems.

Moreover, most other numerical methods can easily be implemented within a spreadsheet simply by making use of the spreadsheet's basic features. Thus, spreadsheets have made it much easier to apply numerical methods to many of the problems arising in engineering analysis. How this is accomplished is one of the major themes of this book.

Problems

Each of the following problems can be solved individually or by a small group of students.

1.1   Suppose you are asked to analyze the truss shown in Fig. 1.2.

(*a*)   What is the overall purpose of the problem?

(*b*)   What information is known?

(*c*)   What information must be determined?

(*d*)   What fundamental engineering principles apply to the problem?

(*e*)   What will be the overall solution strategy?

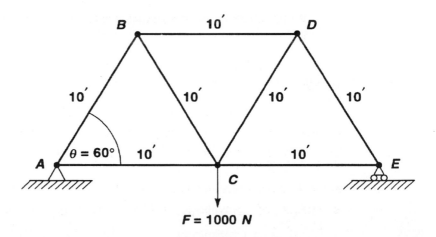

Figure 1.2

1.2 *Fick's law* applies to the diffusion of molecules across a membrane. Fick's law is often written as $q = D(\Delta C/\Delta L)$, where q represents the flow of molecules per unit area per second, D is the diffusion coefficient (or diffusivity), ΔC is the difference in the concentration across the membrane and ΔL is the thickness of the membrane.

(*a*) In this process, what is the flux? What are its units?

(*b*) What is the potential? What are its units?

(*c*) What is the proportionality factor relating flux and potential? What are its units?

1.3 Cite an example of a problem that is based upon force equilibrium. Draw a sketch, indicate what is likely to be known and what is likely to be unknown.

1.4 Cite an example of a problem that is based upon chemical equilibrium. Draw a sketch of the problem. Indicate what fundamental physical and/or chemical principles apply to your problem. Suggest how the problem might be solved (but do not actually try to solve it).

1.5 Cite an example of an industrial process that is based upon conservation of mass. Indicate what other physical and/or chemical processes might apply.

1.6 Cite an example of an industrial process that is based upon conservation of energy. (Choose a different process from that cited in the previous problem.) Indicate what other physical and/or chemical processes might also apply.

1.7 What fundamental engineering principles apply to the launching of a NASA space shuttle?

ADDITIONAL READING

Adams, J. L. *Conceptual Blockbusting.* New York: Norton, 1978.

Florman, S. C. *The Civilized Engineer.* New York: St. Martin's Press, 1987.

Jayaraman, S. *Computer-Aided Problem Solving for Scientists and Engineers.* New York: McGraw-Hill, 1991.

Polya, G. *How to Solve It.* Princeton, NJ: Princeton University Press, 1971.

Rubenstein, M. *Patterns of Problem Solving.* Englewood Cliffs, NJ: Prentice-Hall, 1975.

Scarl, D. *How to Solve Problems. For Success in Freshman Physics, Engineering and Beyond.* 2d ed. Glen Cove, NY: Dosoris Press, 1990.

Starfield, A. M., K. A. Smith, and A. L. Bleloch. *How to Model It: Problem Solving for the Computer Age.* New York: McGraw-Hill, 1990.

CHAPTER 2

EXCEL FUNDAMENTALS

A *spreadsheet* is basically a table containing numerical and/or alphanumeric values. The individual elements within the spreadsheet are known as *cells*. A cell can contain two different kinds of data: a *numerical constant* (a *number*) or a *text constant* (also called a *label* or a *string*). Each cells is referenced by its column heading (typically a single letter) and its row number. This is called a *cell address* or *cell reference*. Thus, B3 refers to the cell in column B, row 3. A tabular collection of cells is referred to as a *worksheet*.

If a cell contains a numerical value, the number may have been entered directly, or it may be the result of a *formula evaluation*. Formulas express interdependencies among cells within a worksheet. For example, suppose the numerical value in cell C7 is generated by the formula =(C3+C4+C5). This formula states that the value in cell C7 is obtained by adding the values in cells C3, C4, and C5. Thus, changing the content of one cell (say C4) will change the contents of other cells (cell C7, and any other cells whose values are generated by formulas that refer to cell C4). This important feature allows the user to carry out *what-if* studies; i.e., *what* happens *if* the content of a particular cell is altered? What will be the effect of this change on other values within the worksheet?

Commercial spreadsheets also provide many other features including formatting, editing, file management and database capabilities, a wide variety of special functions, and the ability to represent data graphically. In addition, the newer versions of most popular spreadsheet programs include special mathematical procedures that are useful for solving problems that arise in engineering analysis. The use of many of these features is discussed in this book.

Excel is a popular spreadsheet program developed and supplied by the Microsoft Corporation. It contains a great many features, including multiple command paths (different ways to do the same thing), capabilities for creating a variety of graphs and drawings, data management capabilities, various short-cut menus, and extensive on-line help. In this chapter we will consider only those bare-bones features that are essential for the material described later in this book. For more information on Excel, you are encouraged to consult the Microsoft Excel User's Guide, Excel's on-line help, or any of the tradebooks describing its use.

Versions of Excel are available both for the Apple Macintosh family of computers and for Intel-compatible (IBM-compatible) computers using various Windows operating environments. This discussion assumes you are familiar with at least one of these environments.

2.1 ENTERING AND LEAVING EXCEL

To enter Excel, select the appropriate icon from the Desktop or from the Microsoft Office Shortcut Bar. (An *icon* is a small graphical symbol.) Typically, this will involve clicking the mouse on an easily recognizable Excel icon. Once you do this, you will see the opening Excel window, consisting of several lines of information surrounding an empty Excel *worksheet*, as shown in Fig. 2.1. The principal items within the window are illustrated in Fig. 2.1 and described below.

Title Bar

The top line is called the *Title Bar*. It includes the spreadsheet name, an icon that closes Excel at the left, and icons that either change the size of the active window or close Excel at the right. Thus, selecting either the left icon (by clicking the mouse pointer on this icon) or the rightmost icon (shaped as a cross) will initiate the steps required to close Excel. We will discuss these icons later, as the need arises. For now, however, note that you can exit from Excel by clicking on the left icon and then selecting Close from the resulting drop-down menu, or by clicking on the rightmost icon.

Menu Bar

The second line is called the *Menu Bar*. Selecting one of the choices (File, Edit, View, Insert, · · ·, Help) causes one of Excel's *drop-down menus* to appear. The File menu, for example, includes selections for opening new or existing worksheet files, closing files, saving files, printing files, and exiting Excel. The Edit menu includes selections for deleting, moving, or copying portions of a worksheet, and for inserting or deleting objects within the worksheet. And so on.

We will discuss the more commonly used menu selections within several of the drop-down menus later in this chapter, as the need arises. For now, however, note that you can leave Excel by selecting Exit from the File menu. (This is an alternative to clicking on one of the icons in the title bar, as described previously.) Or you can close the current worksheet, while remaining in Excel, by clicking on the icon at the left of the menu bar and then selecting Close from the resulting menu. Note also that you can obtain detailed assistance by selecting Help from the menu bar or by pressing function key F1 (more about this later).

Toolbars

The third and fourth lines are called *toolbars*. The icons in these lines duplicate several of the more commonly used menu selections that are available from the Menu Bar. For example, the upper toolbar (the *Standard Toolbar*) contains icons that will open a new workbook, open an existing workbook, save a workbook, print a workbook, etc. (a *workbook* is a collection of one or more worksheets, as described below); it also contains icons that allow you to carry out certain commonly used worksheet operations, such as summing the values in a row or column of numbers. The lower toolbar (the *Formatting Toolbar*) contains icons that allow you to control the appearance of the items shown on the worksheet.

The toolbars also contain drop-down menus whose choices control the portion of the worksheet that is displayed, the current font, and the current font size. These choices also duplicate selections that are available from the Menu Bar. Thus, the toolbars contain repetitions of selections available elsewhere. They are present only for convenience; in fact, they can be removed from the screen if you wish (by selecting Toolbars from the View menu and then adding or removing the available choices from the resulting dialog box).

The Excel Worksheet

Most of the Excel window is occupied by a worksheet consisting of a grid of *cells*. Each cell has its own column letter and row number. These cells will contain the actual problem description; hence, their contents will comprise the detailed worksheet. We will see how information is entered and manipulated within these cells later in this chapter (see Sec. 2.3).

The Formula Bar

The *Formula Bar* appears just above the worksheet. It contains information about the *active worksheet cell* (see Sec. 2.2). The Formula Bar includes the large formula display area on the right, and a smaller *Name Box* on the left. In Fig. 2.1 the formula display area contains a reference to the empty cell A1. We will say more about the Formula Bar later in this chapter (see Sec. 2.3).

Title Bar   Menu Bar   Standard Toolbar   Formula Bar

Currently Active Cell   Formatting Toolbar

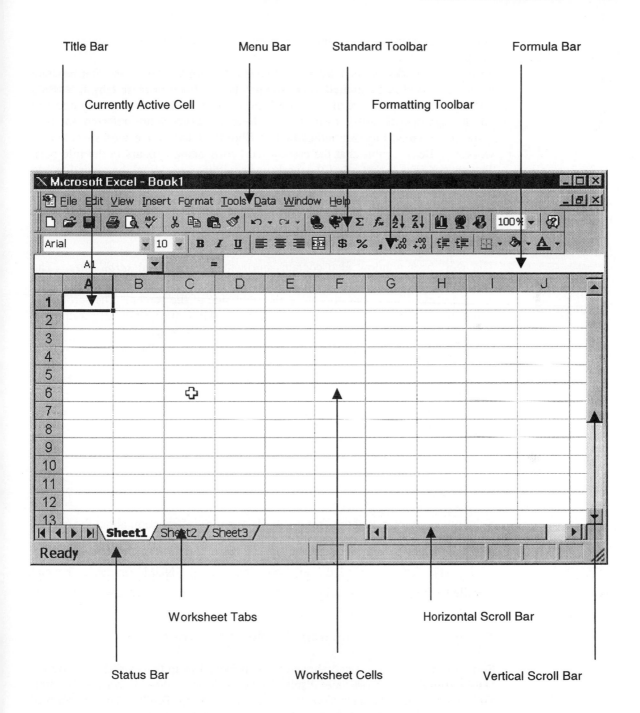

Worksheet Tabs   Horizontal Scroll Bar

Status Bar   Worksheet Cells   Vertical Scroll Bar

Figure 2.1

Worksheet Tabs

Beneath the worksheet is a set of *worksheet tabs,* similar to the tabs that separate the sections within a standard three-ring notebook. Each of these tabs accesses a different worksheet. Any of these worksheets can be selected simply by clicking on the appropriate tab. Collectively, these worksheets are referred to as a *workbook*. Thus, Fig. 2.1 indicates that Sheet1 is the active worksheet in the workbook Book1 (note that the current workbook name appears in the title bar). The general idea is that all of the worksheets within a workbook be related to one another. Each workbook is saved and stored as a single file.

Scroll Bars

The worksheet window also contains horizontal and vertical *scroll bars*. The horizontal scroll bar appears in the second last row, to the right of the sheet tabs, and the vertical scroll bar appears at the right side of the screen. The scroll bars allow rapid movement through the worksheet. They are particularly useful for those worksheets that extend beyond the confines of the screen.

To move through the worksheet, click the mouse on the *scroll button* that appears within each scroll bar and drag the button in the desired direction of movement. For example, to move vertically downward through a long worksheet, drag the vertical scroll button (shown just beneath the upward-pointing arrow in Fig. 2.1) downward. You may also move through the worksheet by clicking on one of the arrows at either end of each scroll bar. Or you may click on the scroll bar itself, on either side of the scroll button.

Status Bar

The bottom line is called the *Status Bar*. It provides information about the current state of the spreadsheet model (e.g., Ready) and the current command or menu selection. It also shows the current input status, with flags such as CAPS (for uppercase), EXT (extended cell selection mode), NUM (numeric keypad activated), and so on.

Adding/Removing Items from the Worksheet Window

Some of the items discussed above can be removed from the worksheet window, resulting in a larger worksheet display (more rows and columns) with less clutter. For example, the Standard Toolbar and the Formatting Toolbar can be removed by selecting Toolbars from the View menu. The Formula Bar, the worksheet tabs, the scroll bars, and the Status Bar can be removed by selecting Options/View from the Tools menu and then deselecting these items from the resulting dialog box, as illustrated in Fig. 2.2. The toolbars can also be removed by selecting **Full Screen** from the **View** menu. These worksheet items can be restored by reversing these procedures. Other items (for example, other toolbars) can also be added in this manner.

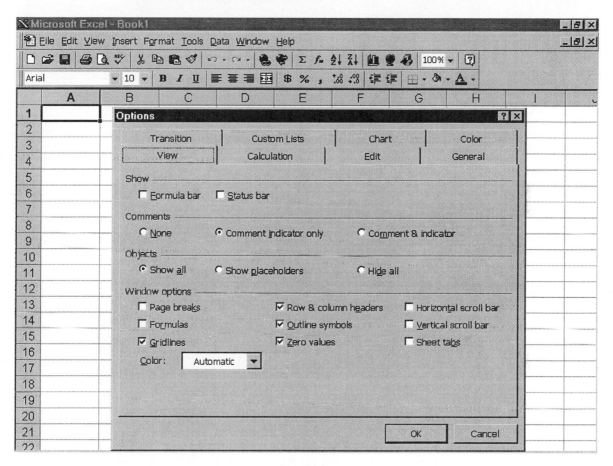

Figure 2.2

Getting Help

Excel provides extensive on-line help in several different forms. One way that is particularly useful is to select **Contents** and **Index** from the **Help** menu, resulting in the dialog box shown in Fig. 2.3. You can then obtain detailed information about any of the major topics included under the **Contents** tab, or the more specific items listed under the **Index** tab, as shown in the figure. There is also a **Find** feature, which allows you to obtain information on many commonly used words or phrases.

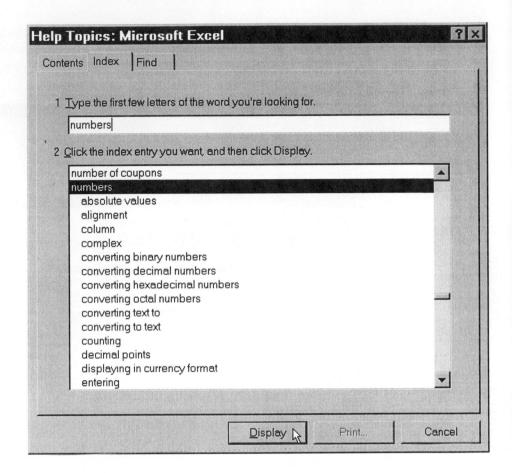

Figure 2.3

Another useful help feature is the **Office Assistant**, which is a cartoon-like dialog box for entering questions in ordinary English (see Fig. 2.4). The **Office Assistant** can be accessed three different ways: by selecting **Microsoft Excel Help** from the **Help** menu, by clicking on the icon shaped as a question mark (?) in the Standard Toolbar, or on most keyboards, by pressing function key F1.

Excel also provides a **What's This?** feature, which explains the meaning of individual items appearing in menus and dialog boxes. To activate this feature, select **What's This?** from the **Help** menu, or on most keyboards, hold down the **Shift** key and press function key F1. Then place the mouse pointer over a specific item, such as a toolbar button or a user-response area within a dialog box, and click the left mouse button. This will generate an information box, typically explaining what the item is and how it can be used.

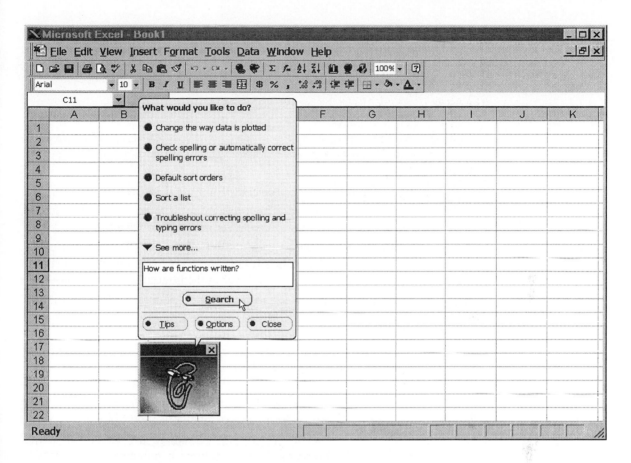

Figure 2.4

Leaving Excel

We have already seen several different ways to leave Excel. To recap, you can double-click on the left icon in the Title Bar, you can click once on the left icon and then select Close from the resulting menu, or you can click on the rightmost icon (containing the ×) in the Title Bar. You may also leave by selecting Exit from the File menu. If your worksheet is new or if it has been altered in any way, an attempt to exit will generate a dialog box, asking you whether or not to save the current version of the worksheet. If you respond by selecting Yes, another dialog box will appear, prompting you for the name and location of the worksheet file. These two dialog boxes safeguard against your inadvertently leaving Excel without first saving your work. (Some other save options, which allow you to save your work periodically before leaving Excel, are described in Sec. 2.8.)

The use of the features found in the Menu Bar and the various toolbars become self-evident with just a little practice. We will discuss certain of these features later in this book, as the need arises.

Exercise

2.1   When working through this and the next few exercises, feel free to experiment on your own after you have worked through the suggested exercises.

(*a*)   Enter Excel from the Macintosh or Windows desktop.

(*b*)   Click on the middle icon within the group of three at the right side of the Title Bar a few times. Be sure you understand what this icon does.

(*c*)   Minimize Excel by clicking on the leftmost icon within the group of three in the Title Bar. Then restore the Excel window by clicking on the Excel button within the *Task Bar* at the bottom of the desktop.

(*d*)   Close Excel (i.e., return to the Desktop) by double-clicking on the left icon in the Title Bar (or by clicking once on the left icon and then selecting Close from the resulting menu).

(*e*)   Reenter Excel. Then select File from the Menu Bar. Examine the selections in the File menu. Then remove the File menu by pressing the Escape key or by clicking on some other portion of the worksheet.

(*f*)   Select Edit from the Menu Bar. Examine the selections in the Edit menu. Then remove the Edit menu using the method described above.

(*g*)   Select Format from the Menu Bar. Examine the selections in the Format menu. Choose the Cells selection, then examine the Number, Alignment, Font, and Border tabs (one at a time). Be sure that you understand what each of these tabs is for. (You may exit from any of the resulting dialog boxes by selecting Cancel.) When you have examined each of these Format selections, remove the Format menu using the method described in (*e*).

(*h*)   Select Help from the Menu Bar. (On most computers you can also obtain the Help menu by pressing function key F1.) Then choose Contents from the Help menu. After examining Contents, choose Index. Be sure you understand how to use each of these features. Open a few specific help items. (Exit from each help item in the usual manner, by double-clicking on the icon in the upper left corner or by clicking on the icon in the upper right corner of the help window.)

(*i*)   Press the CapsLock key on your keyboard. Note what happens in the Status Bar as you turn the CapsLock on and off a few times. Repeat with the NumLock key and with function key F8 (if present).

(*j*)   Close Excel by selecting File/Exit from the Menu Bar. (Note that this is an alternative to the method discussed in (*d*) above).

2.2 MOVING AROUND THE WORKSHEET

We now focus on the worksheet, which is the principal part of the spreadsheet. The worksheet is shown as an empty grid in Fig. 2.1. Each rectangle within the grid represents an empty cell. Cell A1 is the *currently active cell*, as indicated by the rectangle surrounding the cell. The graphical object shown in cell C6 is the *mouse pointer*. This indicates the current mouse position within the worksheet. In Excel the mouse pointer will take on different shapes, depending on the task being performed.

There are many ways to move around the worksheet and change the currently active cell. Cell movement is carried out in any of the ways described below.

Mouse

The easiest way to change the currently active cell is to set the mouse pointer over the desired cell and then click the left button.

If your mouse has a wheel between the buttons (e.g., the Microsoft IntelliMouse), you can turn the wheel to scroll vertically up or down within the worksheet. Or you can move through the worksheet by holding down the wheel and dragging the mouse in the desired direction.

Arrow Keys

The arrow keys are LeftArrow ($\leftarrow$), RightArrow ($\rightarrow$), UpArrow ($\uparrow$), and DownArrow ($\downarrow$). Pressing any of these keys once will result in movement in the indicated direction to an adjacent cell. Holding down one of these keys will result in sustained movement in the indicated direction until the key is released. Once the key is released, whatever cell is selected will be the currently active cell.

PageUp, PageDown

Pressing the PgUp or PgDn keys results in vertical movement within the worksheet. Each move involves several lines. The movement will result in a new currently active cell.

Ctrl-LeftArrow, Ctrl-RightArrow, Ctrl-UpArrow, Ctrl-DownArrow

Holding down the Ctrl key (or the Command key) and pressing one of the arrow keys results in either horizontal or vertical movement to the opposite edge of the worksheet. Each move involves several columns. The movement will result in a new currently active cell.

Scroll Bars

Clicking on an arrow within a scroll bar results in movement one row at a time or one column at a time, depending on the scroll bar. Clicking at an arbitrary location within a scroll bar results in a larger movement. Dragging a scroll button results in a more controlled movement in the indicated direction. This type of movement does *not* result in a new currently active cell.

Home, End

Pressing Home or Ctrl-Home restores the upper left corner of the worksheet and causes cell A1 to be the active cell. (Under some conditions, pressing Home simply moves the active cell to the beginning of the current row.) Pressing Ctrl-End causes movement to the last cell containing data within the worksheet.

GoTo Key

Function key F5 on your keyboard is the GoTo key. Pressing this key generates a dialog box. Respond by entering the address of the cell that you wish to become active at the bottom of the dialog box. (This dialog box can also be generated from the Edit menu.). Note that the dialog box maintains a history of recent moves so that you can easily return to a previous location.

Exercise

2.2 This problem provides practice in moving around the Excel worksheet.

   (*a*) Using the mouse, move the pointer to some arbitrary location within the window and click left. Which cell is now currently active? Repeat several times.

   (*b*) Use the arrow keys to change the location of the currently active cell. Note what happens each time an arrow key is pressed.

   (*c*) Press the PageDown key a few times. What happens each time PageDown is pressed? Which cell is now currently active? Repeat using the PageUp key.

   (*d*) Hold down the Ctrl key (or the Command key) and press the RightArrow key (→). What happens? Which cell becomes currently active? Repeat using Ctrl and each of the other arrow keys.

   (*e*) Using the mouse, drag the vertical scroll button up and down within the vertical scroll bar. Note what happens. Repeat with the horizontal scroll button. Is the currently active cell affected by this type of movement?

(*f*)   Using the mouse, click on the arrows at the ends of each scroll bar and note what happens. Also, try clicking the mouse in different locations within each scroll bar and note what happens.

(*g*)   Does your mouse have a wheel between the buttons? If so, turn the wheel and notice the resulting movement of the worksheet. Press the wheel down and release, then drag the mouse across the worksheet. Notice what happens.

(*h*)   Move the active cell to some arbitrary location within the worksheet. Then press Home and note what happens. Try the same thing with Ctrl-Home.

(*i*)   Press function key F5 and then specify some arbitrary cell address at the bottom of the dialog box. What happens when you press OK?

(*j*)   Select GoTo from the Edit menu. Then specify some arbitrary cell address at the bottom of the dialog box. Note what happens when you press OK. (This is an alternative to the use of function key F5.)

2.3 ENTERING DATA

You may enter either a numerical value (called a *number constant* in Excel) or a string (called a *text constant* in Excel) into any worksheet cell. Remember that errors can always be corrected using the techniques described in Sec. 2.4.

Numerical Values

Numerical values are entered as ordinary numbers, with or without a decimal point. Negative values should be preceded by a minus sign. Scientific notation is also permitted. Thus, any of the following numerical values could be entered into the worksheet:

2	−6	3.33	2.55E−12	−7.08E+6
0.0	0.004	−1.2e−7	4E8	1e10

Numerical values can also be entered in other ways; for example, with embedded commas, preceded by a currency sign ($), or followed by a percent sign (%). You may also enter a numerical date (e.g., 5/24/96) or a time (e.g., 7:20 PM or 19:20:00). However, these formats are generally not used when carrying out technical calculations; hence, we will not discuss them further. Note, however, that the format of any numerical value can be altered by selecting Cells/Number from the Format menu and then making the appropriate selection from the resulting dialog box. Consult Excel's on-line help for more information.

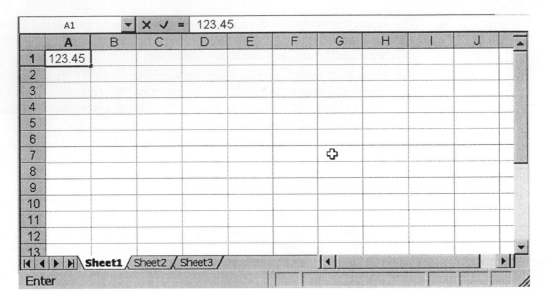

Figure 2.5

As you enter a numerical value in the currently active cell, you will see the value being entered in the Formula Bar as well as in the active cell. When you are finished typing the number, press the Enter key. You may also click on the ✓ symbol in the Formula Bar to accept the number, or click on the ✕ to reject the number. Figure 2.5 shows a number being entered in cell A1. Note the presence of the symbols ✕, ✓, and = in the Formula Bar.

After the number has been entered, it will continue to be visible in the Formula Bar as well as in the active cell. However, the symbols ✕, ✓, and = will no longer appear in the Formula Bar.

Strings (Labels)

A *string* (also called a *label*) is entered as a text constant simply by typing the desired text into the active cell. The text will be visible in the Formula Bar as it is being entered and after it has been entered into the active cell. The ✕, ✓, and = symbols will appear in the Formula Bar while the string is being entered, just as with a numerical value. The string will extend to the right of the active cell if it is wider than the cell width.

Now suppose a long string has been entered into the currently active cell (for example, cell C5) and you wish to enter some other data in the cell to the right of the currently active cell (e.g., cell D5). When the new data is entered, the rightmost portion of the long string will be hidden behind the new data. Hence, only the leftmost portion of the string will appear within the original cell (C5). The entire string will still be stored within the computer's memory, however, and it will still be associated with the original cell (C5). If you again make this cell active, you will see the entire string in the Formula Bar, even though a portion of the string appears overwritten within the worksheet. You may also increase the width of the original cell (by dragging the mouse across the boundary separating the columns in the cell headings or by selecting Column/Width from the Format menu), thus displaying the entire string within the cell.

Exercise

2.3 Enter the following grade report into an empty Excel worksheet. Begin with the heading (Grade Report) in cell A1. (The numerical grades correspond to letter grades, i.e., A = 4.0, B = 3.0, C = 2.0, D = 1.0, F = 0. Fractional values represent pluses and minuses, i.e., A– = 3.7, B+ = 3.3, B– = 2.7, etc.) Be sure that all of the strings are fully visible within the worksheet.

Grade Report
(Your full name)
(Your social security number)

Course	Credits	Grade
Chemistry I	4	3.0
Physics I	4	3.3
English Composition	3	2.0
Intro to Engineering	3	4.0
Calculus I	3	3.7
Seminar	1	2.7

2.4 CORRECTING ERRORS

Errors that occur while entering data can easily be corrected. If you are still in the process of entering data (before pressing the Enter key or before clicking on the ✓ symbol in the Formula Bar), you can use the backspace key to remove unwanted characters, then proceed as usual. Or you can retype the entire data item after a number or a string has been entered incorrectly. The original (incorrect) data item will be replaced by the newer one once you press the Enter key. You can also clear out an active cell (i.e., erase its contents) by pressing the Delete key or by selecting Clear/All or Clear/Contents from the Edit menu.

   If an erroneous data item is lengthy, you may prefer to edit a portion of the data item rather than retype the entire data item. To do so, make the cell active so that the cell content appears in the Formula Bar. Then edit the data item within the Formula Bar or within the cell once you have entered the edit mode (with the three special symbols visible in the Formula Bar). The editing is carried out in the usual manner, by positioning the cursor at the desired edit point and then deleting and/or inserting individual characters, as required. You may drag the cursor across several characters to highlight an area and then delete the entire highlighted area if you wish. You can also double-click on a number or a word within a string and then delete that number or word by pressing the Delete key.

When you are finished editing, you can accept the edited data item by pressing the Enter key or by clicking on the ✓ symbol. You can also cancel the edit by clicking on the ✗ symbol, thus restoring the data item to its original form. (Remember to save your work periodically to avoid unexpected data losses due to power failures, etc. Some save options are explained in Sec. 2.8.)

Exercise

2.4 Create the worksheet described in Exercise 2.3. Then make the following changes within the worksheet:

(*a*) Change the heading from Grade Report to Semester Grade Report.

(*b*) Clear out your social security number.

(*c*) Change Intro to Engineering to Engineering Analysis.

(*d*) Change the number of credits for Physics I from 4 to 3.

(*e*) Change the grade for English Composition from 2.0 to 2.7.

When making these changes, try all of the various cell editing features described above.

2.5 USING FORMULAS

We have already seen that formulas express interdependencies among the values in different cells within a worksheet. Formulas allow you to perform arithmetic operations on numerical values, combine strings, and compare the contents of one cell with another.

In Excel, *formulas begin with an equal sign* (=), followed by a *numerical expression* involving *constants*, *operators*, and *cell addresses*. Consider, for example, the formula =(C3+B2+5). In this formula, the numerical expression is (C3+B2+5), the terms C3 and B2 are cell addresses, 5 is a numerical constant, and the plus sign (+) is an arithmetic operator. If this formula is entered into cell D7, then cell D7 will contain the sum of the numbers in cells C3 and B2, plus 5. You should understand that *if either of the numbers in cells C3 and B2 is changed, the numerical value in cell D7 will change correspondingly.*

Operators

Excel includes *arithmetic operators*, a *string operator*, and *comparison operators*. The arithmetic operators combine numerical values (called *operands*) to produce a numerical value. The string operator is used to combine (i.e., to *concatenate*) two strings. The comparison operators compare one operand with another, resulting in a condition that is either *true* or *false*.

The permissible Excel operators are summarized below. (They are the same as the operators in Microsoft's Visual Basic, QuickBASIC, and QBASIC programming languages.)

To a beginner, the complete list of Excel operators may seem a bit overwhelming. However, most simple engineering applications use only the first five arithmetic operators (addition, subtraction, multiplication, division, and exponentiation).

Arithmetic Operator	*Purpose*	*Example*
+	Addition	A1+B1
−	Subtraction	A1−B1
*	Multiplication	A1*B1
/	Division	A1/B1
^	Exponentiation	A1^3
%	Percentage (divide by 100)	A1%

Note that the % operator requires only one operand (a cell reference or a constant). *Do not confuse this operator with the % icon in the Formatting Toolbar* (see Fig. 2.1).

String Operator	*Purpose*	*Example*
&	Concatenate	A1&B1

Note that the string operator operates only on string operands.

Comparison Operators	*Meaning*	*Example*
>	Greater than	C1 > 100
>=	Greater than or equal to	C1 >= 100
<	Less than	C1 < 100
<=	Less than or equal to	C1 <= 100
=	Equal (equivalent)	C1 = C2
<>	Not equal (not equivalent)	C1 <> C2

Note that the result of a comparison operation is either *true* or *false*.

Operator Precedences

A formula may include several different operators. The question then arises as to which operation is carried out first, which is carried out second, and so on. To answer this question, Excel includes several groups of *operator precedences* that define the order in which the operations are carried out. These operator precedences are listed below, from highest to lowest. Thus, in a complicated formula, the percentage operation would be carried out first, followed by the exponentiation operation, and so on. The relational comparisons would be carried out last.

If multiple operators within the same group appear consecutively, the operations will be carried out from left to right. For example, in the formula =(C1/D2*E3), both operators fall within the same group. Therefore, the division operation will be carried out first, and the resulting quotient will be multiplied by the contents of cell E3.

Operator Precedence	*Operators*
1	Percentage (%)
2	Exponentiation (^)
3	Multiplication and division (* and /)
4	Addition and subtraction (+ and −)
5	Concatenation (&)
6	Comparisons (>, >=, <, <=, =, <>)

These precedences can be altered by introducing pairs of parentheses. For example, the previous formula can be altered to read =(C1/(D2*E3)). Now the multiplication would be carried out first because this part of the formula is included within the innermost pair of parentheses. Then the contents of cell C1 would be divided by the resulting product. You may use these parentheses freely, in accordance with the logic of your particular problem.

We will say more about the use of cell formulas in Section 2.7, when we discuss copying and moving cells within a worksheet.

Example 2.1   A Simple Spreadsheet Application

A small machine shop has the following parts on hand:

Item	*Quantity*
Screws	8000
Nuts	7500
Bolts	6200

Create a worksheet that includes this information, plus the total number of parts on hand. Use a formula to calculate the total.

The desired worksheet is shown in Figure 2.6. Note that the worksheet includes several strings (cells B2 through B5, cell B7, and cell C2), three numerical constants (cells C3, C4, and C5), and one formula (cell C7). The heavy border around cell C7 indicates that it is currently active. Notice that the formula =(C3+C4+C5) used to generate the numerical value in cell C7 is visible in the Formula Bar. Thus, you can distinguish between numerical constants and formula values by examining the content of the Formula Bar.

The sum in cell C7 could also have been obtained simply by selecting cell C7, clicking on the AutoSum icon (Σ) in the Standard Toolbar, and then pressing Enter. The AutoSum feature can be activated beneath a column of numbers or to the right of a row of numbers. Note, however, that the AutoSum feature applies only to summation, whereas the use of a formula is much more general.

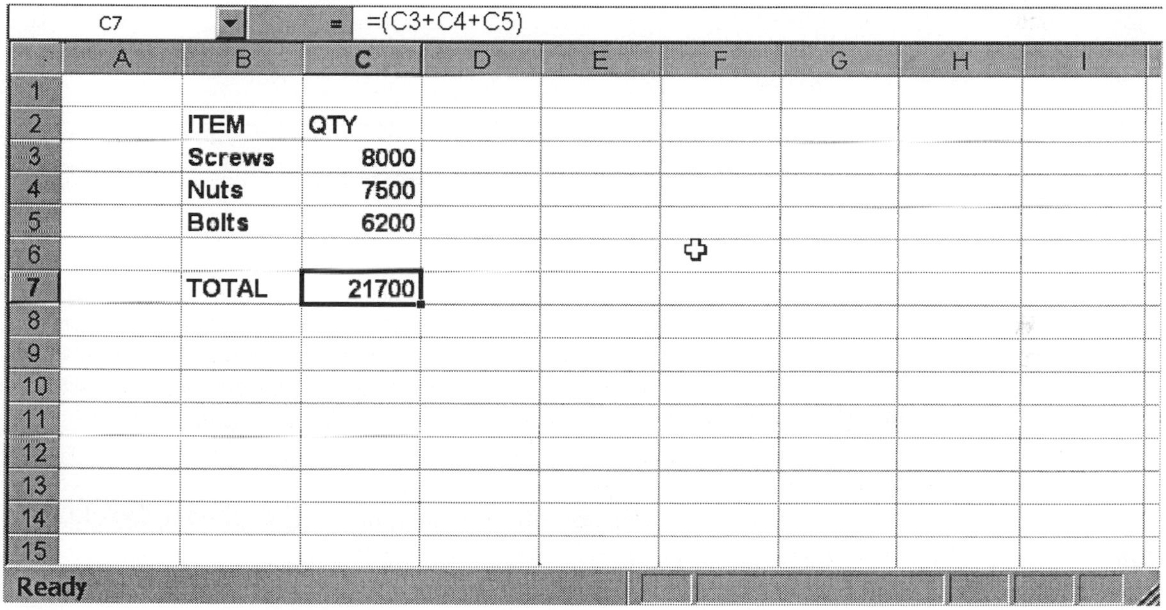

	A	B	C	D	E	F	G	H	I
C7			=(C3+C4+C5)						
1									
2		ITEM	QTY						
3		Screws	8000						
4		Nuts	7500						
5		Bolts	6200						
6									
7		TOTAL	21700						

Figure 2.6

Excel allows you to assign names to individual cells so that you can use cell names rather than cell addresses in formulas. To do so, select a cell and highlight the cell address shown in the Name Box (the leftmost portion of the Formula Bar). Then type the desired cell name in place of the cell address (begin with a letter and avoid blank spaces), and press the Enter key. You may then rewrite any formulas that access these cells so that they use either the cell names or the cell addresses.

Example 2.2 Naming Cells

Alter the worksheet created in Example 2.1 so that cell C3 is called Part1, cell C4 is called Part2, cell C5 is called Part3 and cell C7 is called Part_Total. Then change the formula in cell C7 from =(C3+C4+C5) to =(Part1+Part2+Part3).

Figure 2.7 shows the worksheet with cell C3 active, after the cell has been named Part1. The cell name is shown in the Name Box, in the left portion of the Formula Bar. Note that the *value* within this cell (8000) is not affected by the change in the cell name.

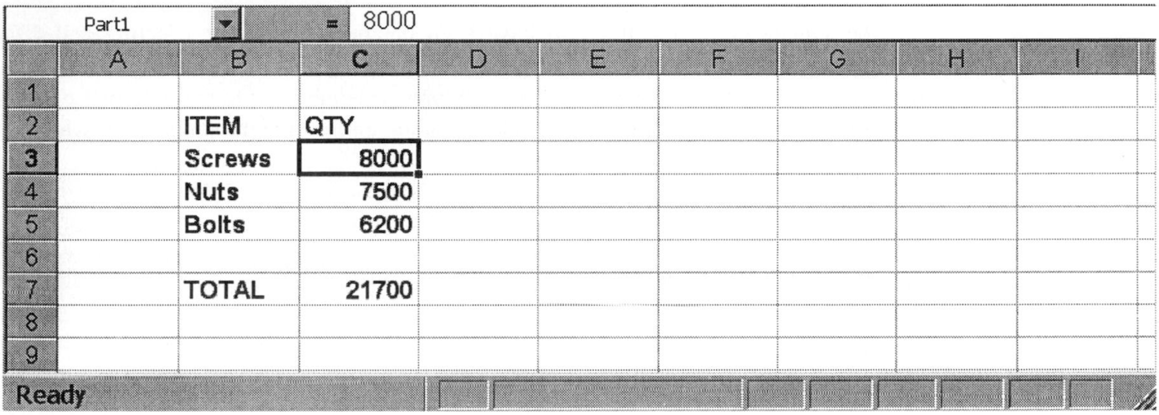

Part1	▼	=	8000						
	A	B	C	D	E	F	G	H	I
1									
2		ITEM	QTY						
3		Screws	8000						
4		Nuts	7500						
5		Bolts	6200						
6									
7		TOTAL	21700						
8									
9									
Ready									

Figure 2.7

Figure 2.8 shows the same worksheet, with cell C7 active. Notice that this cell has been renamed Part_Total, as shown in the Name Box. The formula within this cell has also been changed to =(Part1+Part2+Part3), as shown in the right portion of the Formula Bar. Note that the *value* generated by the formula (21700) is the same as that shown in Example 2.1 (see Fig. 2.6).

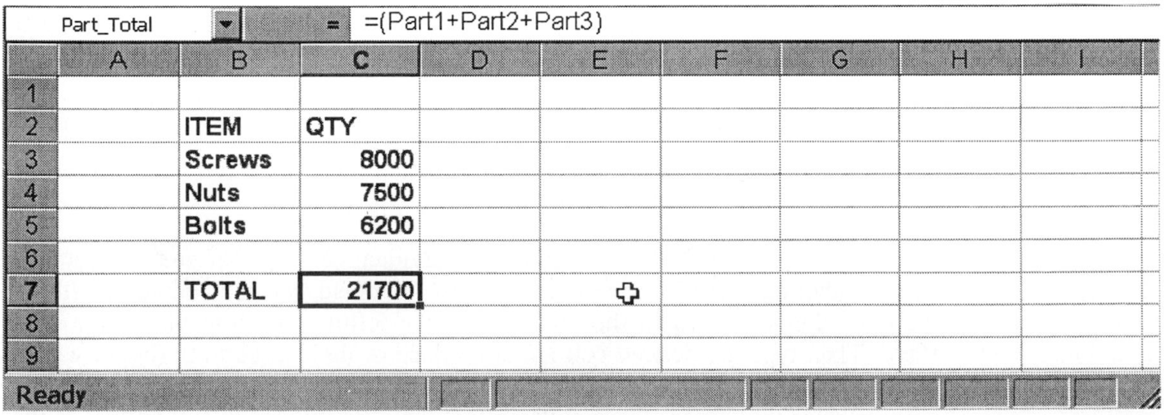

Part_Total	▼	=	=(Part1+Part2+Part3)						
	A	B	C	D	E	F	G	H	I
1									
2		ITEM	QTY						
3		Screws	8000						
4		Nuts	7500						
5		Bolts	6200						
6									
7		TOTAL	21700		⊕				
8									
9									
Ready									

Figure 2.8

Exercises

2.5 Change the numerical values in the worksheet shown in Fig. 2.6, Example 2.1, as follows. Notice what happens to the total as each numerical value is changed.

Item	*Quantity*
Screws	6500
Nuts	9000
Bolts	5400

(*a*) Obtain the sum in cell C7 using a formula, as in Example 2.1.

(*b*) Use the AutoSum feature to obtain the total shown in cell C7.

(*c*) Rename the cells and then obtain the sum using a formula based upon the new cell names, as in Example 2.2.

2.6 Extend the worksheet shown in Figure 2.6, Example 2.1, in the following ways:

(*a*) Add a third column containing the following cost information:

Item	*Cost*
Screws	$0.02 each
Nuts	$0.03 each
Bolts	$0.05 each

(*b*) Add a fourth column containing the value of the screws, the nuts, and the bolts.

(*c*) At the bottom of this fourth column, add a cell containing the total value of all of the parts. Use a formula to obtain the total. Then try using AutoSum to obtain the total.

(*d*) Change the quantity of screws to 6500. What is the effect of this change elsewhere in the worksheet?

(*e*) Change the cost of nuts to $0.04 each. What is the effect of this change elsewhere in the worksheet?

(*f*) Based upon the original values, what is the average cost per part? (Do not distinguish between part types when answering this question.)

2.7 At the end of every term, each student's grade-point average (*GPA*) is calculated as

$$GPA = \left(\sum_i C_i G_i\right) \Big/ \left(\sum_i C_i\right)$$

where C_i refers to the number of credits for the ith course and G_i refers to the numerical grade for that course.

Enter the grade report described in Exercise 2.3 and add a row at the bottom of the worksheet indicating the student's *GPA* for the given semester. Use an Excel formula based upon the above equation to determine the *GPA*.

2.6 USING FUNCTIONS

Excel includes many different *functions* that can be used to carry out a wide variety of operations. For example, there are groups of functions that carry out mathematical and statistical operations, process financial data, process text, and return information about the worksheet. We will use some of these functions in later chapters of this book. For now, however, we will concentrate on general techniques for using functions.

A function consists of a *function name*, followed by one or more *arguments*. The arguments are enclosed in parentheses and separated by commas. Consider, for example, the function specification SUM(C1,C2,C3). This function calculates the sum of the quantities in cells C1, C2, and C3. The function name is SUM, and the arguments are the cell references C1, C2, and C3.

The previous function could also have been written as SUM(C1:C3). The term C1:C3 is called a *range*. It refers to all of the cells between the given extremities (i.e., between cells C1 and C3). The range designation is very convenient when a function requires many contiguous cells as arguments.

Function arguments need not be restricted to cell references. Any formula can be used as a function argument, as long as it is of the proper type (e.g., a numerical formula cannot be used if a string-type argument is required, and so on). A function argument can even include a reference to another function. Consider, for example, the function specification SUM(A1, SQRT(A2/2), 2*B3+5, D7:D12). This function has four arguments: the cell address A1, the function specification SQRT(A2/2), the formula 2*B3+5, and the range D7:D12. The SQRT function returns the square root of its own argument, which in this example is the formula A2/2. (That is, the SQRT function in this example will return the square root of one-half the quantity in cell A2.)

The Paste Function icon (f_x) in the Standard Toolbar provides easy access to all of Excel's functions. Table 2.1 lists the most commonly used functions.

Example 2.3 Student Exam Scores

A group of students obtained the following exam scores in their Introduction to Engineering class. Enter the information into an Excel worksheet. Then determine an overall score for each student, assuming that each exam carries equal weight. Also, determine the class average for each of the exams and determine a class average for the students' overall scores.

Table 2.1   Commonly Used Excel Functions

Function	Purpose
ABS(x)	Returns the absolute value of x.
ACOS(x)	Returns the angle (in radians) whose cosine is x.
ASIN(x)	Returns the angle (in radians) whose sine is x.
ATAN(x)	Returns the angle (in radians) whose tangent is x.
AVERAGE($x1, x2, \cdots$)	Returns the average of $x1, x2, \cdots$.
COS(x)	Returns the cosine of x.
COSH(x)	Returns the hyperbolic cosine of x.
COUNT($x1, x2, \cdots$)	Determines how many numbers are in the list of arguments.
EXP(x)	Returns e^x, where e is the base of the natural (Naperian) system of logarithms.
FV(i, n, x)	Returns the future value of n payments of x dollars each at interest rate i.
INT(x)	Rounds x down to the closest integer.
IRR($x1, x2, \cdots$)	Returns the internal rate of return for a series of cash flows.
LN(x)	Returns the natural logarithm of x ($x > 0$).
LOG10(x)	Returns the base-10 logarithm of x ($x > 0$).
MAX($x1, x2, \cdots$)	Returns the largest of $x1, x2, \cdots$.
MEDIAN($x1, x2, \cdots$)	Returns the median of $x1, x2, \cdots$.
MIN($x1, x2, \cdots$)	Returns the smallest of $x1, x2, \cdots$.
MODE($x1, x2, \cdots$)	Returns the mode of $x1, x2, \cdots$.
NPV($i, x1, x2, \cdots$)	Returns the net present value of a series of cash flows at interest rate i.
PI( )	Returns the value of π. (The empty parentheses are required.)
PMT(i, n, x)	Returns the periodic (e.g., monthly) payment for an n-payment loan of x dollars at interest rate i.
PV(i, n, x)	Returns the present value of a series of n payments of x dollars each at interest rate i.
RAND( )	Returns a random value between 0 and 1. (The empty parentheses are required.)
ROUND(x, n)	Rounds x to n decimals.
SIGN(x)	Returns the sign of x. (Returns +1 if $x > 0$, -1 if $x < 0$.)
SIN(x)	Returns the sine of x.
SINH(x)	Returns the hyperbolic sine of x.
SQRT(x)	Returns the square root of x ($x > 0$).
STDEV($x1, x2, \cdots$)	Returns the standard deviation of $x1, x2, \cdots$.
SUM($x1, x2, \cdots$)	Returns the sum of $x1, x2, \cdots$.
TAN(x)	Returns the tangent of x.
TANH(x)	Returns the hyperbolic tangent of x.
TRUNC(x, n)	Truncates x to n decimals.
VAR($x1, x2, \cdots$)	Returns the variance of $x1, x2, \cdots$.

Student	Exam 1	Exam 2	Final Exam
Davis	82	77	94
Graham	66	80	75
Jones	95	100	97
Meyers	47	62	78
Richardson	80	58	73
Thomas	74	81	85
Williams	57	62	67

An Excel worksheet containing all of the student names and exam scores is shown in Fig. 2.9. The class average for exam 1 (71.6) is shown in active cell B10, beneath the individual scores in cells B2 through B8. Notice the corresponding formula =ROUND(AVERAGE(B2:B8),1) in the Formula Bar. Within this formula, the AVERAGE function is used to obtain the actual average. This function is an argument within the ROUND function, which is used to round the calculated average to one decimal place. Similar formulas are used to generate the averages shown in cells C10 and D10.

Cells E2 through E8 contain the overall scores for the individual students. Each of these rounded averages was obtained using a similar formula. For example, the formula used to obtain the value shown in cell E2 is =ROUND(AVERAGE(B2:D2),1); the formula for cell E3 is =ROUND(AVERAGE(B3:D3),1); and so on. The value in cell E10, which represents the average of the students' overall scores, was obtained in a similar manner (see Exercise 2.8 below.)

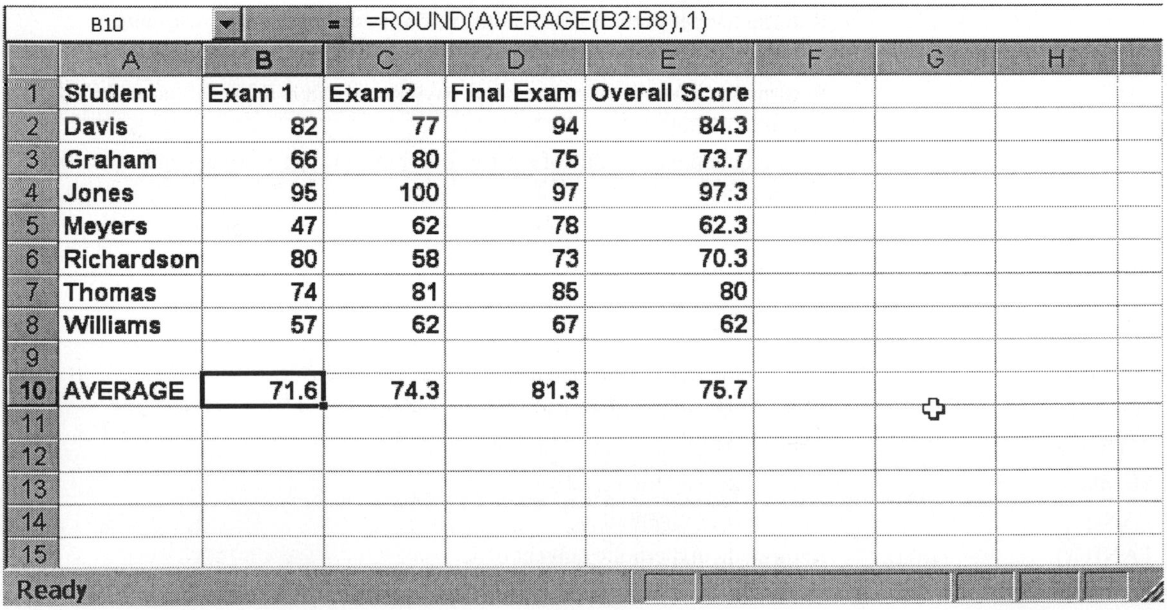

B10	▼	=	=ROUND(AVERAGE(B2:B8),1)				
	A	B	C	D	E	F	G	H
1	Student	Exam 1	Exam 2	Final Exam	Overall Score			
2	Davis	82	77	94	84.3			
3	Graham	66	80	75	73.7			
4	Jones	95	100	97	97.3			
5	Meyers	47	62	78	62.3			
6	Richardson	80	58	73	70.3			
7	Thomas	74	81	85	80			
8	Williams	57	62	67	62			
9								
10	AVERAGE	71.6	74.3	81.3	75.7			
11							⇩	
12								
13								
14								
15								

Ready

Figure 2.9

Exercises

2.8 Answer the following questions for the worksheet shown in Fig. 2.9, Example 2.3:

(*a*) Write the formula used to generate the value shown in cell C10.

(*b*) Write the formula used to generate the value shown in cell E4.

(*c*) Write two different formulas that can be used to generate the value shown in cell E10.

2.9 For the student exam scores given in Example 2.3, suppose that each of the first two exams contributes 30 percent to the student's overall score and the final exam contributes 40 percent.

(*a*) Construct a worksheet similar to that shown in Fig. 2.9 for this situation. Be sure to include the new weighting factors within the formulas, as required.

(*b*) In column G, calculate the difference between each student's overall score and the class average (i.e., the average of the overall scores, shown in cell E10).

(*c*) In cell G10, calculate the average of the differences in cells G2 through G8. What value do you expect to see here, and why?

2.7 EDITING THE WORKSHEET

Many spreadsheet operations are carried out on a *block* of cells (i.e., a group of contiguous cells within a row, a column, or several adjacent rows or columns). For example, you might want to create a graph from a list of numbers, sort a list of names, print a block of data, or move a block of data from one part of a worksheet to another. Excel includes a number of editing commands that allow you to carry out these block-type operations.

Selecting a Block of Cells

To process a block of cells in Excel, you must *first select the cells and then carry out the desired operation*. Multiple-cell selection is best carried out with a mouse. To select a block of cells, move the cursor to one corner of the cell block, hold down the left mouse button, and drag the mouse across the worksheet to the opposite corner. The entire block of cells will then be highlighted.

You can also select the *entire worksheet*, by clicking on the Select All Button (the blank button in the upper left-hand corner, directly to the left of the column headings and directly above the row headings). Or you can hold down the Control and Shift keys and then press the space bar on an Intel-type (IBM-compatible) computer, or Command – Shift – space bar on a Macintosh.

Excel includes an AutoCalculate feature that automatically shows the sum of all numerical values within a selection. This sum appears within the large box in the Status Bar. If you place the mouse cursor over the AutoCalculate box and click the right mouse button, a dialog box will appear, allowing you to display other types of numerical values (e.g., the average value, the max, the min, and so on.).

Example 2.4 Selecting a Block of Cells

In the worksheet created for Example 2.3, select the block of cells extending diagonally from cell B2 to cell E10.

The worksheet shown in Figure 2.9 is again seen in Figure 2.10, with the desired block of cells selected. The selection was obtained by first making cell B2 active and then dragging the mouse diagonally down to cell E10 while holding the left mouse button.

Note that the entire selection is highlighted (the colors are reversed). Cell B2 is currently active, which is why its appearance is different than the remaining cells within the block. Its value (a numerical constant) appears in both the Formula Bar and the cell.

Notice the message Sum=2422.8 within the Status Bar. This is the sum of all numerical values within cells B2 through E10. The value is automatically calculated and displayed in the AutoCalculate Box.

	B2		=	82			
	A	B	C	D	E	F	G	H
1	Student	Exam 1	Exam 2	Final Exam	Overall Score			
2	Davis	82	77	94	84.3			
3	Graham	66	80	75	73.7			
4	Jones	95	100	97	97.3			
5	Meyers	47	62	78	62.3			
6	Richardson	80	58	73	70.3			
7	Thomas	74	81	85	80			
8	Williams	57	62	67	62			
9								
10	AVERAGE	71.6	74.3	81.3	75.7			
11								
12								
13								
14								
15								
Ready					Sum=2422.8			

Figure 2.10

Clearing a Block of Cells

Blocks of cells can be cleared, just as single cells are cleared. (Remember that *any information placed in the cells will be lost when the cells are cleared*.) To clear a block of cells, first select the block of cells and then press the Delete key. Alternatively, you can select the block of cells and then choose Clear/Formulas or Clear/All from the Edit menu.

Copying a Block of Cells

To copy a block of cells from one part of the worksheet to another, you must carry out the following three steps:

1. Select the block of cells to be copied.
2. Select Copy from the Edit menu. The border around the selected cells will then appear to move, indicating that the block is ready to be copied.
3. Move the mouse pointer to the upper left cell of the new location.
4. Press Enter or select Paste from the Edit menu. The original block of cells will then be copied to the new location. Note that the copied block of cells will overwrite whatever was originally in the new location.

You may also copy a block of cells from one location to another by dragging the block of cells from its original location to the new location. To do so, proceed as follows:

1. Select the block of cells to be copied.
2. Move the mouse pointer to any point on the border of the selected cells. (Note that shape of the pointer will change to an arrow.)
3. Hold down the Control key and the left mouse button, and drag the selected block to the new location.
4. Release the mouse button. The original block of cells will then be copied to the new location. Again, note that the copied block of cells will overwrite whatever was originally in the new location.

Moving a Block of Cells

Moving a block of cells from one location to another is similar to copying. Thus, to move a block of cells from one location to another, carry out the following three steps:

1. Select the block of cells to be moved.
2. Select Cut from the Edit menu. The border around the selected cells will then appear to move, indicating that the block is ready to be moved.
3. Move the mouse pointer to the upper left cell of the new location.

4. Press Enter or select Paste from the Edit menu. The original block of cells will then be moved to the new location, overwriting whatever was there originally. The original location will then be empty.

To move a block of cells by dragging, do the following:

1. Select the block of cells to be copied.

2. Move the mouse pointer to any point on the border of the selected cells. (Again, note that shape of the pointer will change to an arrow.)

3. Hold down the left mouse button and drag the selected block to the new location.

4. Release the mouse button. The original block of cells will then be copied to the new location, overwriting whatever was originally in that location.

Undoing Changes

If you change your mind after copying or moving a block of cells from one location to another, you may "undo" the change (i.e., restore the worksheet to its original state) by selecting Undo from the Edit menu or by clicking on the counterclockwise arrow (the Undo icon) in the Standard Toolbar. In fact, Undo can be used to negate many different editing changes. Hence, Undo will always refer to the last editing change (e.g., Undo Paste, Undo Delete, and so on).

Copying Formulas (Relative versus Absolute Cell Addresses)

Special care must be taken when copying formulas. Suppose, for example, that cell C1 contains the formula =A1+B1. This indicates that cell C1 will contain a value that is obtained by adding the values in the two cells to its left. Now suppose that the contents of cell C1 (the formula) is copied to cell C2. The formula will automatically change to =A2+B2, so that cell C2 will contain a value that is the sum of the values in the two cells to *its* left. A cell address that is written in this manner is called a *relative address* because it changes automatically when a formula containing the address is copied to another cell.

Now suppose the formula in cell C1 is written differently, as =$A$1+$B$1. When a cell address is preceded by dollar signs in this manner, it *does not* change when a formula containing the cell address is moved to another location. Thus, if the contents of cell C1 is copied to cell C2, the formula in cell C2 will be the same as the formula in cell C1; i.e., =$A$1+$B$1. A cell address that is preceded by dollar signs in this manner is called an *absolute address* because it remains unchanged when a formula containing the address is copied to another cell.

Cell formulas can include both relative and absolute addresses; e.g., =$A$1+B1. If such formulas are copied to another cell, the relative addresses will change automatically, but the absolute addresses will remain unchanged.

Thus, if the formula =$A$1+B1 is copied from cell C1 to cell E1, the formula in cell E1 will be =$A$1+D1. Such formulas containing mixed address types may be desirable in certain types of situations.

Moving Formulas

If a formula is *moved* rather than copied, all of the cell addresses within the formula remain unchanged. Thus, when moving a formula, it does not matter whether the cell addresses are written as relative addresses or absolute addresses.

On the other hand, if an *object* cell (i.e., a cell that is *referred to* in a formula) is moved, the formula will automatically change to accommodate that move, whether the cell reference is relative or absolute. For example, suppose that cell C1 contains the formula =A1+B1. Now suppose that the content of cell A1 is moved to cell B5. Then the formula in cell C1 will automatically change to =B5+B1. Moreover, if the original formula had been =$A$1+B1, the formula would become $B$5+B1. Thus, absolute addresses and relative addresses *both* change automatically if a cell *reference* changes within a formula.

Working with cell formulas is not as complicated as it may appear. Remember that relative cell addressing is *usually* (but not always) appropriate in most elementary applications. Be especially careful when copying formulas from one cell location to another.

Inserting and Deleting Rows and Columns

Sometimes a need arises to insert one or more rows or columns into an existing worksheet without destroying any of the information currently in the worksheet. To insert a single row into a worksheet, select any cell in the area where the new row is to be inserted. Then choose Rows from the Insert menu. The new row will then be inserted in the location of the selected cell. The row that initially occupied this location and all of the rows beneath it will be "pushed down" so that none of the information originally in the worksheet will be lost. Cell addresses appearing in formulas will automatically be adjusted for the new location of any rows that were moved.

A *block* of rows is inserted in the same way. First select several contiguous cells within a *column*, indicating where the new rows will appear. Then choose Rows from the Insert menu, causing the new rows to be inserted into the worksheet. The rows originally in this location and all of the rows beneath them will be "pushed down" under the insertion. Cell addresses appearing in formulas will automatically be adjusted for the new location of the rows that were moved.

Figures 2.11 and 2.12 illustrate the appearance of a worksheet before and after inserting two new rows. Notice the manner in which room has been made for the new rows, by moving the displaced rows (and the rows beneath them) below the insertion point.

B2	▼	=	82				
	A	B	C	D	E	F	G	H
1	Student	Exam 1	Exam 2	Final Exam	Overall Score			
2	Davis	82	77	94	84.3			
3	Graham	+ 66	80	75	73.7			
4	Jones	95	100	97	97.3			
5	Meyers	47	62	78	62.3			
6	Richardson	80	58	73	70.3			
7	Thomas	74	81	85	80			
8	Williams	57	62	67	62			
9								
10	AVERAGE	71.6	74.3	81.3	75.7			
11								
12								
13								
14								
15								
Ready			Sum=148					

Figure 2.11

B2	▼	=					
	A	B	C	D	E	F	G	H
1	Student	Exam 1	Exam 2	Final Exam	Overall Score			
2								
3		+						
4	Davis	82	77	94	84.3			
5	Graham	66	80	75	73.7			
6	Jones	95	100	97	97.3			
7	Meyers	47	62	78	62.3			
8	Richardson	80	58	73	70.3			
9	Thomas	74	81	85	80			
10	Williams	57	62	67	62			
11								
12	AVERAGE	71.6	74.3	81.3	75.7			
13								
14								
15								
Ready								

Figure 2.12

Column insertions are made in the same manner as row insertions. To insert a single column, select any cell in the area where the column is to be inserted. Then choose Columns from the Insert menu, causing the new column to be inserted in the location of the selected cell. The column that initially occupied this location and all of the columns to its right will be "pushed to the right," so that all of the information originally in the worksheet will be retained. All cell addresses appearing in formulas will automatically be adjusted, as required.

To insert a *block* of columns, the procedure is much the same as the insertion of a single column. First select several contiguous cells within a *row*, indicating where the new columns will appear. Then choose Columns from the Insert menu. The new columns will then be inserted into the worksheet. The columns originally occupying this location and all of the columns to their right will be "pushed to the right" of the insertion.

Rows and columns can be *deleted* from a worksheet in much the same manner as they are inserted. Thus, to delete a row, select any cell within the row and then choose Delete/Entire Row from the Edit menu. The worksheet will "close up" following the deletion. To delete a *block* of rows, select a group of adjacent cells within the unwanted rows and choose Delete/Entire Row from the Edit menu.

Column deletion works the same way. A single column is deleted by selecting any cell within the column and then choosing Delete/Entire Column from the Edit menu. Similarly, a *block* of columns is deleted by selecting a group of adjacent cells within the unwanted columns and then choosing Delete/Entire Column from the Edit menu.

Remember that the deletion of a row or column will result in a loss of information within the worksheet. If you inadvertently make an unwanted deletion, however, you can recover the deleted information (providing the undo is carried out *immediately* after the deletion) by selecting Undo Delete from the Edit menu or by clicking on the Undo icon in the Standard Toolbar.

Any remaining cell formulas that refer to a deleted row or a deleted column will result in an error message.

Inserting and Deleting Cells

Individual cells can also be inserted and deleted within a worksheet. To insert one or more adjacent cells, first select the location of the insertion on a cell-for-cell basis. Then choose Cells/Shift cells right or Cells/Shift cells down from the Insert menu. The cells displaced by the insertion will then be shifted to the right or down, as requested. Note that Shift cells right causes the insertion to occur within existing *rows*, whereas Shift cells down causes the insertion within existing *columns*.

Cell *deletions* work the same way. Simply select the block of cells to be deleted and choose Delete/Shift cells left or Delete/Shift cells up from the Edit menu. The first menu selection (Delete/Shift cells left) will cause the cells to the right of the deleted cells (in the same *rows*) to be shifted to the left, thus filling the "hole" formed by the deleted cells. Similarly, the second menu selection (Delete/Shift cells up) will result in the cells beneath the deleted cells (in the same *columns*) being shifted up, filling the vacancy left by the deleted cells.

You should realize that *deleting a cell is not the same as clearing a cell*. The distinction is important when the deleted cell is referenced in a formula that resides elsewhere. Suppose, for example, that cells A1 and B1 contain the numerical constants 10 and 20, and cell C1 contains the formula =A1+B1. The value 30 (10+20=30) will appear in cell C1. Now suppose cell A1 is *cleared*. Its numerical value will then become zero so that the value displayed in cell C1 will be 20 (0+20=20). If cell A1 is *deleted*, however, the formula in cell C1 will indicate an error because the reference to cell A1 will no longer be valid.

Remember that all remaining cell formulas respond to insertions or deletions of individual cells in the same manner as insertions or deletions of entire rows or columns. Thus, the formulas will automatically be adjusted for any existing cells that are displaced as a result of an insertion. An error message will result from any cell formula that refers to a deleted cell.

Adjusting Column Widths

You may improve the appearance of a worksheet by changing the width of one or more columns, particularly if a column contains very short numbers (e.g., two or three digits), very long numbers, or lengthy strings. If a cell is not wide enough to display a number, several *pound signs* (e.g., ###) will appear in place of the number. You can change the width of the columns individually, or you can change the width of a block of columns collectively.

The easiest way to change the width of a single column is by dragging. To do so, position the mouse pointer on the column heading (the row containing A, B, C, etc.) at the right edge of the desired column. The pointer will then change its shape to a cross. You may then change the width of the column by holding down the left mouse button and dragging the column edge in the desired direction.

The width of a column can also be changed via a menu selection and dialog box. To do so, select any cell within the column and then choose Column/Width from the Format menu. This will result in a dialog box requesting a numerical value for the column width. You may also choose Column/AutoFit Selection, which customizes the column width to the data item within the currently active cell. (AutoFit can also be activated by double-clicking on the right edge of a column heading.) Or you may choose Column/Standard Width from the Format menu. This will restore the column width to its original (default) size.

You can change the width of a *block* of columns by first selecting the appropriate cells within any row and then choosing Column/Width (or Column/Standard Width or Column/AutoFit Selection) from the Format menu. Then proceed in the same manner as when resizing a single column. Note, however, that the AutoFit Selection will size the columns separately, to accommodate each individual data item.

Formatting Data Items

You can also improve the appearance of a worksheet by *formatting* the data items within the individual cells. Formatting refers to the appearance of the numerical values, the alignment of the data items within the cells, the choice of fonts, etc. To format a data item, you must first select the data item and then select Cells from the Format menu. Then choose the appropriate tab from the resulting dialog box (Number, Alignment, Font, etc.) and respond to the items that are of interest.

Of particular interest are the different ways that numerical values can be represented. The various selections can be seen by choosing the Number tab from the Format Cells dialog box. This results in a display of several different categories of numbers, most of which have multiple format codes. A detailed discussion of all of these choices is beyond the scope of this chapter. For most technical applications, however, the Number category is the most frequently used. This feature allows you to control the number of decimal places that are displayed in a calculated value. (Generally, the total number of significant figures in a calculated result should not exceed the number of significant figures in the associated input data. This usually limits the number of decimal places.) Try formatting some numbers in a worksheet using several of these format codes.

Cell alignment is another useful formatting feature. Here we are concerned primarily with the horizontal and/or vertical alignment of data items within their cells. The cell alignment can be altered by first selecting one or more cells and then choosing the Alignment tab from the Format Cells dialog box. The desired horizontal and vertical alignment features can then be selected from the choices that are displayed. The orientation of the text can also be selected from this dialog box. (Note that the horizontal cell alignment can also be controlled from the Formatting Toolbar.)

Excel permits other types of formatting, including the choice of several *fonts* and *font sizes*, the placement of *borders* around various cells, and the use of *cell patterns*. These features can be accessed by selecting Cells from the Format menu. Some can also be accessed from the icons on the Formatting Toolbar. The details are straightforward and can be determined by simple experimentation. More information can be obtained by accessing Excel's on-line help.

Editing Shortcuts

Many of the features in the Edit and Format menus can be accessed more rapidly by selecting one or more cells and then pressing the right mouse button on IBM-compatible computers, or by pressing the Command+Options keys and then pressing the mouse button on a Macintosh. Thus, you can access Edit features such as Cut, Copy, Paste, Insert, Delete, and Clear Contents. You can also access the Paste Special feature, which allows you to paste selectively (e.g., formulas only, values only, etc.). The multitab Format Cells dialog box can also be obtained in this manner.

Example 2.5 Editing a Worksheet

Enhance the appearance of the worksheet shown in Fig. 2.6, which was originally developed in Example 2.1.

The edited worksheet is shown in Fig. 2.13. Notice that a title and several blank rows have been added, and columns B and C have been widened (by selecting cells within the columns and then choosing Column/Width from the Format menu). In addition, some of the fonts have been changed, the data items have been centered within their cells, a border has been placed around the original block of cells, horizontal lines have been added beneath the headings and above the total, and the numbers are displayed differently (note the commas). All of these changes were made from the Format Cells dialog box, obtained by selecting Cells from the Format menu. Finally, note that the grid lines and row/column headings have been removed (by selecting Options/View from the Tools menu).

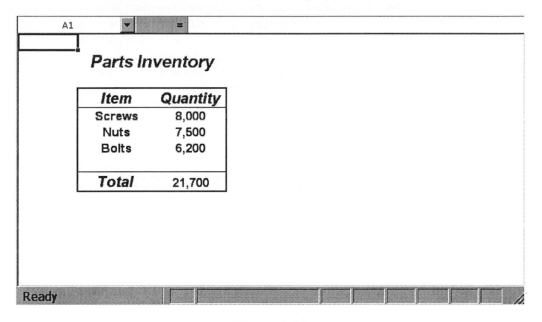

Figure 2.13

2.8 SAVING AND RETRIEVING THE WORKSHEET

There are several ways to save a worksheet in Excel. The most common is to click on the Save icon in the Standard Toolbar (see Fig. 2.14) or to select Save from the File menu. Either of these methods will cause the current version of the worksheet to be saved under its current file name, thus replacing any earlier version saved under the same name. If the worksheet has not been named previously, a dialog box will appear, prompting you for a file name.

You may also save a worksheet under a *different* name by selecting **Save As** from the **File** menu. A dialog box will appear whenever this option is chosen, prompting you for a file name.

It is a good idea to save your work frequently when building or editing a worksheet. This will minimize the damage caused by an unexpected power failure or a careless reply to a **Save** prompt (e.g., answering **No** rather than **Yes**).

If a worksheet has been changed in any manner since the last save and you attempt to exit from Excel, you will first see a dialog box asking you whether or not to save the current version of the worksheet before exiting. This feature is intended to prevent you from leaving Excel without saving your most recent work, thereby losing whatever changes you may have made since the last save.

To *retrieve* a worksheet that has been saved previously, either click on the Open icon in the Standard Toolbar (see Fig. 2.14), or select **Open** from the **File** menu. In either case a dialog box will appear, prompting you for a file name within the current folder (directory). You may also change folders if you wish.

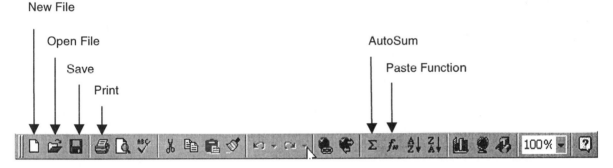

Figure 2.14

2.9 PRINTING THE WORKSHEET

Once you have built and edited a worksheet, you will want to print it out on paper. In Excel you can print either the entire workbook, selected worksheets within the workbook, or a selected portion of one worksheet. To print an *entire* worksheet, select **Print** from the **File** menu. Then choose **Active sheet(s)** from the resulting dialog box. (Note that there are several options available in this dialog box, most of which are self-explanatory.) The entire worksheet will then be printed out onto one or more pages, as required. (This assumes, of course, that Excel has been set up to recognize your particular printer, that the printer has been turned on and loaded with paper, etc.) You can also print the worksheet by simply clicking on the **Print** icon in the Standard Toolbar (see Fig. 2.14).

To print a *part* of a worksheet, you must first select the block of cells to be printed, choose **Print** from the **File** menu, and then choose **Selection** from the resulting dialog box. The selected block of cells will then be printed.

You can also "preview" the appearance of the printed sheet on your screen by selecting Print Preview from the File menu. Sometimes it is a good idea to do this before actually printing, to verify that you like the appearance of the fonts, column widths, page breaks, and so on. Print Preview includes some very useful features, such as the ability to print your worksheet either vertically (*portrait*) or horizontally (*landscape*) or to fit an entire worksheet on a single sheet of paper. To use these features, click on the Setup button in Print Preview, then select the desired features from the dialog box beneath the Page tab.

Exercises

2.10 Recreate the worksheet containing student exam scores as shown in Fig. 2.9. Then enhance its appearance by editing it in the following ways:

(*a*) Add some empty space at the top of the worksheet. Place a worksheet title (e.g., **Engineering Analysis 100 Semester Grade Report**) at some convenient location within this space. Try a single-line title and a two-line title, and use the one you like the best.

(*b*) Format the numerical values so that they all show two decimal places.

(*c*) Adjust the column widths as required to make the worksheet more readable.

(*d*) Align the numerical values and their headings so that they are centered horizontally within their respective cells.

When you are finished editing, print the edited version of the worksheet. Fill the printed page with the worksheet, using a landscape layout.

2.11 In Exercise 2.7 you were asked to prepare a worksheet containing a student's first-semester grade report including a calculation of the student's grade-point average (GPA) for that semester, using the formula given in Exercise 2.3. This problem requires you to expand that worksheet by adding a second-semester grade report, thus showing a full academic year (two semesters). The second-semester grade report is shown below:

Semester 2

Course	Credits	Grade
Chemistry II	4	3.7
Physics II	3	3.0
Economics	3	3.3
Engineering Analysis	3	4.0
Calculus II	3	3.3
Seminar	1	3.0

To expand the worksheet, proceed as follows

(*a*) Add two blank lines between the social security number and the heading line (**Course Credits Grade**). Then add the title **Semester 1** within this space.

(*b*) To the right of the first-semester grade report, add the second-semester grade report. Remember to include a grade point average for the second semester beneath the grade report, as described in Exercise 2.7. (You can enter this information easily by copying the first-semester grade report and then editing as required.)

(*c*) Add a calculation of the overall (first-year) grade point average at the bottom of the worksheet.

(*d*) Enhance the appearance of the spreadsheet by adding blank rows and columns, adjusting column widths, formatting numerical values, underlining headings, and so on.

(*e*) Try changing the text by selecting different typefaces, selecting larger fonts for the headings, and using boldface fonts selectively.

When you are finished adding information, print the new version of the worksheet. Fill the printed page with the worksheet, using a landscape layout.

2.12 Rearrange the worksheet created in Exercise 2.11 in the following ways:

(*a*) Place the second-semester grade report beneath rather than to the right of the first-semester grade report.

(*b*) Place the grade point average for each semester to the right of the individual course listings and grades.

(*c*) Place the overall (first-year) grade point average beneath the semester grade point averages.

(*d*) Adjust the overall appearance of the worksheet by adding or removing blank rows and columns, readjusting column widths, repositioning the headings, and so on.

When you are finished, print the rearranged version of the worksheet. Use a portrait layout to fill the printed page with the worksheet.

2.10 DISPLAYING CELL FORMULAS

When a formula is being entered into a cell, it is displayed both within the cell and within the Formula Bar. Once the formula has been completed and the Enter key has been pressed, however, the formula is no longer visible within the cell. Instead, the *value* generated by the formula is displayed within the cell (though the formula may still be seen within the Formula Bar of the *currently active* cell).

There are times when it may be desirable to see the actual cell formulas within their respective cells, rather than the values generated by the formulas. If you hand in a homework assignment, for example, your instructor may want to verify that you have been using cell formulas rather than typing in numbers. Or you may wish to examine the cell formulas within a worksheet yourself, simply to verify that they are correct.

To view the cell formulas within their respective cells, choose Options from the Tools menu. When the Options dialog box appears, select the View tab and then select Formulas under the Window options heading. (This will cause the cells to increase in width somewhat.) You can remove the formula display at a later time by deactivating the Formulas selection from the Options dialog box.

Figure 2.15 shows a simple worksheet in which the cell formulas are displayed (notice cell C7). This is the same worksheet that was originally shown in Fig. 2.6. Note that the cell widths are now wider, to accommodate the cell formulas.

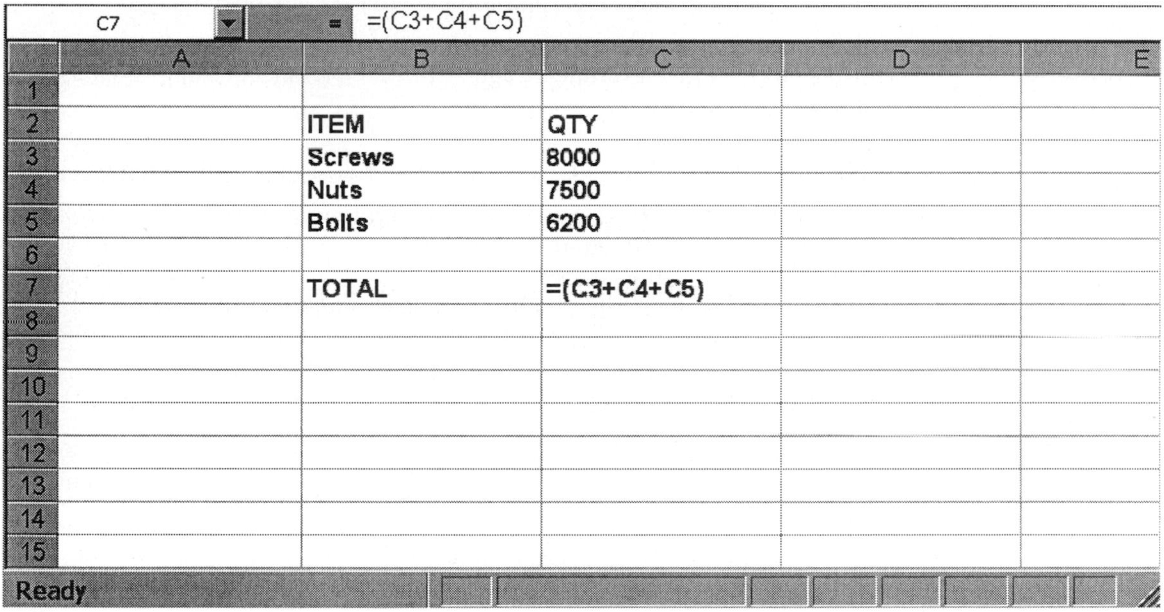

C7	▼	=	=(C3+C4+C5)		
	A	B	C	D	E
1					
2		ITEM	QTY		
3		Screws	8000		
4		Nuts	7500		
5		Bolts	6200		
6					
7		TOTAL	=(C3+C4+C5)		
8					
9					
10					
11					
12					
13					
14					
15					

Ready

Figure 2.15

Exercise

2.13 Display the cell formulas in the student exam worksheet created in Exercise 2.9. Then print the worksheet showing the cell formulas.

2.11 CREATING AND RUNNING MACROS

A *macro* is a series of consecutive keystokes and/or mouse actions that is recorded for later playback. Each macro is stored under its own name and has a *shortcut key* (such as Ctrl-P, Ctrl-X, etc.) associated with it. Once a macro has been created, its instructions can be executed automatically simply by pressing its shortcut key. Thus, a macro allows you to automate a series of instructions within an Excel spreadsheet. Generally speaking, the more complicated the series of instructions, the more useful the macro.

To record a macro, select Macro/Record New Macro from the Tools menu. A dialog box will then appear, as shown in Fig. 2.16 (see Example 2.6 below). You must then provide the name for the macro (do not include blank spaces), a shortcut key (a single letter, either upper- or lowercase), a storage designation (typically the current workbook), and a description of the macro. Then press OK. From this point on, any instructions (i.e., keystrokes or mouse movements) that are *completed* (i.e., that are not cancelled) will be recorded within the macro. The Status Bar will display the message Recording during this time (see Fig. 2.17 below). Once all of the instructions have been completed, the recording is terminated by clicking on the Stop button within the pop-up Stop Recording menu or by selecting Macro/Stop Recording from the Tools Menu. The macro will then be saved in the designated manner.

Once a macro has been saved, its instructions can be executed simply by pressing the shortcut key (or by selecting Macro/Macros from the Tools menu, selecting a macro from the list shown in the resulting dialog box, and then selecting Run). The instructions will then be executed within the cells specifically designated by the macro. It is also possible to execute the instructions beginning from whatever cell is currently active; to do so, the macro must be recorded using *relative references* (i.e., by activating the Relative Reference button in the Stop Recording menu while the macro is being recorded).

The instructions that comprise the macro can be viewed by selecting Macro/Macros from the Tools menu and double-clicking on the macro name shown in the dialog box or by selecting the macro and then selecting Edit. These instructions are written in a special language developed by Microsoft, called *Visual Basic*. This language uses a command structure based upon classical BASIC, though Visual Basic includes many enhancements designed specifically for the Microsoft Windows environment. A detailed discussion of Visual Basic is, unfortunately, beyond the scope of this text.

Example 2.6 Creating a Macro

In Example 2.3 we built a worksheet to record and average a set of student exam grades (see Fig. 2.9). Let us now create and execute a macro that will compare each student's overall score with the class average. In other words, let us determine the difference between each student's overall score (shown in column

E) and the class average (cell E10), to determine the number of points by which the student is above or below the class average.

The process is initiated by selecting Macro/Record New Macro from the Tools menu. This results in the dialog box shown in Fig. 2.16. Within this dialog box, we see that the macro will be named Class_Standing, its shortcut key will be Ctrl-s, and it will be stored within the current worksheet. A description of the macro is also provided at the bottom of the dialog box.

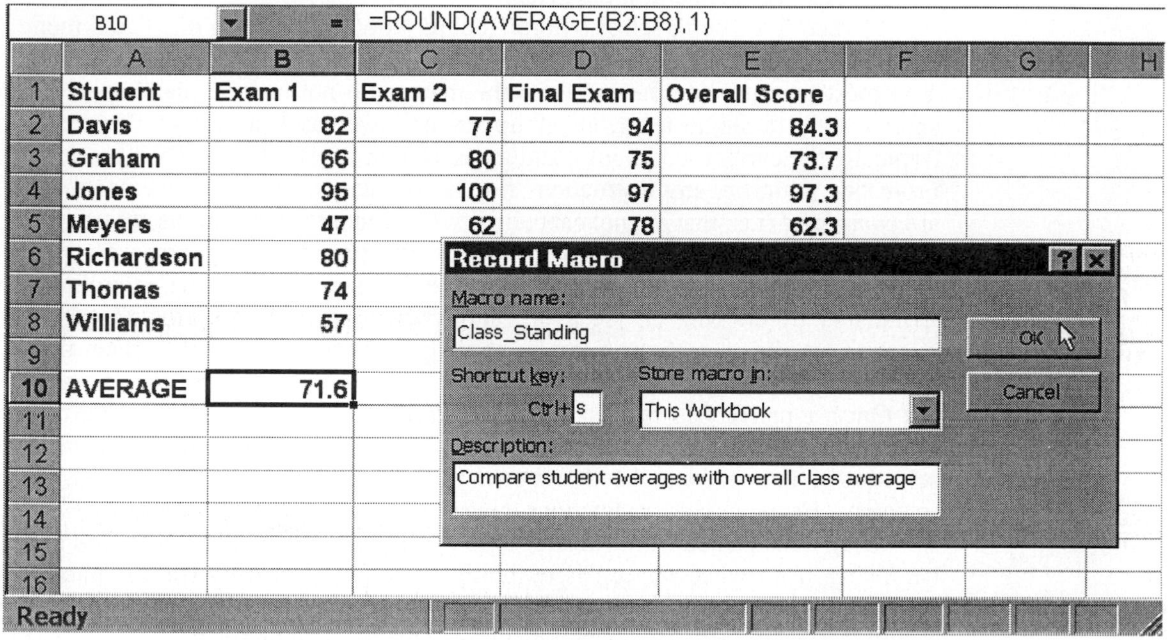

B10		=	=ROUND(AVERAGE(B2:B8),1)				
	A	B	C	D	E	F	G	H
1	Student	Exam 1	Exam 2	Final Exam	Overall Score			
2	Davis	82	77	94	84.3			
3	Graham	66	80	75	73.7			
4	Jones	95	100	97	97.3			
5	Meyers	47	62	78	62.3			
6	Richardson	80						
7	Thomas	74						
8	Williams	57						
9								
10	AVERAGE	71.6						

Record Macro

Macro name: Class_Standing

Shortcut key: Ctrl+s   Store macro in: This Workbook

Description: Compare student averages with overall class average

Ready

Figure 2.16

Once we press OK, the recording process begins and the message Recording appears within the Status Bar, as shown in Fig. 2.17. At this point, we will carry out the following actions:

1. Enter the title Comparison in cell F1.
2. Enter the formula =E2-E$10 in cell F2.
3. Copy this formula into cells F3 through F8.
4. Widen column F.
5. Select cells F1 through F8 and apply a bold font.
6. Adjust the size of the worksheet, eliminating some excess empty space.

When all of these actions have been completed, we select Macro/Stop Recording from the Tools menu. The macro will then be saved with the worksheet, under the name Class_Standing.

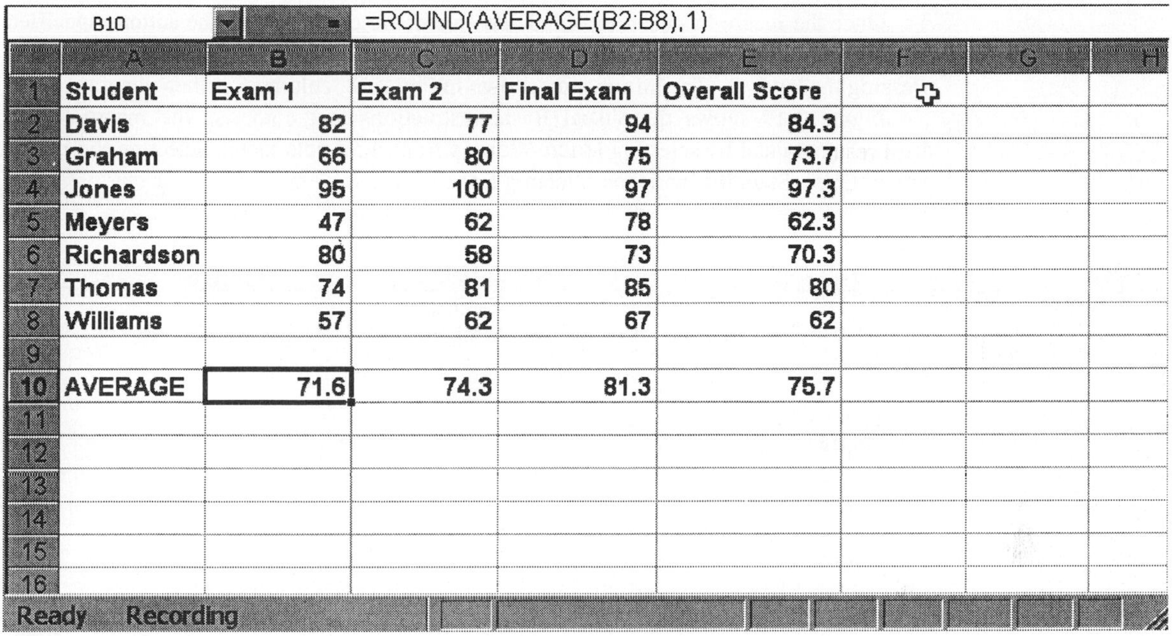

B10   =ROUND(AVERAGE(B2:B8),1)

	A	B	C	D	E	F	G	H
1	Student	Exam 1	Exam 2	Final Exam	Overall Score	✛		
2	Davis	82	77	94	84.3			
3	Graham	66	80	75	73.7			
4	Jones	95	100	97	97.3			
5	Meyers	47	62	78	62.3			
6	Richardson	80	58	73	70.3			
7	Thomas	74	81	85	80			
8	Williams	57	62	67	62			
9								
10	AVERAGE	71.6	74.3	81.3	75.7			
11								
12								
13								
14								
15								
16								

Ready   Recording

Figure 2.17

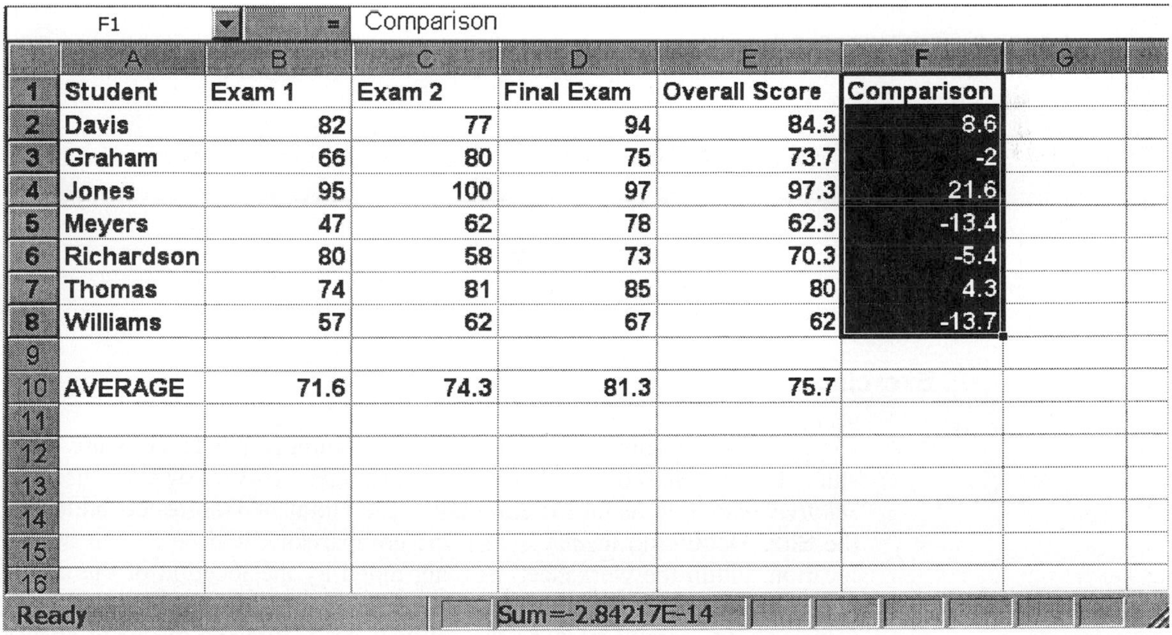

F1   Comparison

	A	B	C	D	E	F	G
1	Student	Exam 1	Exam 2	Final Exam	Overall Score	Comparison	
2	Davis	82	77	94	84.3	8.6	
3	Graham	66	80	75	73.7	-2	
4	Jones	95	100	97	97.3	21.6	
5	Meyers	47	62	78	62.3	-13.4	
6	Richardson	80	58	73	70.3	-5.4	
7	Thomas	74	81	85	80	4.3	
8	Williams	57	62	67	62	-13.7	
9							
10	AVERAGE	71.6	74.3	81.3	75.7		
11							
12							
13							
14							
15							
16							

Ready   Sum=-2.84217E-14

Figure 2.18

Once the macro has been recorded and stored, we can carry out the actions specified by the macro simply by pressing Ctrl-s (i.e., by holding down the Ctrl key and then pressing the letter s). The information shown in Fig. 2.18, column F is then generated.

Figure 2.19 shows the Visual Basic instructions that comprise the macro. This display is obtained by selecting Macro/Macros from the Tools menu, selecting the macro named Class_Standing, and then selecting Edit.

```
Ex2-6.xls - Module2 (Code)
(General)                                    Class_Standing

Sub Class_Standing()
'
' Class_Standing Macro
' Compare student averages with overall class average
'
' Keyboard Shortcut: Ctrl+s
'
    Range("F1").Select
    ActiveCell.FormulaR1C1 = "Comparison"
    Range("F2").Select
    ActiveCell.FormulaR1C1 = "=RC[-1]-R10C[-1]"
    Selection.AutoFill Destination:=Range("F2:F8"), Type:=xlFillDefault
    Range("F2:F8").Select
    Columns("F:F").ColumnWidth = 10.89
    Range("F1:F8").Select
    Selection.Font.Bold = True
    Application.Left = 106.6
    Application.Top = 81.4
End Sub
```

Figure 2.19

Exercises

2.14 Create a macro similar to that shown in Example 2.6 to compare each student's overall score with the class average. Now, however, use the *relative reference* feature (i.e., activate the Relative Reference button in the Stop Recording menu) so that the comparisons will be shown in *any* location within the worksheet, as determined by the location of the active cell when the macro is initiated. Save the macro under the name Comparisons, using Ctrl-A as the shortcut key. Run the macro to ensure that it executes correctly.

2.15 For the worksheet described in Example 2.1, create a macro that will transform the appearance of the worksheet from that shown in Fig. 2.6 to that shown in Fig. 2.13. Save the macro with the worksheet, using the name Fancy. Use Ctrl-f as a shortcut key. Run the macro, and compare the resulting output with that shown in Fig. 2.13.

2.12 CLOSING REMARKS

In this chapter we have focused only on some of the most frequently used features of Excel, in preparation for the various engineering applications to be discussed in the following chapters. Excel includes many other features, most of which are simple to use but beyond the scope of our present discussion. You are encouraged to learn more about Excel from the excellent on-line help and tutorials included with Excel and the many detailed Excel texts that are currently on the market.

ADDITIONAL READING

Dodge, M., C. Kinata, and C. Stinson. *Running Microsoft Excel 97.* Microsoft Press, 1997.

Jacobson, R. *Microsoft Excel 97/Visual Basic Step by Step.* Microsoft Press, 1997.

Microsoft Excel 97 Worksheet Function Reference. Microsoft Corp., 1997.

Microsoft Office 97/Visual Basic Programmer's Guide. Microsoft Corp., 1997.

CHAPTER 3

GRAPHING DATA

Graphing data is one of the most common tasks carried out with a spreadsheet program. Excel includes the capability to generate a wide variety of graphs. (In Excel, graphs are referred to as *charts*.) All graphs are relatively easy to create in Excel, though some types of graphs are more easily generated than others. Our interest will be directed primarily toward the use of *bar graphs* (called *Column charts* in Excel) and *x-y graphs* (called *Scatter charts* or *XY charts* in Excel), as these are the types of graphs that are most commonly used in engineering applications.

Engineers generally refer to the horizontal axis (the *x*-axis) as the *abscissa* and the vertical axix (the *y*-axis) as the *ordinate*. In Excel, however, the horizontal axis is called either the *Category Axis* or the *Value Axis*, depending on the chart type. (*Category Axis* is used in most charts, though *Value Axis* is used for XY charts.) The vertical axis is always called the *Value Axis*.

3.1 CREATING A GRAPH IN EXCEL

The easiest way to generate a graph in Excel is to use the so-called *Chart Wizard*, which is found in the Standard Toolbar (see Fig. 3.1). Or, you can select Chart from the Insert menu.

In either case, the general procedure is as follows:

1. Select a block of contiguous cells containing the data to be plotted (the *source data*). The selection may include column headings or row labels.

Chart Wizard

Figure 3.1

2. Click on the *Chart Wizard* icon within the Standard Toolbar or select Chart from the Insert menu.

3. Select a chart type from either the Standard Types dialog box or the Custom Types dialog box. If one of the standard types has been selected (which will generally be the case), choose a chart subtype from the selections appearing to the right. Then press either Next or Finish. (Pressing Next allows you to specify additional information about the final appearance of the chart.) You can also press Cancel if you wish to start over again.

4. If you pressed Next in step 3, a second dialog box will appear, allowing you to specify information about the source data selection. Then press either Next of Finish. (You can also press Back, to change something entered earlier, or Cancel, to terminate the entire process.)

5. If you pressed Next in step 4, a third dialog box will appear, allowing you to specify additional information, such as a title, labels along the axes, gridlines, a data legend, data labels (i.e., labels along the axes), and so on. Then press either Next or Finish.

6. If you pressed Next in step 5, a fourth and final dialog box will appear, asking you whether to display the graph within an existing worksheet or a separate worksheet. (The worshkeet containing the data is the default.) Press Finish once this information has been supplied.

When these steps have been completed, the desired graph will appear somewhere within the indicated worksheet. The graph can then be moved by dragging it by one edge, or its size or shape can be altered by dragging one of the eight small squares located along its edges.

The Chart Toolbar shown in Fig. 3.2 will also appear somewhere within the worksheet. This toolbar allows you to select and format various parts of a graph, change the type of graph, or change the appearance of the graph. You can relocate this toolbar by dragging it to a new location, and you can change its shape by pulling one of its edges either toward or away from the center.

Figure 3.2

When you generate a graph, remember that the data selected must be appropriate to the type of graph desired. Thus, a single set of data is required for a bar graph, whereas a set of *x*-values and a corresponding set of *y* values are required for an *x-y* graph. The data should be entered in adjacent cells, in either a column or a row format. The data selection can include adjacent labels (e.g., column headings or row labels) but not empty cells.

Once an embedded graph has been created within a worksheet, it can easily be moved to its own worksheet by clicking on the graph and then selecting Location/As new sheet from the Chart menu. Or the following procedure can be used (providing more control over the size and location of the graph):

1. Click on the graph so that it becomes the active object within the worksheet. Then choose Cut from the Edit menu.

2. Select another worksheet from the tabs displayed at the bottom of the screen. (The Sheet Tabs feature found in the View tab under Tools/Options must be enabled so that the tabs are visible, as in Fig. 2.1.)

3. Select a cell whose location will define the upper left corner of the chart in the new worksheet. Then choose Paste from the Edit menu.

4. If the graph is to be *copied* rather than *moved* from the original worksheet, select Edit/Copy rather than Edit/Cut in Step 1.

3.2 BAR GRAPHS (EXCEL *COLUMN CHARTS*)

A *bar graph* consists of a series of vertical rectangles (bars) that represent *single-valued* data. For example, the bar graph shown in Fig. 3.3 represents the inventory level of several different parts within a small machine shop. Each vertical bar represents the inventory level for a particular item.

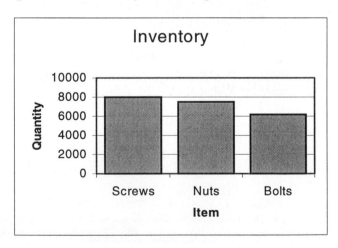

Figure 3.3

Example 3.1 Creating a Bar Graph in Excel

Modify the worksheet developed in Example 2.1 by embedding a bar graph within the worksheet (see Fig. 2.4). Include the individual part quantities, but not the total, within the bar graph.

Figure 3.4 shows a highlighted portion of the worksheet (cells B2 through C5) that will be used to create the graph. The mouse pointer is pointing to the *Chart Wizard* icon in the Standard Toolbar.

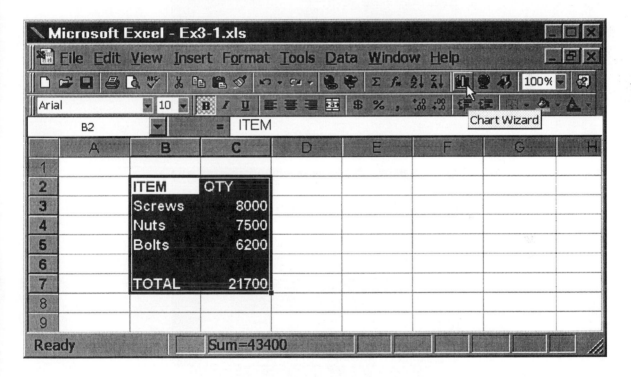

Figure 3.4

Once the *Chart Wizard* icon has been selected the first of four dialog boxes appears, as shown in Fig. 3.5. This is where the type of graph is selected. In Fig. 3.5, a *Column chart* is shown highlighted. This is the graph type that will be used in this example.

Figure 3.6 shows the second dialog box, where the corresponding data range is specified. Note that a sample bar chart is included as a part of this dialog box.

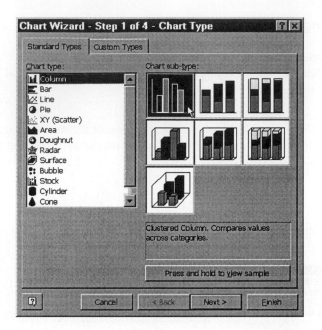

Figure 3.5

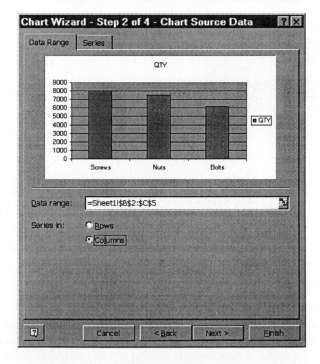

Figure 3.6

The third dialog box is shown in Fig. 3.7. It allows us to specify additional information, such as a title, labeling of the axes, appearance of gridlines, and so on.

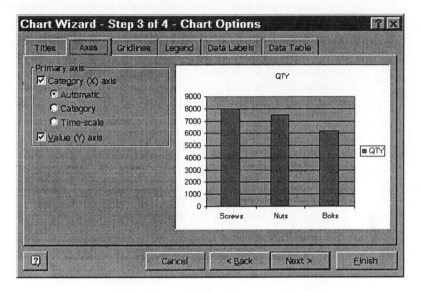

Figure 3.7

The fourth dialog box, shown in fig. 3.8, allows us to specify whether the final graph will appear in the original worksheet, alongside the given data, or in a separate worksheet. In this example, the graph will be placed within the original worksheet.

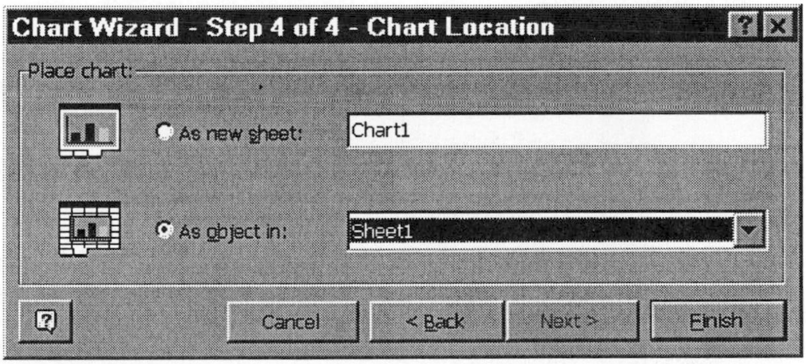

Figure 3.8

The resulting worksheet containing the bar chart is shown in Fig. 3.9. This worksheet has been rearranged somewhat, by dragging the original chart to a more convenient location and by expanding the size of the chart. (A chart can be relocated by clicking on it, then dragging it to the desired location. To resize a chart, click on it and then drag one of the eight peripheral *resizing boxes* in the desired direction.)

Notice the numerical scale that is automatically placed along the *y*-axis. Also note that the labels in column B are automatically copied to the bar graph, beneath their corresponding bars. These labels would not appear if the selection had been based solely on the numerical data in cells C3 through C5.

The legend QTY, shown to the right of the actual graph, and the title QTY, centered above the graph, were taken from cell C2 and automatically placed within the chart area. These items could have been edited (i.e., altered or deleted) while the chart was being created, in the third dialog box (see Fig. 3.7). However, they can also be edited after the graph has been completed, as explained below.

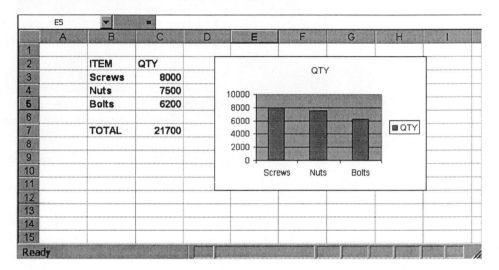

Figure 3.9

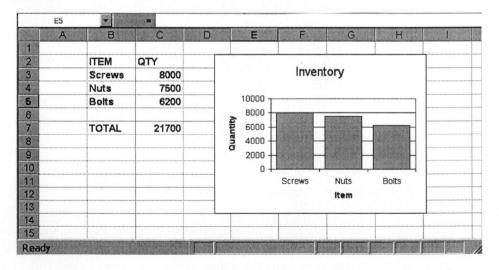

Figure 3.10

Figure 3.10 shows the same worksheet after the graph has been edited. In particular, note that the legend has been deleted, the chart title has been changed to Inventory, titles (Item and Quantity) have been added to the *x*- and *y*- axes, the colors (shown as different shades of gray) of the background and the vertical bars have been changed, and the bars have been widened.

To alter the graph in this manner, we first click on the legend and remove it by pressing the Delete key. We then change the title by clicking on it, typing the new title in the resulting editing box, and selecting a new font size from the Formatting Toolbar. The titles along the axes were obtained by clicking on the chart area to make it active, selecting Chart Options/Titles from the Chart menu, and then responding to the resulting dialog box. (The chart title could also have been changed in this manner. Note, however, that the Chart menu appears in the Menu Bar only when the graph is the active object within the worksheet.)

The background color was changed by clicking on the background area and selecting Selected Plot Area from the Format menu. Similarly, the color of the vertical bars was changed by clicking on one of the bars and selecting Selected Data Series/Patterns from the Format menu. Finally, we widened the vertical bars by clicking on one of the bars and then selecting Selected Data Series/Options from the Format menu. Note that we could have completely removed the space between the bars if we had wished.

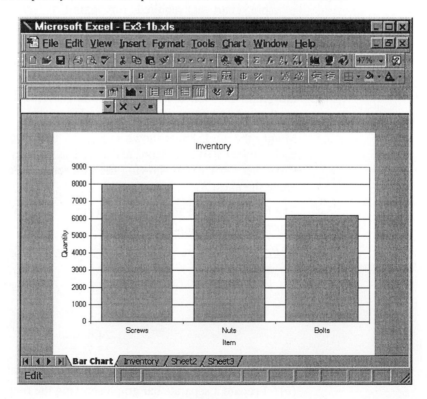

Figure 3.11

Figure 3.11 shows the same graph that appeared in Fig. 3.10, but the graph is now located within a separate sheet. To move the graph from its original location we first clicked on it to make it an active object, then selected Location/As new sheet from the Chart menu. The new sheet name (Inventory) was then entered into the dialog box. Note that the Chart Toolbar now appears beneath the Format Toolbar (compare with Fig. 3.4).

Notice also that the the original worksheet has been renamed Inventory. This was accomplished by choosing Sheet/Rename from the Format menu, and then retyping the new name within the tab.

Exercises

3.1 Reconstruct the worksheet shown in Fig. 3.9 on your own computer. Then modify the embedded bar graph in the following ways:

(*a*) Edit the graph by adding the chart title *Parts in Stock*. In addition, add the title *Number* along the *y*-axis and the title *Part* along the *x*-axis. Then print the entire worksheet.

(*b*) Transform the bar graph into a *Pie chart*, showing the distribution of screws, nuts, and bolts within the entire inventory of parts. Edit the pie chart as required for legibility. Then print the worksheet containing the pie chart.

3.2 Reconstruct the worksheet shown in Fig. 2.7 containing student exam scores. Construct a bar graph of the overall scores for the individual students. Do not include the overall class average in the graph. Construct the graph in such a manner that the individual vertical bars touch one another. Add a chart title and a title along each of the axes.

3.3 Modify the worskheet shown in Fig. 2.7 to include the difference between each student's overall score and the class average, as shown in Fig. 2.16. Construct a bar graph showing the differences. Add a chart title and titles along the axes.

3.4 The following data show the repair times for two machines that have been experiencing frequent breakdowns:

Breakdown No.	Machine 1 (min)	Machine 2 (min)
1	12.1	22.5
2	27.8	15.1
3	18.5	8.2
4	6.5	11.9
5	24.6	7.7
6	33.7	19.4

(*a*) Prepare a bar graph showing the repair times for machine 1.

(*b*) Prepare a separate bar graph showing the repair times for machine 2.

(*c*) Prepare a single bar graph showing the repair times for both machines.

3.3 X-Y GRAPHS (EXCEL *XY CHARTS*, OR *SCATTER CHARTS*)

An *x-y graph* is created by passing a line or a curve through a series of *paired data points*. Each paired data point consists of an *x*-value and a *y*-value; hence, the name *x-y* graph. Most engineering and scientific data are displayed in the form of *x-y* graphs.

Note that Excel supports another type of graph, called a *Line chart,* which looks like an *x-y* graph. However, a line chart is *not* a graph of paired data points but is a graph of *single-valued* data points that are *uniformly spaced* along the *x*-axis, regardless of the *x*-values. Graphs of this type are not widely used in engineering. Hence, we will not make use of Excel Line charts in this book. *Do not confuse x-y graphs (called XY charts or Scatter charts in Excel) with Excel Line charts.*

Arithmetic (Cartesian) Coordinates

The most common type of *x-y* graph is one in which the *x*-axis is subdivided into a series of equally spaced intervals, and the *y*-axis is subdivided into another set of equally spaced intervals. When the axes are subdivided in this manner, we refer to the graph as having *arithmetic coordinates* (also called *cartesian coordinates*).

Figure 3.12 shows an x-y graph with arithmetic coordinates. The graph is a plot of voltage as a function of time, based upon the formula

$$V = 10e^{-0.5t}$$

where *V* represents voltage and *t* represents time, in seconds.

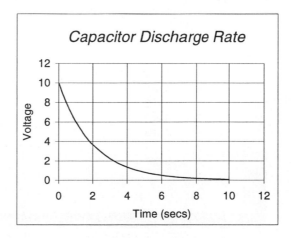

Figure 3.12

The easiest way to create an arithmetic *x-y* graph in Excel is to enter the *x*- and *y*-values in adjacent rows or columns, with the *x*-values in the top row or the leftmost column. Then select a single block of cells, containing both the *x*- and *y*-values. You can also enter the data in nonadjacent rows or columns, selecting the *x*-data in the usual manner and then holding down the Control key (Command key) while selecting the *y*-data.

After selecting the data, click on the *Chart Wizard* icon or select Chart from the Insert menu, as before. Now choose an *XY (Scatter) chart* from the first dialog box (*do not* select a *Line chart!*), and select the desired type of *x-y* graph from the available subtypes. The selections allow you to display or hide the individual data points, pass line segments or a curve through the data points, etc.

Once the chart type has been selected, finish the graph by providing the information requested in the remaining dialog boxes, as described earlier. (Note that both axes are called *Value Axes* in an XY chart.)

Example 3.2 Creating an X-Y Graph in Excel (Arithmetic Coordinates)

The voltage within a capacitor varies with time in accordance with the formula

$$V = 10e^{-0.5t}$$

where *V* represents voltage and *t* represents time, in seconds. Prepare an *x-y* graph of the voltage as the time varies from 0 to 10 seconds. Display the data to three-decimal precision using arithmetic coordinates. Label the graph so that it is legible and attractive.

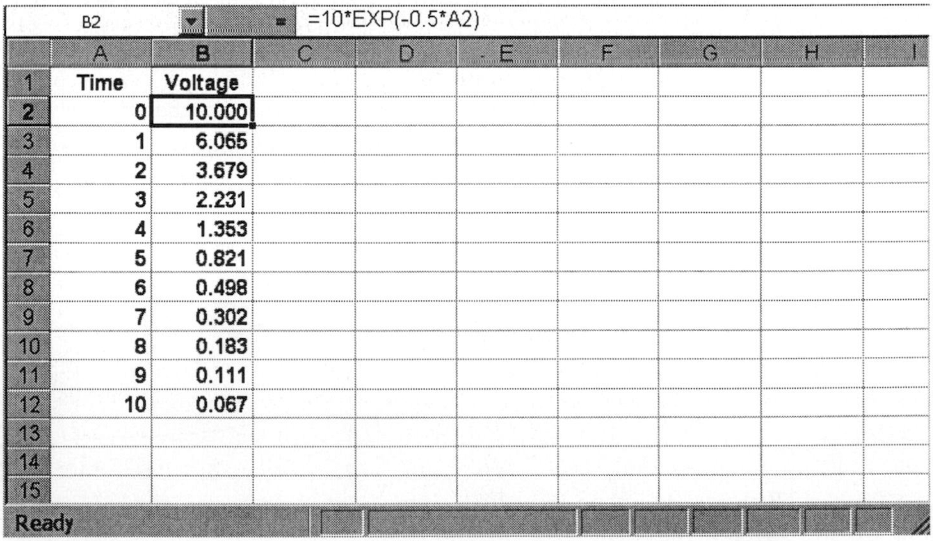

	A	B	C	D	E	F	G	H	I
	Time	Voltage							
2	0	10.000							
3	1	6.065							
4	2	3.679							
5	3	2.231							
6	4	1.353							
7	5	0.821							
8	6	0.498							
9	7	0.302							
10	8	0.183							
11	9	0.111							
12	10	0.067							

B2 =10*EXP(-0.5*A2)

Ready

Figure 3.13

Figure 3.13 shows a worksheet containing several values of *V* vs *t* spanning the 10-second time interval. The independent variable (time) is tabulated in the first column (column A), and the dependent variable (voltage) is tabulated in the second column. The dependent variables are generated by a formula, as seen in the formula bar for cell B2.

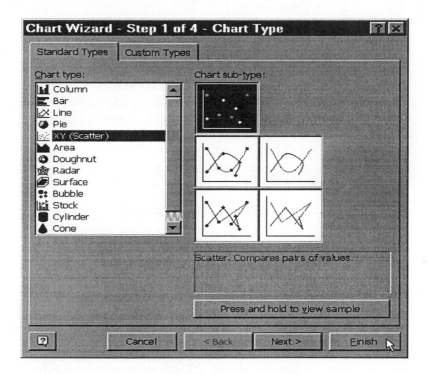

Figure 3.14

To create an *x-y* graph, we first select the data in columns A and B (cells A1 through B12). We then click on the *Chart Wizard* icon in the Standard Toolbar, resulting in the dialog box shown in Fig. 3.14. Notice that an XY (Scatter) chart has been selected from the left column, and the first sub-type, showing the individual data points without any interconnection, was selected from the right display area. We then select Finish, resulting in the graph shown within Fig. 3.15.

The graph was then edited to improve its appearance and legibility. The editing was initiated by clicking on the graph, causing it to be activated for editing. Figure 3.16 shows the worksheet after the editing was completed.

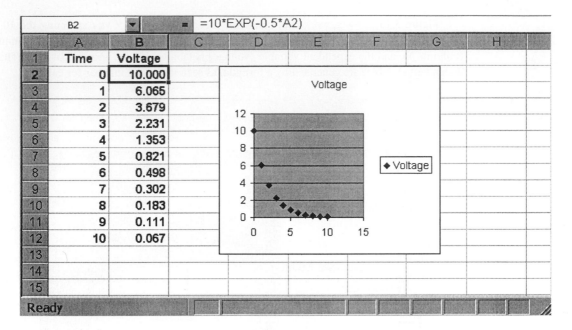

Figure 3.15

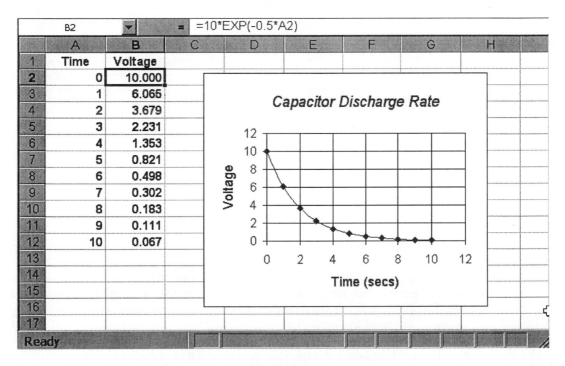

Figure 3.16

The following changes were made to the graph:

1. The graph was relocated by dragging it to a new location within the worksheet and was then enlarged by dragging its edges in the desired directions.

2. The chart subtype was changed by clicking on the chart, selecting Chart Type from the Chart menu, and then choosing a different subtype from the available selections.

3. The data legend was removed by clicking on it and then pressing the Delete key.

4. The chart title was changed from Voltage to Capacitor Discharge Rate. This was accomplished by selecting Chart Options/Titles from the Chart menu and then entering the new title into the Chart title area near the top of the dialog box.

5. The new title was then made larger and formatted in bold italic, by selecting the font size from the Formatting Toolbar and then clicking on the *Bold* and *Italic* icons.

6. Labels were added along the axes. To do so, we again selected Chart Options/Titles from the Chart menu and then entered the labels into the indicated areas for the x- and y-axes. (Note that each axis is referred to as a Value Axis in an x-y chart.)

7. The background color within the plot area was changed by clicking on the plot background and choosing Selected Plot Area from the Format menu. A new color (white) was then selected from the resulting dialog box.

8. The scale along the x-axis was changed and vertical grid lines added by clicking on the chart and then choosing Selected Axis/Scale from the Format menu. Within the resulting dialog box, the Auto check marks were removed from the Maximum, Major unit, and Minor unit selctions, and the values were changed to Maximum 12, Major unit 2, and Minor unit 2.

Note that there are other ways to perform many of these editing changes. For example, we could have used the shortcuts resulting from clicking the right mouse button (on a Macintosh, holding down the Command+Options keys and pressing the mouse button). Or we could have clicked directly on certain objects (such as the chart title) and then changed them as desired.

Exercises

3.5 The following measurements of voltage versus time were obtained for the capacitor described in Example 3.2. (Note that there is some scatter in the data since these are *measured* rather than *calculated* values.)

Seconds	Volts	Seconds	Volts
0	9.8	6	0.6
1	5.9	7	0.4
2	3.9	8	0.3
3	2.1	9	0.2
4	1.0	10	0.1
5	0.8		

(a) Place the data in a worksheet using a row-oriented layout. (Place the values of the independent variables in the top row and the values of the dependent variables in the bottom row.)

(b) Construct an *x-y* graph of the data. *Do not* interconnect the data points. Add appropriate titles and edit the overall appearance of the graph, as in Example 3.2.

(c) Change the location of the *x-y* graph from the original worksheet to a separate worksheet. Test the various editing features of the graph after it has been relocated.

3.6 A polymeric material contains a solvent that dissolves as a function of time. The concentration of the solvent, expressed as a percentage of the total weight of the polymer, is shown in the following table as a function of time.

Solvent Concentration (weight percent)	Time (sec)
55.5	0
44.7	2
38.0	4
34.7	6
30.6	8
27.2	10
22.0	12
15.9	14
8.1	16
2.9	18
1.5	20

Enter the data into an Excel worksheet and plot the data as an *x-y* graph with time as the independent variable. Show the individual data points.

3.7 The relationship between pressure, volume, and temperature for many gases can be approximated by the ideal gas law, which is written as

$$PV = RT$$

where *P* is the absolute pressure, *V* is the volume per mole, *R* is the ideal gas constant (0.082054 liter atm / mole °K), and *T* is the absolute temperature.

(a) Construct an Excel worksheet containing a table of pressure versus absolute temperature for absolute temperatures ranging from 0 to 800 °K, and specific volumes of 20, 35, and 50 liters/mole.

Note that the worksheet should contain four columns. The values of the absolute temperature should be placed in the first column, and the second column should contain the corresponding pressure values for a specific volume of 20 liters/mole. The third and fourth columns should contain the pressure values for specific volumes of 35 and 50 liters/mole, respectively.

(b) Plot all of the data (i.e., all three pressure vs temperature curves) on the same x-y graph. Edit the graph so that it is legible and attractive. Include a legend, indicating the specific volume associated with each curve.

3.8 The force exerted by a spring is given by

$$F = -kx^2$$

where F is the force, in newtons; x is the displacement of the spring from the equilibrium position, in centimeters; and k is a spring constant. Prepare an Excel worksheet, including an x-y graph, showing force as a function of distance for two springs whose spring constants are 0.1 newtons/cm² and 0.5 newtons/cm², respectively. Plot both curves on the same graph. For each spring, consider displacements ranging from 0 to 20 cm.

3.9 Several engineering students have built a wind-driven device that generates electricity. The following data have been obtained with the device:

Wind velocity, mph	Power, watts
0	0
5	0.26
10	2.8
15	7.0
20	15.8
25	28.2
30	46.7
35	64.5
40	80.2
45	86.8
50	88.0
55	89.2
60	90.3

Enter the data into an Excel worksheet and plot the electrical power as a function of wind velocity. Show the individual data points.

3.10 The following data describe the current, in milliamps, passing through an electronic device as a function of time.

Seconds	Milliamps	Seconds	Milliamps
0	0	9	0.77
1	1.06	10	0.64
2	1.51	12	0.44
3	1.63	14	0.30
4	1.57	16	0.20
5	1.43	18	0.14
6	1.26	20	0.091
7	1.08	25	0.034
8	0.92	30	0.012

Enter the data into an Excel worksheet and plot the current as a function of time. Show the individual data points.

3.4 SEMI-LOG GRAPHS

Sometimes it is advantageous to plot the *log* of y against x. This is known as a *semi-log* graph. Semi-log graphs are commonly used in many diverse fields, including engineering, chemistry, physics, biology, and economics.

Figure 3.17 shows a semi-log graph of the same equation that was plotted in Fig. 3.12; that is,

$$V = 10e^{-0.5t}$$

over the interval $0 \le t \le 10$. Notice that the curve now appears as a straight line. Also, note that the lower limit of the y-axis is 0.01 rather than 0 since the log of 0 is undefined. (*Remember that Fig. 3.17 is actually a plot of the $\log_{10}$ Voltage versus time, even though the y-axis is labeled simply Voltage.*)

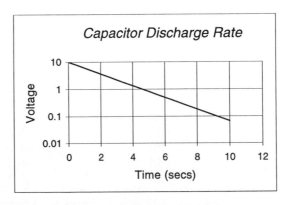

Figure 3.17

Engineers plot data on a semi-log graph for two reasons: First, the range of y-values can be much greater, often spanning several orders of magnitude (i.e., several powers of 10). For example, the y-axis in Fig. 3.17 ranges from 0.01 to 10, which is three orders of magnitude. Accordingly, Fig. 3.17 is known as a *three-cycle* semi-log graph.

Second, the *exponential equation*

$$y = ae^{bx} \tag{3.1}$$

appears as a straight line when it is plotted on a semi-log graph (i.e., when the log of y is plotted against x). Many phenomena in science and engineering are governed by this equation or some variation, such as

$$y = a(1 - e^{bx}) \tag{3.2}$$

Hence, if a data set shows up as a straight line when plotted on a semi-log graph, we can conclude that an exponential-type equation will best represent the data (or, more significantly, we can conclude that the data were generated by a *process* governed by an exponential-type equation).

To see why Equation (3.1) plots as a straight line on a semi-log graph, let us take the $\log_{10}$ of each side of the equation. We obtain

$$\log_{10} y = \log_{10} a + \log_{10}(e^{bx}) \tag{3.3}$$

which becomes

$$\log_{10} y = \log_{10} a + bx(\log_{10} e) \tag{3.4}$$

Since $\log_{10} e = 0.434294$, we can rewrite Equation (3.4) in a simpler form as

$$\log_{10} y = 0.434294\, bx + \log_{10} a \tag{3.5}$$

A similar result can be obtained using natural logarithms rather than base-10 logarithms, resulting in the equation

$$\ln y = bx + \ln a \tag{3.6}$$

This equation is somewhat simpler than Equation (3.5). However, Equation (3.5) corresponds directly to the straight lines plotted on semi-log graphs because semi-log graphs are displayed in terms of multiples of 10 rather than multiples of e (recall that $e = 2.7182818 \ldots$ is an irrational number that represents the base of the natural system of logarithms).

In general terms, we can write the equation for a straight line as

$$ordinate = slope \times abscissa + constant \tag{3.7}$$

Thus, we can interpret Equation (3.5) as a special type of straight line; that is,

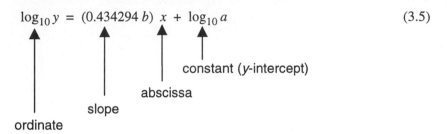

$$\log_{10} y = (0.434294\, b)\ x + \log_{10} a \tag{3.5}$$

constant (*y*-intercept)

abscissa

slope

ordinate

In Excel, a semi-log graph can easily be generated from an ordinary (arithmetic) *x-y* graph, provided the *y*-values are all positive. To do so, click on the *y*-axis and choose Selected Axis/Scale from the Format menu (or double-click on the *y*-axis). Then check the box labeled Logarithmic scale in the resulting dialog box.

A semi-log graph can also be generated directly by selecting Custom Types/Logarithmic within the Chart Type dialog box. (Recall that the Chart Type dialog box appears in step 1 of the *Chart Wizard* when generating a graph from scratch. It can also be obtained by clicking on an existing graph and then selecting Chart Type from the Chart menu.) Remember, however, that *the Logarithmic chart selection in Excel refers to a semi-log graph, not a log-log graph* (see below).

In either case you may want to relocate the labels along the *x*-axis so that they appear at the bottom of the graph. To do so, click on the *y*-axis and choose Selected Axis/Scale from the Format menu. Then enter the lowest *y*-value in the area labeled Value (X) axis Crosses at:. Be sure that the Auto box in front of this selection is not checked.

Example 3.3 Creating a Semi-Log Graph in Excel

Convert the arithmetic *x-y* graph developed in Example 3.2 (see Fig. 3.16) into a semi-log graph.

We begin with the Excel worksheet shown in Fig. 3.16. To convert the graph into a semi-log graph, we click on the *y*-axis and choose Selected Axis/Scale from the Format menu. This results in the Format Axis dialog box shown in Fig. 3.18. Within this dialog box, we check the first four Auto boxes in the Value (Y) axis scale area and the Logarithmic scale box near the bottom. We also enter the value 0.01 (the minimum *y*-value) in the area labeled Value (X) axis Crosses at:. Figure 3.18 shows all of the new selections. Finally, we move to the Number tab in the Format Axis dialog box and select the General category, to automatically display the correct number of decimal places along the y-axis.

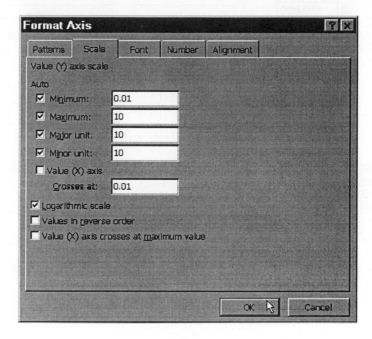

Figure 3.18

The resulting graph is shown within the worksheet in Fig. 3.19. Notice that the chart now appears as a semi-log graph, and the exponential equation appears as a straight line.

Exercises

3.11 Using Excel, generate an arithmetic x-y plot of $\log_{10}V$ vs t, based upon the equation that appeared in Example 3.2; that is,

$$V = 10e^{-0.5t}$$

Compare the resulting graph with the semi-log graph shown in Fig. 3.19. What can you conclude about the relationship between the two graphs?

3.12 A chemical reaction is being carried out in a well-stirred tank. The concentration of the substance being created can be calculated as a function of time using the formula

$$C = a(1 - e^{-bt})$$

where C is the concentration in moles per liter and t is the time in seconds.

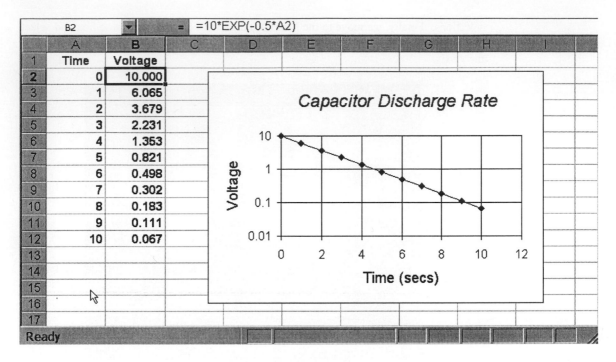

Figure 3.19

(a) Construct a table of concentration versus time within an Excel worksheet for the case where $a = 8$ and $b = 0.25$. Include individual cells that contain the current values of a and b. Select a long enough time period so that the concentration approaches its equilibrium value.

(b) Create an *x-y* graph of concentration versus time from the tabulated values. Connect the individual data points with line segments. Add an appropriate title and label the axes.

(c) Change the values of a and b to 12 and 0.5, respectively. Notice what happens to the tabulated values and the graph. Note the comparative ease with which these changes are carried out.

(d) Restore the original values of a and b. Then alter the type of graph by replacing the arithmetic coordinates with semi-logarithmic coordinates. Can you explain the shape of the resulting graph?

3.13 Enter the data given in Exercise 3.6 into an Excel worksheet and plot the data as a semi-log graph with time as the independent variable. Show the individual data points. Compare the resulting graph with the graph obtained in Exercise 3.6.

3.5 LOG-LOG GRAPHS

A *log-log* graph has logarithmic coordinates along both axes. Thus, it is actually a plot of log *y* against log *x*. Log-log graphs are useful for plotting scientific and technical data because they allow the data to span several orders of magnitude and because a *power equation* of the form

$$y = ax^b \tag{3.8}$$

will appear as a straight line.

For example, Figure 3.20 shows a log-log graph of the well-known equation for the volume of a sphere; that is,

$$V = \frac{4}{3}\pi r^3$$

within the interval $0 \le r \le 10$. Notice that the resulting curve appears as a straight line on the log-log graph. In addition, note that the lower limit of each axis is some small positive number rather than 0 since the log of 0 is undefined. (*Remember that Fig. 3.20 is really a plot of $\log_{10}$ Volume vs $\log_{10}$ Radius, even though the axes are labeled simply Volume and Radius.*)

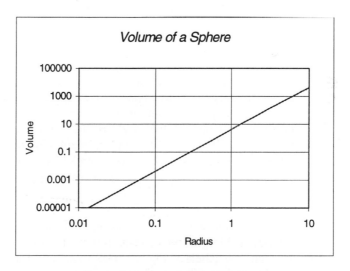

Figure 3.20

To see why Equation (3.8) plots as a straight line on a log-log graph, we take the $\log_{10}$ of each side of the equation, resulting in

$$\log_{10} y = \log_{10} a + b \log_{10} x \tag{3.9}$$

Rearranging Equation (3.9) slightly, we obtain

$$\log_{10} y = b\,(\log_{10} x) + \log_{10} a \qquad\qquad (3.10)$$

which is the equation for a straight line in which $\log_{10} y$ is the dependent variable and $\log_{10} x$ is the independent variable. In other words,

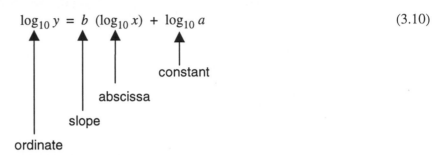

$$\log_{10} y = b\,(\log_{10} x) + \log_{10} a \qquad\qquad (3.10)$$

Note that the last term, $\log_{10} a$, is not called the y-intercept since the value of the abscissa will always exceed 0 in a log-log graph.

Equation (3.10) can also be obtained by taking the *natural* log of each term in Equation (3.8), resulting in the expression

$$\ln y = b \ln x + \ln a \qquad\qquad (3.11)$$

We then multiply each term in this equation by $\log_{10} e = 0.434294$, thus converting each term to a base-10 logarithm.

A log-log graph cannot be created directly in Excel, but it can easily be constructed from an ordinary (arithmetic) x-y graph or from a semi-log graph, provided all values of the independent and dependent variables are positive. (Remember that the log of zero is undefined, as is the log of a negative number. Hence, Excel will generate an error message if any of the x- or y-values are either zero or negative.)

To create a log-log graph from an ordinary x-y graph, simply click on the y-axis and choose Selected Axis/Scale from the Format menu. Then check the box labeled Logarithmic scale in the resulting dialog box. You may also want to relocate the labels along the x-axis by entering the lowest y-value in the area labeled Value (X) axis Crosses at: (be sure that the Auto box in front of this selection is not checked).

The process is then repeated for the x-axis; that is, click on the x-axis and choose Selected Axis/Scale from the Format menu and check the box labeled Logarithmic scale. You can also relocate the labels along the y-axis by entering the lowest x-value in the area labeled Value (Y) axis Crosses at: (again, be sure that the Auto box in front of this selection is not checked).

If you begin with a semi-log graph rather than an ordinary (arithmetic) *x-y* graph, then only the *x*-axis must be altered, using the method described in the preceding paragraph.

Example 3.4 Creating a Log-Log Graph in Excel

Create a log-log graph of the area and volume of a sphere as a function of the radius within the interval $0 \leq r \leq 10$, using the following two formulas.

$$A = 4\pi r^2 \qquad \text{and} \qquad V = \frac{4}{3}\pi r^3$$

Let us begin with a worksheet containing tabular values and an ordinary (arithmetic) *x-y* graph of *A* and *V*, as shown in Fig. 3.21. The *x-y* graph was created by first selecting the data in columns A, B, and C, and then following the steps described in Sec. 3.1. In order to convert this graph into a log-log graph, we first change the value of *r* shown in cell A2 from 0 to some small positive number, say 0.001. We then click on the *x*-axis and choose Selected Axis from the Format menu. If we then select the Scale tab, we obtain the dialog box shown in Fig. 3.22. Notice the values that have been entered for the minimum, the maximum, and the *y*-axis crossing point. Also note that the Logarithmic scale box has been checked.

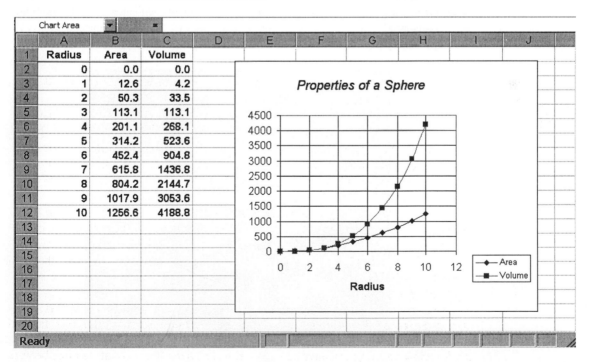

Figure 3.21

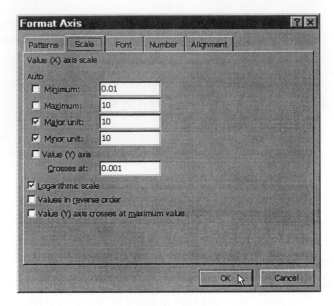

Figure 3.22

We then click on the *y*-axis and repeat the entire process. This results in the dialog box shown in Fig. 3.23. Again, notice the values that have been entered for the minimum, the maximum, and the *x*-axis crossing point, and note that the Logarithmic scale box has been checked.

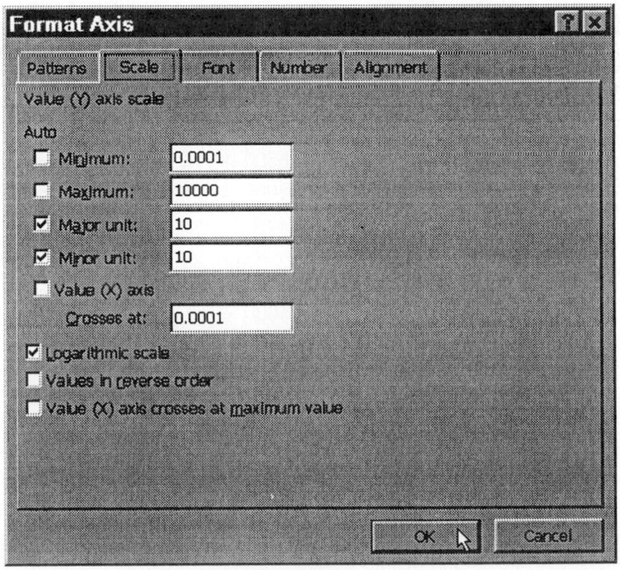

Figure 3.23

The resulting worksheet, containing the desired log-log graph, is shown in Fig. 3.24. Notice that the two curves shown in Fig. 3.21 are now represented as straight lines. In addition, notice that the y-axis has been subdivided into eight cycles and the x-axis into three cycles, where each cycle represents an order of magnitude (factor of 10) increase. Hence, this is an 8 × 3 log-log graph.

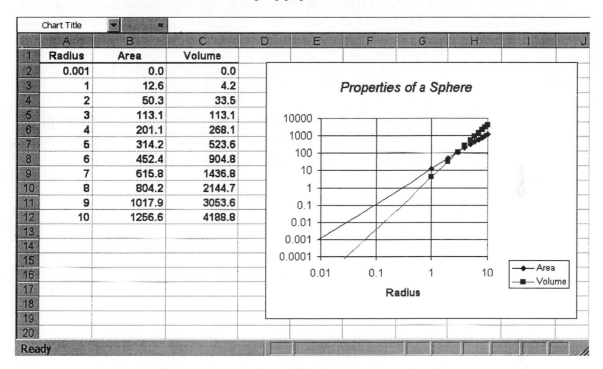

Figure 3.24

Exercises

3.14 Construct a worksheet containing values of y versus x generated by the equation

$$y = 1.5 \, x^{0.8}$$

over the interval $0 \leq x \leq 100$. Plot the data in the following three ways:

(a) Plot y versus x using arithmetic coordinates.

(b) Plot $\log_{10} y$ versus $\log_{10} x$ using arithmetic coordinates.

(c) Plot y versus x using logarithmic (log-log) coordinates.

Compare the resulting graphs and explain any similarities in the shape of the curves.

3.15 The force exerted by a spring is given by

$$F = -kx^2$$

where F is the force, in newtons; x is the displacement of the spring from the equilibrium position, in centimeters; and k is a spring constant. Prepare an Excel worksheet, including a log-log graph, showing force as a function of distance for two springs whose spring constants are 0.1 newtons / cm^2 and 0.5 newtons/cm^2, respectively. Plot both curves on the same log-log graph. For each spring, consider displacements ranging from 0.001 to 20 cm. Compare with the results obtained in Exercise 3.8.

3.16 A group of students have measured the quantity of water discharged from a tank as a function of time. The following data have been obtained:

Time, min	Volume, gal	Time, min	Volume, gal
0	0	35	70.2
5	17.2	40	77.2
10	27.8	45	83.7
15	38.4	50	89.9
20	46.5	55	96.9
25	54.3	60	103.4
30	62.5		

Enter the data into an Excel worksheet and plot the volume as a function of time using semi-log coordinates. Then plot the same data in a separate graph, using log-log coordinates. Show the individual data points in both graphs. What, if anything, can you conclude about the equation that might be used to represent the data?

3.17 An environmental engineer has obtained a bacteria culture from a municipal water sample and allowed the bacteria to grow within a petri dish. The following data were obtained:

Time, min	Bacteria Concentration, ppm
0	6
1	9
2	15
3	19
4	32
5	42
6	63
7	102
8	153
9	220
10	328

Enter the data into an Excel worksheet and plot the data several different ways. Use the resulting graphs to solve the following problem:

Suppose the growth of Type A bacteria is governed by a process described by the equation

$$C_A = ae^{bt}$$

and the growth of Type B bacteria is governed by a process described by the equation

$$C_B = ax^t.$$

What type of bacteria is the engineer dealing with?

3.6 CLOSING REMARKS

In this chapter we have confined our attention to bar graphs (i.e., Excel Column charts) and x-y graphs (Excel Scatter charts) since they are the types of graphs most commonly used in science and engineering. You should understand, however, that Excel is able to generate many other types of graphs. Some are used for business or financial applications, while others are intended simply to add pizazz to displays of numerical data. You are encouraged to explore them on your own.

ADDITIONAL READING

Dodge, M., C. Kinata, and C. Stinson. *Running Microsoft Excel 97*. Microsoft Press, 1997.

Eide, A. R., R. D. Jenison, L. H. Mashaw, and L. L. Northrup. *Engineering Fundamentals and Problem Solving*. 3d ed. New York: McGraw-Hill, 1997.

CHAPTER 4

ANALYZING DATA

Engineering analysis generally begins with the analysis of data. Engineers often gather data to measure *variability* or *consistency*. For example, a car manufacturer may want to determine the precise dimensions of several manufactured parts that are all of the same type, such as the diameters of several ball bearings coming off an assembly line. This information will provide an indication of the consistency with which the ball bearings are being manufactured. Similarly, a clothing manufacturer may want to study the variation in sizes among potential customers so that the company will know how many items of each size should be manufactured.

Measured data can tell you a great deal if you know how to interpret the results. Usually, however, a great deal of tedious arithmetic must be carried out in order to interpret the results, particularly if the data set is large. Excel can help in this regard by carrying out the tedious arithmetic for you, allowing you to concentrate on the interpretation of the results. We will see how this is done in this chapter.

4.1 DATA CHARACTERISTICS

There are several commonly used parameters that allow us to draw conclusions about the characteristics of a data set. They are the *mean*, *median*, *mode*, *min*, *max*, *variance*, and *standard deviation*. Let us discuss each of them individually.

Mean

The *mean* is the most commonly used characteristic of a data set. It is also referred to as the *average* or the *arithmetic average*. It is an indication of the *expected behavior* of a data set.

The mean is determined by the well-known formula

$$\bar{x} = \frac{(x_1 + x_2 + \cdots + x_n)}{n} = \frac{1}{n}\sum_{i=1}^{n} x_i \tag{4.1}$$

where $\bar{x}$ represents the mean, x_i represents an individual data value, and n represents the total number of data values. (Note that the subscript i ranges from 1 to n.)

In Excel, the AVERAGE( ) function is used to determine the mean. The argument contained in parentheses indicates a block of cells containing the values to be averaged. Thus, the expression =AVERAGE(B1:B12) will determine the mean of the values stored in cells B1 through B12. Within the indicated block of cells, zeros are included in the calculation but blank cells are ignored.

Median

The *median* is a value such that half the data values lie above and half lie below. If the number of data values is odd, the median coincides with one of the data values. For example, 3 is the median for the data set (2, 0, 8, 3, 5). If the number of data values is even, however, the median is usually taken as the average of the two centermost values. Thus, 4 is the median for the data set (2, 8, 3, 5).

In Excel, the MEDIAN( ) function is used to determine the median. It is used in the same manner as the AVERAGE function described above. Thus, the expression =MEDIAN(B1:B12) will determine the median of the values stored in cells B1 through B12. The numerical values within the cells need *not* be sorted.

Mode

The *mode* is the value that occurs with the greatest frequency within a data set. Not all data sets have a mode. On the other hand, some data sets have multiple modes. The data set (1, 2, 3, 4, 5), for example, does not have a mode because no value occurs more frequently than any other. However, in the data set (1, 2, 2, 4, 5), the mode is 2. The data set (1, 2, 2, 3, 4, 4, 5) has two modes, 2 and 4.

In Excel, the mode can be determined with the MODE( ) function. Again, the arguments indicate the block of cells containing the data. Thus, the expression =MODE(B1:B12) will determine the mode of the values stored in cells B1 through B12. The MODE function returns an error message (#N/A) if the data set does not have a mode.

If multiple modes are present, the MODE function will return the value that occurs with the greatest frequency, or the first value encountered if all of the modes occur with the same frequency. Thus, the MODE function will return 4 for the data set (1, 2, 2, 3, 4, 4, 4, 5), but it will return 2 for the data set (1, 2, 2, 3, 4, 4, 5).

Min and Max

The *min* and the *max* (i.e., the minimum and the maximum) simply indicate the extremities of the data set. In Excel, the MIN( ) and MAX( ) functions return these values. The arguments again indicate the block of cells containing the data. Thus, the expression =MIN(B1:B12) returns the smallest value within the cells B1 through B12, whereas =MAX(B1:B12) returns the largest value. Blank cells are ignored.

Note that the MIN and MAX functions return the values that are the smallest and the largest *algebraically*. They do *not* return the values that are the smallest and the largest in magnitude. Thus, for the data set (−5, −2, 1), the MIN function would return −5 (which is algebraically the smallest value), and the MAX function would return 1 (which is algebraically the largest).

Variance

The *variance* provides an indication of the degree of *spread* in the data. The greater the variance, the greater the spread.

The variance is determined by the following formula:

$$s^2 = \frac{1}{n-1} \sum_{i=1}^{n} (x_i - \overline{x})^2 \tag{4.2}$$

where s^2 represents the variance, x_i represents an individual data value, $\overline{x}$ represents the mean, and n represents the number of data values. Note that the equation involves summing the square of the difference between each data value and the calculated mean.

To see why the variance provides a measure of the spread in the data, let us consider the summation term more carefully. Each of the terms being added will always be greater than or equal to zero since it is the square of another number. If there is little spread in the data, each of the individual data values will be close to the calculated mean. Hence, each of the terms being added will be relatively small, and the variance will therefore be relatively small.

On the other hand, if there is considerable spread in the data, some of the data values will be relatively far from the calculated mean. Therefore some of the terms in the summation will be relatively large positive numbers, and the variance will also be relatively large. Thus, the greater the spread in the data, the larger the variance.

Excel includes the VAR( ) function, which is used to determine the variance. As before, the argument indicates the block of cells containing the data. Thus, the expression =VAR(B1:B12) returns the variance of the numerical values in cells B1 through B12.

Standard Deviation

The *standard deviation* also provides a measure of spread in the data set. In fact, the standard deviation is simply the square root of the variance. Thus,

$$s = \sqrt{s^2} = \sqrt{\frac{1}{n-1}\sum_{i=1}^{n}(x_i - \bar{x})^2} \qquad (4.3)$$

where s represents the standard deviation and all other symbols are as defined for the variance.

Clearly, the standard deviation will be large if the variance is large, and the standard deviation will be small if the variance is small. Why, then, are *both* of these parameters used to indicate the degree of spread? The answer is that the variance is the more fundamental quantity, though its dimensions are expressed in units *squared* rather than the same units used for the mean, median, mode, min, and max. For example, if the individual data values have the units of inches, the resulting mean, median, mode, min, and max will also be expressed in inches, but the variance will be expressed in square inches. The standard deviation, on the other hand, will be expressed in inches, which is consistent with the other parameters. By selecting the standard deviation to indicate spread, we therefore obtain consistent units among all of the parameters. To determine a value for the standard deviation, however, you must first calculate the variance. Hence, you should have some understanding of both parameters and their interrelationship.

In Excel, the function STDEV( ) returns the standard deviation of its arguments. Again, the arguments indicate the block of cells containing the data. Hence the expression =STDEV(B1:B12) will return the standard deviation of the data in cells B1 through B12. We could, of course, first determine the variance and then calculate its square root, but the direct calculation is easier.

Example 4.1 Analyzing a Data Set

A car manufacturer wishes to determine how accurately the cylinders are being machined in several engine blocks. The design specifications call for a cylinder diameter of 3.500 inches, with a tolerance of ±0.005 inches.

To determine the accuracy of the cylinders, several engine blocks were taken from the assembly line during manufacture and one cylinder was measured in each block. For consistency, the measurement was always perpendicular to the axis of the engine block (i.e., perpendicular to the straight line connecting the centers of the cylinders). The following data were obtained:

Sample	Diameter (in)	Sample	Diameter (in)
1	3.502	11	3.497
2	3.497	12	3.504
3	3.495	13	3.498
4	3.500	14	3.499
5	3.496	15	3.501
6	3.504	16	3.500
7	3.509	17	3.503
8	3.497	18	3.494
9	3.502	19	3.499
10	3.507	20	3.508

Note that four of the individual values exceed the allowable tolerance of ± 0.005 inches (three above, one below).

Analyze the data by placing them in an Excel worksheet and then calculating the mean, median, mode, min, max, and standard deviation.

Figure 4.1 shows the Excel worksheet containing the data and the calculated parameters. The calculated values are located to the right of the actual data. Note that cell E5, which contains the numerical value for the mean, is highlighted. The expression used to obtain this value, =AVERAGE(B4:B23), is shown in the formula bar at the top of the figure. Similar formulas were used to obtain the other five values.

The numerical values, other than the standard deviation, were *formatted* to display three decimal places by selecting Cells from the Format menu, selecting the Number tab, then the Number category, and finally, the value 3 in the data area labeled Decimal places within the resulting dialog box.

From the worksheet, we obtain the following values (all units are in inches):

mean:	3.501	median:	3.500	mode:	3.497
min:	3.494	max:	3.509	std dev:	0.00427

Notice that the mean (3.501) is slightly higher than the median (3.500), whereas the mode (3.497) is somewhat lower. Also note that the standard deviation (0.00427 inches) is only 0.122 percent of the mean.

The values obtained for the mean and the standard deviation can provide useful statistical information about the likelihood that the individual cylinder diameters will exceed the specified tolerance, though the details are beyond the scope of this book.

E5	▼	■	=AVERAGE(B4:B23)							
	A	B	C	D	E	F	G	H	I	J	K
1		**Engine Cylinder Data**									
2											
3	**Sample**	**Diameter**									
4	1	3.502	inches								
5	2	3.497		Mean =	3.501	in					
6	3	3.495									
7	4	3.500		Median =	3.500	in					
8	5	3.496									
9	6	3.504		Mode =	3.497	in					
10	7	3.509									
11	8	3.497		Min =	3.494	in					
12	9	3.502									
13	10	3.507		Max =	3.509	in					
14	11	3.497									
15	12	3.504		Std Dev =	0.00427	in					
16	13	3.498									
17	14	3.499									
18	15	3.501									
19	16	3.500									
20	17	3.503									
21	18	3.494									
22	19	3.499									
23	20	3.508									
24											

Figure 4.1

Problems

4.1 Using your own version of Excel, recreate the worksheet shown in Fig. 4.1 showing the mean, median, etc. for the data given in Example 4.1. Change the values of some of the diameters and observe what effect this has on the mean, the median, etc.

4.2 The heights of 20 engineering students are given in the following table. Use Equations (4.1), (4.2), and (4.3) to determine the mean, median, mode, min, max, variance, and standard deviation. Carry out the calculations by hand, using only a calculator, a pencil, and a piece of paper. (Do *not* use a spreadsheet to obtain a solution. Also, do *not* use any of the built-in statistical functions that may be present in your calculator.)

Student	Height (in)	Student	Height (in)
1	70.6	11	70.2
2	71.1	12	72.0
3	73.3	13	70.0
4	72.6	14	69.8
5	70.0	15	69.0
6	71.6	16	69.4
7	66.5	17	68.3
8	71.1	18	73.8
9	67.0	19	66.9
10	68.8	20	71.6

4.3   Enter the data given in Problem 4.2 into an Excel worksheet and determine the mean, median, mode, min, max, variance, and standard deviation. Be sure the worksheet is legible and clearly labeled.

4.2 HISTOGRAMS

Though the mean, median, mode, min, max, and standard deviation provide useful information about a data set, it is often desirable to plot the data in a manner that illustrates how the values are distributed within their range. This type of plot is called a *histogram*, or a *relative frequency* plot. From the relative frequency plot, we can obtain a plot of the *cumulative distribution*, which allows us to estimate the likelihood that a data value associated with an item drawn at random is less than or greater than some specified value. The details associated with each plot are discussed below.

Histogram Fundamentals

To create a histogram, you must first subdivide the range of the data into a series of adjacent, equally spaced intervals. The first interval must begin at or below the smallest data value (the min), and the last interval must extend to or beyond the largest data value (the max). Each interval will have some lower bound, x_i, and an upper bound, x_{i+1}, where $x_{i+1} = x_i + \Delta x$ and Δx represents the fixed interval width. These intervals are sometimes referred to as *class intervals*.

Once the intervals have been defined, you must determine how many data values fall within each interval. The *relative frequencies* are then obtained as

$$f_i = \frac{n_i}{n} \tag{4.4}$$

where f_i represents the relative frequency for the ith interval, n_i represents the number of data values in the ith interval, and n represents the total number of data values. Note: if k is the number of intervals, then

$$\sum_{i=1}^{k} n_i = n \tag{4.5}$$

and

$$\sum_{i=1}^{k} f_i = 1 \tag{4.6}$$

The histogram is usually expressed as a bar graph showing the values of the interval counts (n_i) or the relative frequencies (f_i). From this bar graph, it is easy to see how the data are distributed.

Example 4.2 Constructing a Histogram

Construct a histogram of the data given in Example 4.1. Choose 10 equally spaced intervals ranging from $x = 3.490$ to $x = 3.510$ inches. If a data value falls on an interval boundary, assign the data value to the lower interval. (*Note*: This boundary rule is arbitrary. Many authors suggest that a data value falling on an interval boundary be assigned to the *upper* interval. However, our choice of the lower interval is consistent with the rule employed in Excel.)

The first interval will extend from 3.490 to 3.492 inches. The second interval will extend from 3.492 to 3.494 inches though it will actually include only those values that *exceed* 3.492 inches because a value of exactly 3.492 inches will be assigned to the first interval, and so on. For simplicity, however, we will write the interval bounds to four significant figures. This practice is customarily followed when constructing histograms.

The class intervals and their corresponding values are shown in the following table:

Interval Number	Interval Range	Number of Values	Relative Frequency
1	3.490 – 3.492	0	0
2	3.492 – 3.494	1	0.05
3	3.494 – 3.496	2	0.10
4	3.496 – 3.498	4	0.20
5	3.498 – 3.500	4	0.20
6	3.500 – 3.502	3	0.15
7	3.502 – 3.504	3	0.15
8	3.504 – 3.506	0	0
9	3.506 – 3.508	2	0.10
10	3.508 – 3.510	1	0.05

Thus, we see that 5 percent of the data values fall within the interval 3.492 − 3.494, 10 percent fall within the interval 3.494 − 3.496, and so on. (More precisely, 5 percent of the data exceed 3.492 but are less than or equal to 3.494, 10 percent exceed 3.494 but are less than or equal to 3.496, etc.)

Histogram Generation in Excel

When working with real data, the number of values within a data set is often quite large. Thus, it may be very tedious to count the number of values within each interval, as we have done in the previous example. Fortunately, however, most spreadsheets will do this counting for us once we have entered the data.

In Excel, histograms can be constructed very easily using the Histogram feature found in the Analysis ToolPak. To install the Analysis ToolPak, choose Add-Ins from the Tools menu. Then select Analysis ToolPak from the resulting Add-Ins dialog box. (Once the Analysis ToolPak has been installed, it will remain installed unless it is removed, by reversing the above procedure.)

To construct a histogram in Excel, proceed as follows:

1. Enter the basic data and the interval bounds into the worksheet. (The intervals are called *bins* in Excel.) Typically, the data are entered in a single column (or a single row), and the interval bounds are entered in another column (or another row). Only the *right* interval bounds are entered. Hence, the last interval (rightmost interval) will be open-ended; that is, it will contain all data values that exceed the largest bound.

2. Choose Data Analysis/Histogram from the Tools menu. Then provide the required information in the resulting dialog box. In particular, you must specify the cell blocks that contain the data (called the Input Range) and the interval bounds (the Bin Range). You can specify each range either by typing it directly into the dialog box or by dragging the mouse over the block of cells containing the data (hold down the mouse button as you move the mouse). If the location of the Histogram dialog box interferes with your dragging the mouse over the desired block of cells, you can move the dialog box by placing the mouse pointer in the dialog box title bar (i.e., Histogram) and then dragging the dialog box to a more convenient location.

3. If the histogram is to be embedded within the current worksheet, you must also select the Output Range option and enter the cell address of the upper corner of the block that will contain the output data.

4. When building a conventional histogram, the Pareto box at the bottom of the dialog box should *not* be selected. You *may* select either of the remaining two boxes (Cumulative Percentage and Chart Output), though they should be left blank for now. We will say more about these two boxes later in this chapter.

Example 4.3 Generating a Histogram in Excel

Modify the worksheet developed in Example 4.1 (see Fig. 4.1) by adding a histogram of the data. Use the same class intervals as in Example 4.2.

We begin by creating some additional space in the worksheet, so that there is ample room for the histogram. To do so, we will shorten the label in column C of Fig. 4.1 and then decrease the width of columns C and F. We then add the upper interval bounds in column G, as shown in Fig. 4.2. The next step is to select Data Analysis from the Tools menu and then select Histogram from the Data Analysis dialog box.

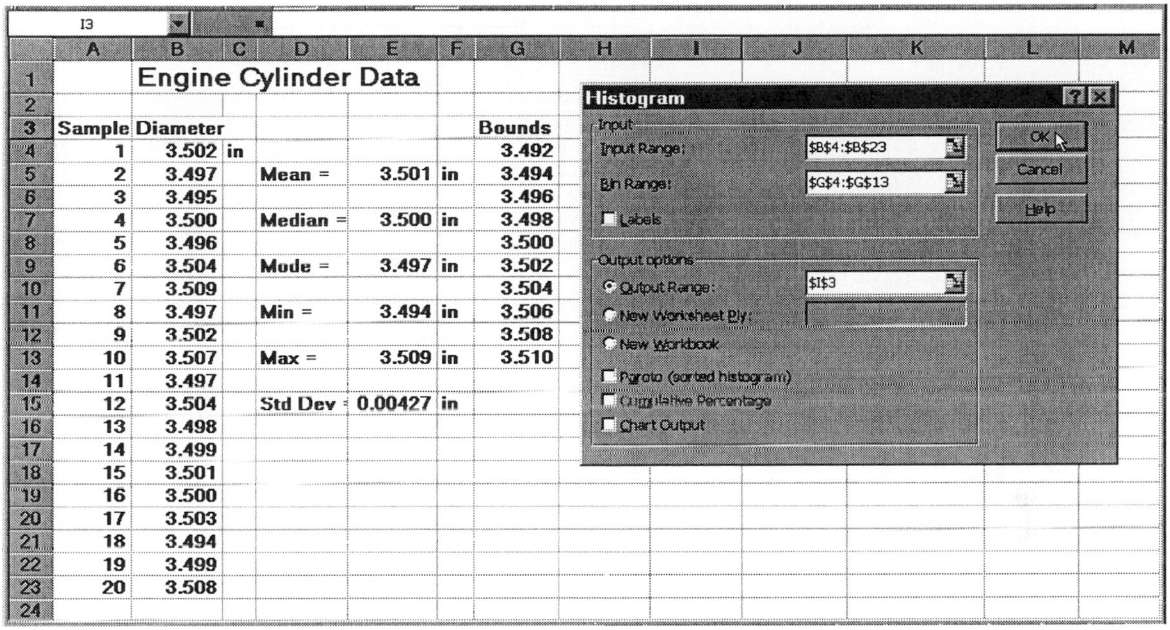

Figure 4.2

Once the Histogram selection is made, the Histogram dialog box will appear, as shown in Fig. 4.2. (In Fig. 4.2, the Histogram dialog box has been moved to one side of the worksheet so that the requested cell ranges can be entered by dragging the mouse pointer over the appropriate cell blocks.) We then proceed to fill in the requested information. The cells containing the data are specified by first clicking on the Input Range data area and then dragging the mouse pointer across cells B4 through B23. Similarly, the interval boundaries are specified by clicking on the Bin Range data area and then dragging the mouse pointer across cells G4 through G13. Finally, the upper left corner of the output area is specified by selecting Output Range as an output option and then selecting cell I3. Figure 4.2 shows these selections having been made. Note that none of the three boxes at the bottom of the dialog box (Pareto, Cumulative Percentage, and Chart Output) have been selected.

When the required ranges have been specified, we click on the OK button, resulting in the table of frequencies shown in Fig. 4.3. Note that the intervals in column I are labeled *Bin*, and the interval values in column J are labeled *Frequency*. The values tabulated in the *Frequency* column are integer sums, not the relative frequencies (i.e., fractional values) described earlier. We can, of course, obtain a set of relative frequencies in the next column (column K) by dividing each of the cell values in column J by the total number of data points, as in Equation (4.4) above.

	A	B	C	D	E	F	G	H	I	J	K	L	M
	K17												
1		Engine Cylinder Data											
2													
3	Sample	Diameter					Bounds		Bin	Frequency			
4	1	3.502	in				3.492		3.492	0			
5	2	3.497		Mean =	3.501	in	3.494		3.494	1			
6	3	3.495					3.496		3.496	2			
7	4	3.500		Median =	3.500	in	3.498		3.498	4			
8	5	3.496					3.500		3.500	4			
9	6	3.504		Mode =	3.497	in	3.502		3.502	3			
10	7	3.509					3.504		3.504	3			
11	8	3.497		Min =	3.494	in	3.506		3.506	0			
12	9	3.502					3.508		3.508	2			
13	10	3.507		Max =	3.509	in	3.510		3.510	1			
14	11	3.497							More	0			
15	12	3.504		Std Dev =	0.00427	in							
16	13	3.498											
17	14	3.499											
18	15	3.501											
19	16	3.500											
20	17	3.503											
21	18	3.494											
22	19	3.499											
23	20	3.508											
24													

Figure 4.3

A histogram is customarily plotted as a bar graph (i.e., an Excel *Column chart*.). This can be carried out manually, using the techniques described in Chapter 3, or it can be carried out automatically when the histogram is created. To create a bar graph automatically, simply select Chart Output at the bottom of the Histogram dialog box (see Fig. 4.2). When this option is selected, the bar graph will appear as an embedded bar graph, as shown in Fig. 4.4.

Figure 4.4 differs from a traditional histogram in two ways. First, the numerical labels shown beneath every other interval (the *bin* values) actually correspond to the interval *right boundaries* (which are shown as tick marks) rather than the intervals themselves. Thus, the first interval extends from a value of 3.490 to 3.492, and so on. When interpreting the histogram, each of these values should therefore be associated with the "tick" mark to its right.

The graph can be clarified somewhat by rotating the labels so that they are vertical, centered beneath each interval. To do so, click on the *x*-axis and choose

Selected Axis/Alignment from the Format menu. Then either drag the alignment indicator so that it is in the 12 AM position or enter the value 90 in the area labeled Degrees. It will then be easier to associate each numerical label with the right boundary beneath each interval, as shown in Fig. 4.5.

The other difference concerns the manner in which the vertical bars are drawn within the bar graph. Since the vertical bars represent contiguous intervals, they should be drawn adjacent to each other, without any intervening space. This is easy to accomplish, as follows: Click on any of the vertical bars and choose Selected Data Series/Options from the Format menu. Then choose a value of 0 for the Gap width. The resulting histogram will then appear as shown in Fig. 4.5.

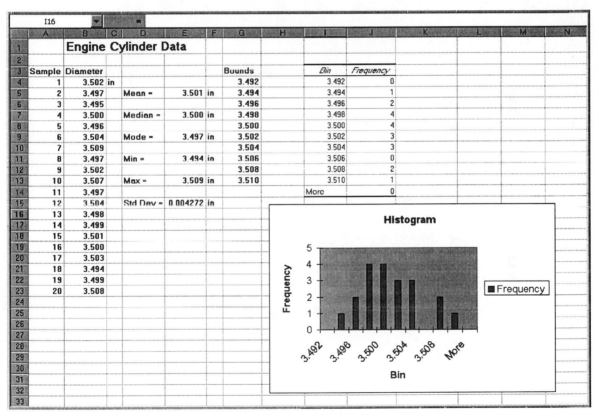

Figure 4.4

When constructing a histogram, another important consideration concerns the choice of the number of intervals. Too few intervals results in a histogram that lacks detail and consequently provides little information about how the data are distributed. On the other hand, too many intervals results in gaps within the histogram, which detracts from the overall shape of the histogram. As a rule, 10 to 15 intervals works well with most data sets. Fewer intervals might be better, however, if the number of data values is relatively small.

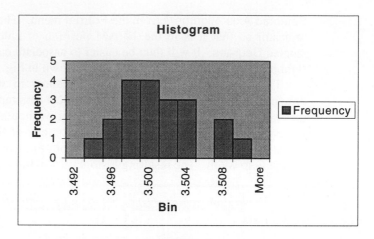

Figure 4.5

Problems

4.4 Using your own copy of Excel, generate the worksheets and figures shown in Figs. 4.3, 4.4, and 4.5 (see Example 4.3). Edit the graphs in Figs. 4.4 and 4.5 so that they fill the screen and are clearly labeled.

4.5 A class of 30 students obtained the following exam scores in their Introduction to Engineering class:

Student No.	Exam Score	Student No.	Exam Score
1	87	16	71
2	64	17	41
3	74	18	77
4	56	19	74
5	95	20	56
6	74	21	79
7	76	22	90
8	67	23	47
9	82	24	44
10	67	25	79
11	91	26	96
12	64	27	69
13	71	28	66
14	41	29	50
15	78	30	77

Enter the data into an Excel worksheet and carry out the following operations:

(a) Determine the mean, median, min, and max. Explain the difference between the mean and the median.

(b) Construct a 10-interval histogram spanning the range from 0 to 100.

(c) Based upon this histogram, how many students have exam scores ranging from 71 to 80? How many have exam scores above 90? How many have exam scores of 50 or less?

4.6 Use the data given in Problem 4.5 to construct the following histograms:

(a) A five-interval histogram that spans the range of the data. (Note that this histogram need not necessarily begin at 0 and extend to 100. Choose any convenient values for the lower and upper bounds.)

(b) A 15-interval histogram that spans the same range.

(c) A 30-interval histogram that spans the same range.

(d) Based upon these results, what would you suggest is the best number of intervals for this data set, and why?

4.3 CUMULATIVE DISTRIBUTIONS

We have seen that a histogram provides a graphical illustration of how a data set is distributed. Of equal importance is the *cumulative distribution*, which provides another way to view the manner in which the data are distributed. The cumulative distribution allows us to determine the likelihood that a particular value drawn at random is less than or greater than some specified value. For example, we might wish to determine the likelihood that a cylinder diameter within a randomly selected engine block is greater than some specified value, say 3.500 inches.

To construct a cumulative distribution, we must first construct a histogram and determine the relative frequency associated with each of the intervals, using Equation (4.4). We then determine the following cumulative values:

$$F_1 = f_1$$
$$F_2 = f_1 + f_2$$
$$F_3 = f_1 + f_2 + f_3$$

. . . . .

$$F_j = f_1 + f_2 + \cdots + f_j = \sum_{i=1}^{j} f_i$$

. . . . .

$$F_k = f_1 + f_2 + \cdots + f_k = \sum_{i=1}^{k} f_i \tag{4.7}$$

Note that the last term, F_k, will always equal 1 because of Equation (4.6). This provides a convenient means to check the accuracy of the individual f_is. (Add up the f_is and see if they sum to 1.)

Example 4.4 Constructing a Cumulative Distribution

Construct a set of numerical values for the cumulative distribution corresponding to the histogram developed in Example 4.2.

We have already determined the class intervals and their corresponding relative frequencies. In order to determine the cumulative distribution, we must determine the partial sums of the relative frequencies. Thus, we see that

$$F_1 = 0$$

$$F_2 = 0.05$$

$$F_3 = 0.05 + 0.10 = 0.15$$

$$F_4 = 0.05 + 0.10 + 0.20 = 0.35$$

$$F_5 = 0.05 + 0.10 + 0.20 + 0.20 - 0.55$$

and so on. The results are summarized below. (Note that $F_{10} = 1.00$, as required.)

Interval Number	Interval Range	Number of Values	Relative Frequency	Cumulative Distribution
1	3.490 – 3.492	0	0	0
2	3.492 – 3.494	1	0.05	0.05
3	3.494 – 3.496	2	0.10	0.15
4	3.496 – 3.498	4	0.20	0.35
5	3.498 – 3.500	4	0.20	0.55
6	3.500 – 3.502	3	0.15	0.70
7	3.502 – 3.504	3	0.15	0.85
8	3.504 – 3.506	0	0	0.85
9	3.506 – 3.508	2	0.10	0.95
10	3.508 – 3.510	1	0.05	1.00

Sometimes the cumulative distribution is expressed in terms of percentages rather than decimal quantities. The percentages are obtained by multiplying the decimal quantities by 100. Thus, in the previous example, we could have written $F_2 = 5$ percent, $F_3 = 15$ percent, and so on.

Cumulative Distributions in Excel

In Excel, the cumulative distribution can be obtained at the same time the histogram is generated. To do so, we simply select Cumulative Percentage in the Histogram dialog box. The distribution will be expessed in terms of percentages. Note that it is possible to generate the histogram, the cumulative percentages, and the associated graphs all at the same time. The graph of the cumulative percentages will be shown as an *x-y* graph superimposed over the histogram bar graph, as illustrated in the next example.

Example 4.5 Generating a Cumulative Distribution in Excel

Modify the worksheet developed in Example 4.3 (see Figs. 4.3 and 4.4) to include the cumulative distribution, expressed as percentages, in addition to the histogram frequencies. Generate a combined graph of the frequencies and the cumulative percentages.

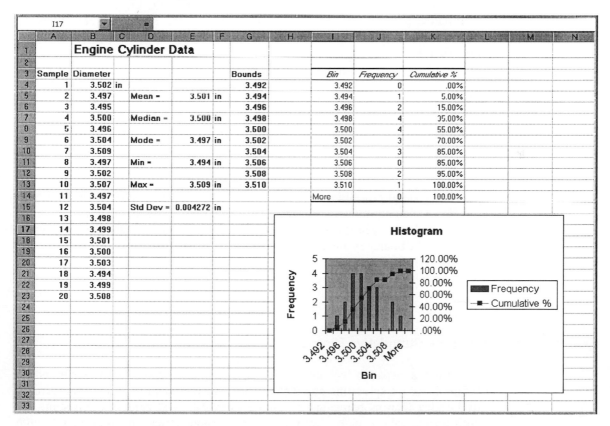

Figure 4.6

We begin in the same manner as in Example 4.3, with the worksheet shown in Fig. 4.2. (Recall that the Histogram dialog box was obtained by selecting Data Analysis/Histogram from the Tools menu). Now, however, we select both the Cumulative Percentage and the Chart Output features at the bottom of the Histogram dialog box. Clicking on the OK button produces the histogram shown in Fig. 4.6. The resulting worksheet is identical to that shown in Fig. 4.3 except that column K now contains the cumulative percentages in addition to the information generated earlier. The graphical display now shows a bar graph of the histogram and a plot of the cumulative percentages superimposed as a line graph. (You may have to move around in the worksheet in order to see the graphical display.)

Note that the graph in Fig. 4.6 contains *two* y-axis scales, one to the left of the graph and the other to the right. The axis on the left, labeled Frequency, applies to the bar graph of the frequencies, whereas the axis on the right, labeled with percent signs, applies to the superimposed x-y graph of the cumulative percentages.

There are two problems associated with histograms that include a cumulative distribution, such as the graph shown in Fig. 4.6. First, as discussed earlier, the bar graph intervals do not touch each other. Second, the x-y graph of the cumulative percentages connects points drawn at the *centers* of the intervals, whereas we would prefer to draw this line graph from the right interval boundaries.

We have already discussed a solution to the first problem; namely, click on a vertical bar, choose Selected Data Series/Options from the Format menu, and specify a value of 0 for the Gap width. The only way to correct the second problem, however, is to plot the cumulative distribution as a separate x-y graph, using the methods described in Sec. 3.3. This permits us much greater flexibility in the appearance of the final graph. In particular, it allows us to plot the cumulative distribution with the cumulative values associated with the right interval bounds rather than the centers of the intervals.

The general procedure is to assign a y-value of zero to the left boundary of the first nonempty interval and the first cumulative value as the y-value for the right boundary. Each successive cumulative value is then assigned as the y-value for the right boundary of its corresponding interval. The last cumulative value (either 1.0 or 100 percent) will then be assigned as the y-value corresponding to the right boundary of the last interval. The method is illustrated in Example 4.6.

Example 4.6 Plotting the Cumulative Distribution

Construct a separate x-y graph of the cumulative distribution generated in Example 4.5. Express the cumulative values as fractional values ranging from 0 to 1 rather than as percentages ranging from 0 to 100 percent. Associate the cumulative values with the right interval boundaries, in accordance with traditional procedures for displaying a cumulative distribution.

We begin with the worksheet shown in Fig. 4.6. We will proceed by first deleting the original graph and then copying the interval bounds (the *Bin* values) into a new location. Let us choose column I, beneath the original interval bounds (cells I17 through I27). These values will form the labels for the *x*-axis of the new graph. We then place the corresponding cumulative values within adjoining cells in column J (cells J17 through J27). Note that the cumulative values are now shown as fractions rather than percentages (choose Cells/Number from the Format menu, then select the desired numerical format.)

It is now quite easy to generate the desired *x-y* graph, using the methods described in Sec. 3.3. Figure 4.7 shows the resulting worksheet, containing the new *x-y* graph. Note that the titles, labels, and background color have been changed, grid lines have been added, and the legend has been removed. In addition, the entire graph has been enlarged so that it is more legible.

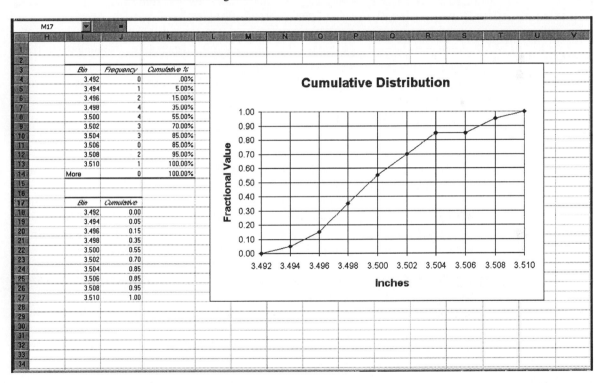

Figure 4.7

We can obtain some very useful information from the cumulative distribution once it is plotted in the form shown in Fig. 4.7. In particular, if we choose any value along the *x*-axis, the corresponding *y*-value indicates the likelihood that a single sample selected at random from the given population will have a value less than or equal to the *x*-value. Information of this type is of great importance to engineers involved in manufacturing processes or other processes in which quality control is an important consideration.

Example 4.7 Drawing Inferences from the Cumulative Distribution

From the cumulative distribution plotted in Fig. 4.7, estimate the likelihood that an arbitrary cylinder diameter within a randomly selected engine block will not exceed 3.503 inches.

To solve this problem, locate the value 3.503 along the *x*-axis in Fig. 4.7. The corresponding *y*-value is approximately 0.77. (The value of the gridlines should now be readily apparent.) Hence, we conclude that there is a 77 percent likelihood that the diameter of an arbitrarily selected cylinder does not exceed 3.503 inches. Conversely, there is a 23 percent likelihood (1 − 0.77) that the cylinder diameter *does* exceed 3.503 inches. The manufacturing engineer must now exercise some judgment in deciding whether or not the 77 percent value is sufficiently high.

The information obtained from the graph would probably be more accurate if a smooth curve were drawn through the aggregate of the data points (assuming that a larger number of data points would result in a smoother and more accurate graph). This can, of course, always be done by hand. However, Excel can also pass a smooth curve (a *trendline*) through a set of data without going through the individual data points. We will see how this is accomplished in the next chapter.

The subject of data analysis can be studied much more extensively, though the material is beyond the scope of our present discussion. You might consider taking a course in probability and statistics if you wish to pursue this topic further. For now, however, let us recognize that Excel includes many other features that are statistical in nature. You might wish to explore some of them by examining the entries listed in the Data Analysis dialog box under the Tools menu.

Problems

4.7 Using your own version of Excel, verify the solution to each of the following problems:

(*a*) Recreate the cumulative distribution and the corresponding graph shown in Fig. 4.6 (see Example 4.5).

(*b*) Modify the worksheet developed in Prob. 4.7(*a*) to obtain the worksheet and the graph shown in Fig. 4.7.

4.8 Using Equations (4.4) through (4.7), determine the relative frequencies and the cumulative distribution for the data given in Problem 4.2. Carry out the calculations by hand, using only a calculator, a pencil, and a piece of paper. (Do *not* use a spreadsheet to obtain a solution.) Do not plot the data.

4.9 Using the Analysis Toolpak/Histogram features included in Excel, carry out the following operations on the data given in Problem 4.2.

(a) Enter a series of interval (*bin*) bounds based upon one-inch interval spacing.

(b) Determine the frequencies associated with the intervals.

(c) Determine the cumulative percentages associated with the intervals.

(d) Create a bar graph of the frequencies with adjacent intervals, as shown in Fig. 4.5.

4.10 Using the histogram data developed in Problem 4.5, construct a graph of the cumulative distribution in each of the following ways:

(a) Using the Analysis Toolpak/Histogram features included in Excel, develop a combined bar graph/line graph with separate *y*-axes similar to that shown in Fig. 4.6.

(b) Develop a separate *x-y* graph of the cumulative distribution, as described in Example 4.6. Include a set of grid lines and appropriate labels, as shown in Fig. 4.7.

4.11 Plot the cumulative distribution data developed in Prob. 4.9(*c*). Then use the graph to answer the following questions:

(a) What is the likelihood that the height of an engineering student selected at random will not exceed 5 feet 10 inches?

(b) What is the likelihood that an engineering student selected at random will be at least six feet tall?

(c) Suppose the data used for Prob. 4.9(*c*) applies only to male students and that female students are assumed to be 10 percent shorter than male students. What is the likelihood that a female engineering student selected at random will be at least 5 feet 4 inches tall?

(d) Suppose you were selecting one male and one female engineering student to pose as a couple for a picture in the student newspaper. If the students are selected at random, what is the likehood that the height of the female student will not exceed 5 feet 5 inches and the height of the male student will be at least 5 feet 9 inches? (*Hint*: Obtain the product of the individual probabilities.)

4.12 The U.S. Environmental Protection Agency has tested the average fuel efficiency of 24 late-model cars equipped with V-6 engines and automatic transmissions. The results obtained are shown below.

Sample	Mileage (mpg)	Sample	Mileage (mpg)
1	22.9	13	25.5
2	23.9	14	22.2
3	21.4	15	21.7
4	25.4	16	23.5
5	23.9	17	27.1
6	24.4	18	23.0
7	23.1	19	23.9
8	22.0	20	23.6
9	25.4	21	19.2
10	20.7	22	22.7
11	21.4	23	26.0
12	22.8	24	21.3

Enter the data into an Excel spreadsheet and then analyze the data as follows:

(a) Determine the mean, median, mode, min, max, and standard deviation.

(b) Construct a histogram, based upon a reasonable interval width.

(c) Construct a cumulative distribution. Show the cumulative distribution in the form of an x-y graph.

4.13 Use the results of the previous problem to answer the following questions:

(a) Explain the difference between the mean and the median.

(b) From an examination of the histogram, what would you conclude about the data?

(c) If a car of this type is chosen at random, what is the likelihood that the fuel efficiency of this car will not exceed 20 mpg? What is the likelihood that the fuel efficiency will not exceed 22 mpg? What is the likelihood that the fuel efficiency *will* exceed 25 mpg?

4.14 An engineer is responsible for monitoring the quality of a batch of 1000-ohm resistors. To do so, the engineer must accurately measure the resistance of a number of resistors within the batch, selected at random. The results obtained are shown below.

Sample No.	Resistance, ohms	Sample No.	Resistance, ohms
1	1006	16	960
2	1006	17	976
3	978	18	954
4	965	19	1004
5	988	20	975
6	973	21	1014
7	1011	22	955
8	1007	23	973
9	935	24	993
10	1045	25	1023
11	1001	26	992
12	974	27	981
13	987	28	991
14	966	29	1013
15	1013	30	998

Enter the data into an Excel spreadsheet and then analyze the data as follows:

(a) Determine the mean, median, mode, min, max, and standard deviation.

(b) Construct a histogram, based upon a reasonable interval width.

(c) Construct a cumulative distribution. Show the cumulative distribution in the form of an *x-y* graph.

(d) Based upon the cumulative distribution for this random sample, how likely is it that a resistor selected at random will deviate from the target value of 1000 ohms by more than two percent, either above or below?

ADDITIONAL READING

Akai, T. J. *Applied Numerical Methods for Engineers.* New York: Wiley, 1994.

Barnes, J. W. *Statistical Analysis for Engineers and Scientists: A Computer-Based Approach.* New York: McGraw-Hill, 1994.

Devore, J. L. *Probability and Statistics for Engineers and the Sciences.* 3d ed. Pacific Grove, California: Brooks/Cole Publishing Co., 1991.

Eide, A. R., R. D. Jenison, L. H. Mashaw, and L. L. Northrup. *Engineering Fundamentals and Problem Solving.* 3d ed. New York: McGraw-Hill, 1997.

Hogg, R. V. and J. Ledolter. *Engineering Statistics.* New York: Macmillan, 1987.

Mendenhall, W. and T. Sincich. *Statistics for the Engineering and Computer Sciences.* 2d ed. San Francisco: Dellen Publishing Co., 1988.

Milton, J. S. and J. C. Arnold. *Introduction to Probability and Statistics: Principles and Applications for Engineering and the Computing Sciences.* 2d ed. New York: McGraw-Hill, 1990.

CHAPTER 5

FITTING EQUATIONS TO DATA

In the last chapter, we were concerned with the analysis of sets of single-valued data (i.e., $x_1, x_2, x_3, \cdots$, and so on.). We now turn our attention to the analysis of *paired* data values (i.e., $P_1 = (x_1, y_1)$, $P_2 = (x_2, y_2), \cdots$, and so on.).

Engineers frequently collect paired data in order to understand the characteristics of an object or the behavior of a system. The data may indicate a *spatial profile* (for example, temperature versus distance) or a *time history* (voltage versus time). Or the data may indicate *cause-and-effect relationships* (for example, force as a function of displacement) or *system output as a function of input* (yield of a chemical reaction as a function of temperature). Such relationships are often developed graphically, by plotting the data in a particular way. Mathematical expressions that capture the relationships shown in the data can then be developed.

Data that represent measured values usually show some *scatter*, which is due to fluctuations or errors in the measurements. Therefore, when fitting a curve through the data, we pass a curve through the *aggregate* of the data rather than the individual data points (though it is common practice to disregard "outliers"; that is, occasional isolated data points that are far removed from the main cluster, possibly because of erroneous measurements). This procedure allows us to capture the overall trend reflected by the entire data set.

Our goal, then, is to determine the equation of a curve that represents the aggregate of the data. To do so, we will make use of the *method of least squares* —a commonly used procedure for fitting a straight line or a curve to a set of data. In this chapter, we will see how the method can be applied to several different types of curves.

5.1 THE METHOD OF LEAST SQUARES

To understand the method of least squares, consider the graph shown in Fig. 5.1. This graph contains four data points and a curve passing through the aggregate of the data. The individual data points are represented as $P_1 = (x_1, y_1)$, $P_2 = (x_2, y_2)$, $P_3 = (x_3, y_3)$, and $P_4 = (x_4, y_4)$, and the equation for the curve is represented in general terms as $y = f(x)$.

For each data point, $P_i = (x_i, y_i)$, we can define an *error*, e_i, as the difference between y_i, the actual y-value, and $f(x_i)$, the corresponding calculated y-value. Thus, we can write

$$e_i = y_i - f(x_i) \tag{5.1}$$

for each of the data points; that is, for $i = 1, 2, \cdots, n$, where n is the total number of data points. In Fig. 5.1, for example, we can see that e_2 is defined as the difference between y_2, the actual data point, and $f(x_2)$, the corresponding point on the curve.

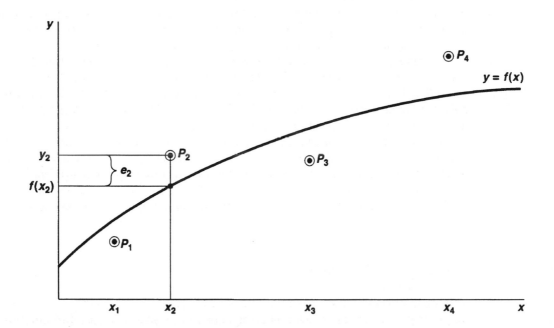

Figure 5.1

Remember that the curve that we refer to in general terms as $y = f(x)$ is actually represented by some specific equation. The exact equation depends upon the specific curve. For example, if we wish to fit the straight line $y = ax + b$ to the data, our goal is to determine the values of a and b that will result in a good curve fit. Similarly, if we wish to fit the polynomial $y = c_0 + c_1x + c_2x^2 + c_3x^3$ to the data, we must determine the values of c_0, c_1, c_2, and c_3 that will result in a good fit. Thus, *the overall strategy in fitting a curve to a set of data points is to determine an appropriate set of values for the coefficients in the equation of the chosen curve.*

Intuitively, we would like to select the unknown coefficients in such a manner that the error terms, e_i, will be as small *in magnitude* as possible (see Fig. 5.1). One approach might be to select the coefficients in such a manner that the sum of the errors is minimized. The problem with this approach, however, is that the individual errors might be large in magnitude but opposite in sign (i.e., several *positive* errors that are large in magnitude and several *negative* errors that are large in magnitude), so that the errors would tend to cancel each other when added together. This could result in some very bad curve fits, even though the *sum* of the errors might be very small.

A better approach is to select the coefficients in such a manner that the sum of the *squares* of the errors is minimized. Keep in mind that the square of the error will always be positive, regardless of whether the individual error is positive or negative. Hence, by summing the squares of the errors, we will always be summing positive numbers, thus eliminating the possibility that large errors might cancel one another because of differing signs. In other words, the only way the sum of the *squares* of the errors can be minimized is by making the individual errors as small in magnitude as possible, thus assuring us of a good fit. This is the idea behind the method of least squares.

The procedure, then, is to form an equation for the sum of the square errors in terms of the unknown coefficients. We can then use calculus to determine the values of the coefficients that minimize the sum of the errors. We will see how this is accomplished in Sec. 5.3. First, however, let us see exactly how the method is carried out, both by hand and using Excel, when fitting a straight line to a data set.

5.2 FITTING A STRAIGHT LINE TO A SET OF DATA

The most common type of curve fitting involves fitting a straight line to an aggregate of data. Situations of this type arise frequently, both in business and in technical applications.

To summarize, we wish to determine the values of the coefficients a and b in the straight-line equation $y = ax + b$ so that the line will pass through n data points. The coefficients are obtained by satisfying the *least squares criteria*. In particular, the coefficients are determined by solving the following two simultaneous equations:

$$a\sum_{i=1}^{n} x_i + bn = \sum_{i=1}^{n} y_i \tag{5.2}$$

$$a\sum_{i=1}^{n} x_i^2 + b\sum_{i=1}^{n} x_i = \sum_{i=1}^{n} x_i y_i \tag{5.3}$$

We will see how these two equations are obtained in Sec. 5.3. For now, let us simply observe that the unknown quantities in these equations are a and b. All of the other terms are either known (n is known in advance) or can be calculated from the given data (the various summation terms). The method is illustrated in the following example.

Example 5.1 Fitting a Straight Line to a Set of Data

An engineer has measured the force exerted by a spring as a function of its displacement from its equilibrium position. The following data have been obtained:

Data Point No.	Distance (cm)	Force (N)
1	2	2.0
2	4	3.5
3	7	4.5
4	11	8.0
5	17	9.5

Pass a straight line through the data points using the method of least squares. Then plot the data points and the resulting straight line.

Note that distance is the independent variable (x) and force is the dependent variable (y). Thus, we are seeking an equation for force as a function of distance. Note that there are five data points; hence, $n = 5$ in the least squares equations.

In order to apply the method of least squares, we must expand the above table as follows:

i	x_i	y_i	x_i^2	$x_i y_i$
1	2.0	2.0	4.0	4.0
2	4.0	3.5	16.0	14.0
3	7.0	4.5	49.0	31.5
4	11.0	8.0	121.0	88.0
5	17.0	9.5	289.0	161.5
Sums:	41.0	27.5	479.0	299.0

If we substitute these values into Equations (5.2) and (5.3), we obtain the following two simultaneous equations for a and b:

$$41a + 5b = 27.5$$

$$479a + 41b = 299$$

These equations can be solved by a variety of techniques, such as direct substitution or the use of Cramer's rule. They can also be solved using Excel, as explained in Chap. 8. For now, however, we simply state the solution as $a = 0.514706$ and $b = 1.279412$. (You can verify that these solutions are valid by substituting them into the equations and recalculating the right-hand values.) Thus, the equation of the line that passes through the data points is

$$y = 0.514706x + 1.279412$$

Figure 5.2 shows a plot of the given data and the line.

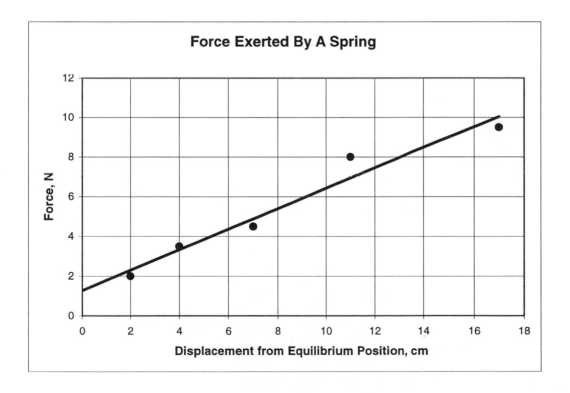

Figure 5.2

Once a line passing through the data has been determined, it is generally helpful to evaluate each of the error terms and then compute a numerical value for the sum of the squares of the errors (*SSE*), using the formula

$$SSE = \sum_{i=1}^{n} [y_i - f(x_i)]^2 \tag{5.4}$$

This quantity is an indication of the quality of the curve fit — the smaller the sum of the square errors, the better the fit. This procedure is particularly useful when fitting several different curves to the same set of data since the curve resulting in the smallest value for the sum of the square errors will generally provide the best fit.

Another indication of the quality of the curve fit is the so-called *r-squared* value, which is defined as

$$r^2 = 1 - \frac{SSE}{SST} \tag{5.5}$$

where *SST* represents the sum of the squares of the deviations about the mean, given by

$$SST = \sum_{i=1}^{n} [y_i - \bar{y}]^2 \tag{5.6}$$

The *r-squared* value varies between 0 and 1. Note that r^2 will equal 1 when *SSE* equals zero. Hence, an r^2 value close to 1 (which means that the sum of the square errors is small) generally indicates a good fit.

Example 5.2 Assessing a Curve Fit

Assess the quality of the curve fit obtained in the last example by computing the individual error terms and the sum of the square errors.

From the last example, the equation of the straight line that best represents the aggregate of the given data is $y = 0.514706x + 1.279412$. Knowing this equation, it is now a simple procedure to determine each of the error terms. For example, we can write

$$e_1 = y_1 - [0.514706\, x_1 + 1.279412]$$

$$e_1 = 2.0 - [(0.514706)(2.0) + 1.279412] = -0.308824$$

Similarly,

$$e_2 = y_2 - [0.514706 x_2 + 1.279412]$$

$$e_2 = 3.5 - [(0.514706)(4.0) + 1.279412] = 0.161764$$

and so on. The results are summarized in the table below. Note that the value of y determined from the least squares equation (column 4) is expressed as $y(x_i)$.

i	x_i	y_i	$y(x_i)$	e_i	e_i^2
1	2.0	2.0	2.308824	−0.308824	0.095372
2	4.0	3.5	3.338236	0.161764	0.026168
3	7.0	4.5	4.882354	−0.382354	0.146194
4	11.0	8.0	6.941178	1.058822	1.121104
5	17.0	9.5	10.029414	−0.529414	0.280279
	Sum:	27.5		*Sum:*	1.669117

If we sum the values in the last column, we determine that the sum of the squares of the errors (*SSE*) is 1.669117. This value is not particularly meaningful by itself, but if we were to fit several different curves to the same set of data, the sum of the square errors could be compared, thus providing a measure of the quality of each fit.

From the sum of the y-values, we can also determine the mean y value as

$$\bar{y} = 27.5 / 5 = 5.5$$

Hence, we can determine the values of *SST* and r^2 as

$$SST = [(2.0 - 5.5)^2 + (3.5 - 5.5)^2 + (4.5 - 5.5)^2 + (8.0 - 5.5)^2 + (9.5 - 5.5)^2] = 39.5$$

$$r^2 = 1 - 1.669117 / 39.5 = 0.957744$$

Note that the r^2 value is close to 1, which suggests a good fit.

Least Squares Curve Fitting in Excel

If all of this seems rather complicated, don't despair. Excel will do most of the work for you. In fact, in order to fit a straight line to a data set using Excel, you need only enter the data into a worksheet, plot the data in the usual manner, and then generate a *trendline* through the data. The method of least squares will automatically be applied to the data set, and the results displayed both graphically and algebraically. Thus, you need not concern yourself with constructing extended tables, solving simultaneous equations, or evaluating error terms.

To fit a straight line to a set of data in Excel, proceed as follows:

1. Open a new worksheet and enter the x-data (the independent variable) in the leftmost column.
2. Enter the y-data (the dependent variable) in the next column.
3. Plot the data as an x-y graph (i.e., an *XY Chart*) with arithmetic coordinates. Do not interconnect the individual data points.

4. Click on one of the plotted data points, thus selecting the data set as the active editing object. The data points will appear highlighted if this step is carried out correctly.

5. Choose Add Trendline from the Chart menu. Then specify the type of curve and request any pertinent options when the resulting dialog box appears. (Generally, you should request that the equation of the curve and its associated r^2 value be displayed. You may also wish to force the curve through a specified intercept or extrapolate the curve fit forward, i.e., beyond the rightmost data point, or backward, beyond the leftmost data point.)

6. Press the OK button. The curve fitting will then be carried out and the results displayed automatically.

The procedure is illustrated in the following example.

Example 5.3 Fitting a Straight Line to a Set of Data in Excel

Fit a straight line through the data given in Example 5.1 by first plotting the data and then using the Trendline feature in Excel. Extend the curve fit backward, to the point $x = 0$. Display the equation of the line passing through the data and its associated r^2 value.

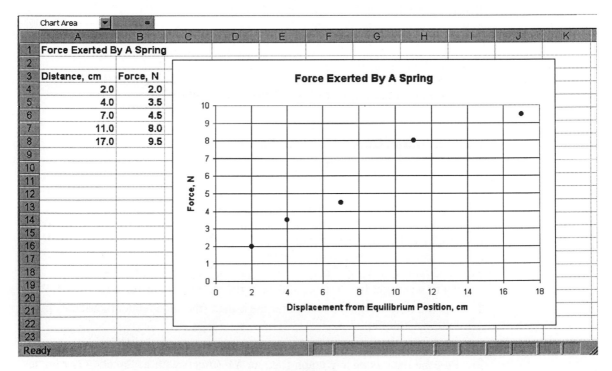

Figure 5.3

Following the procedure outlined above, we begin by entering and plotting the data within a worksheet, as shown in Fig. 5.3. The data are plotted as an *x-y* graph against a background of horizontal and vertical grid lines. Note that the individual data points are not interconnected.

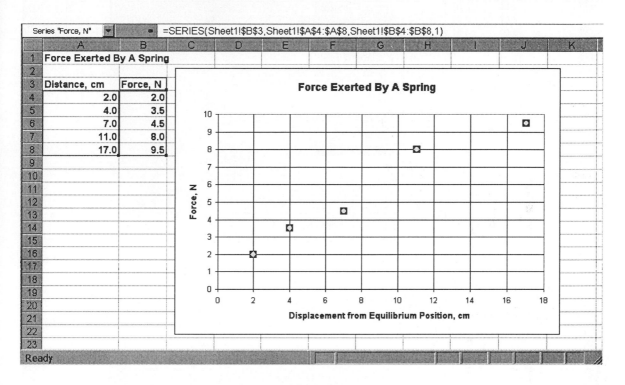

Figure 5.4

Figure 5.4 illustrates the appearance of the graph after the data set has been selected as the active editing object. We then select Add Trendline from the Chart menu, resulting in the dialog box shown in Fig. 5.5.

Notice that the Trendline dialog box shown in Fig. 5.5 contains two tabs. Initially, the dialog box associated with the Type tab is active, allowing us to select a linear (straight line) trendline. (Several of the remaining types of curves will be discussed in Sec. 5.4.)

We then select the Options tab. This results in another dialog box, from which we can specify that the trendline equation and the r^2 value will be displayed on the graph. The selection of these features is indicated at the bottom of Fig. 5.6. Also, note that 2 units are indicated in the backward direction under Forecast. This causes the trendline to be extended back two units beyond the leftmost data point ($x = 2$), to the point $x = 0$. We could also have forced the trendline to intercept the *y*-axis at a specified location if we had wished, by selecting the Set Intercept feature and entering a *y*-value in the data entry area.

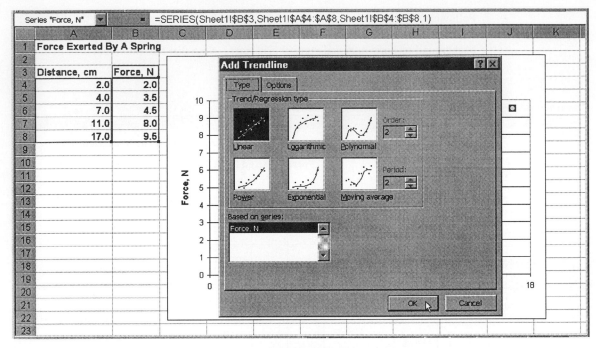

Figure 5.5

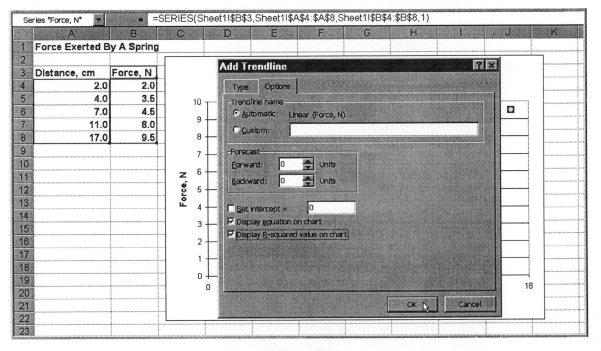

Figure 5.6

Clicking the OK button results in the graph shown in Fig. 5.7. Notice the straight line that has been drawn through the data set, ranging from the y-axis ($x = 0$) to the last data point ($x = 17$). This is the desired trendline. We also see that the equation of the trendline is

$$y = 0.5147x + 1.2794$$

as determined in Example 5.1. (The resulting equation for the trendline and the r^2 value have been moved above the graph in order to improve their legibility.) Moreover, we see that the calculated r^2 value is

$$r^2 = 0.9577$$

as determined in Example 5.2. *Note that the determination of the trendline equation and the r^2 value is entirely automatic.*

Unfortunately, the Trendline feature in Excel does not display the sum of the squares of the errors (*SSE*) associated with the curve fit. This value can, of course, be obtained manually within the worksheet, by determining the square of each of the error terms (once the trendline equation is known) and then summing their values.

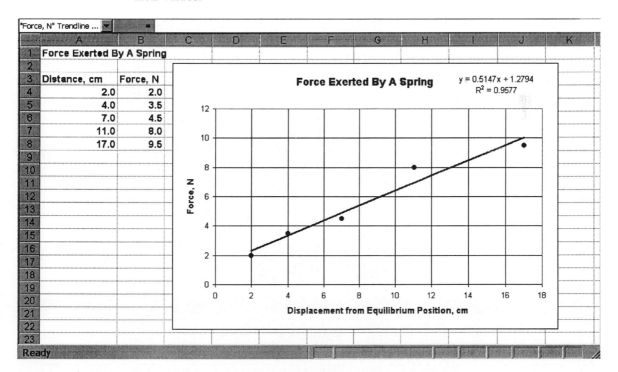

Figure 5.7

Excel also provides a Regression feature within the Analysis Toolpak. This feature will fit a straight line (called a *regression line*) to a set of data using the method of least squares. The resulting output includes the coefficients of the regression line equation, the sum of the squares of the errors, and the r^2 value. In addition, optional output can be requested, which includes a listing of the individual error terms (called *residuals*) and a plot of the data. From an engineer's perspective, however, the Regression feature may provide too much statistical information in the output summary. Much of this information is extraneous and may be confusing, as seen in the example presented below. Thus, the trendline feature discussed earlier provides a more straightforward approach to the problem of fitting a straight line through a set of data.

Example 5.4 Use of the Regression Feature in Excel

Fit a straight line through the data given in Example 5.1 using the Regression feature found in the Excel Analysis Toolpak. Include a list of the residuals in the output and generate a plot of the data.

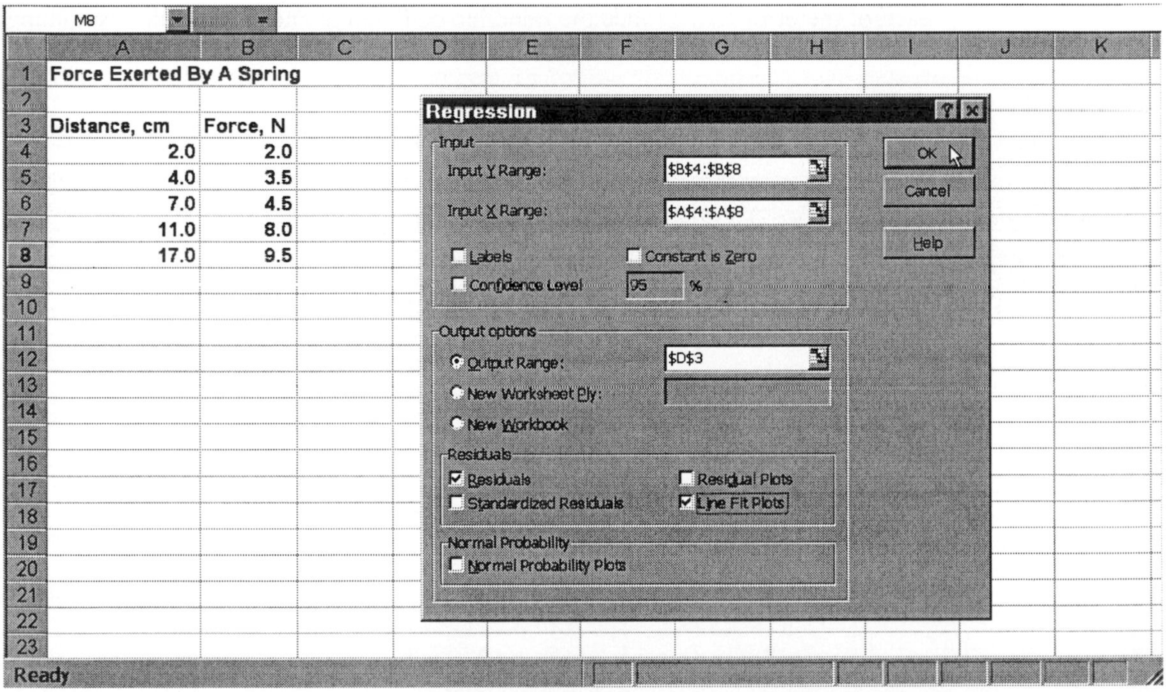

Figure 5.8

We begin by entering the data into a worksheet, as shown in the earlier figures (e.g., Fig. 5.3). We then select Data Analysis/Regression from the Tools menu. This results in the Regression dialog box shown in Fig. 5.8. Within this dialog box, the *y*-values are designated as the contents of cells B4 through B8, and the *x*-values as the contents of cells A4 through A8. The upper left corner of the output will appear in cell D3, as indicated in the address space corresponding to Output Range. Notice also that Residuals and Line Fit Plots have been selected as options.

Figure 5.9 shows the resulting output, once OK has been selected in the Regression dialog box. Much of this information is extraneous, of interest only to a trained statistician. Within the computed output, however, we see that the r^2 value (0.957744) is shown in cell E7, under the heading *Regression Statistics*. Similarly, the sum of the squares of the errors (1.669118) is shown in cell F15, under the heading *SS*. This value is labeled *Residual*.

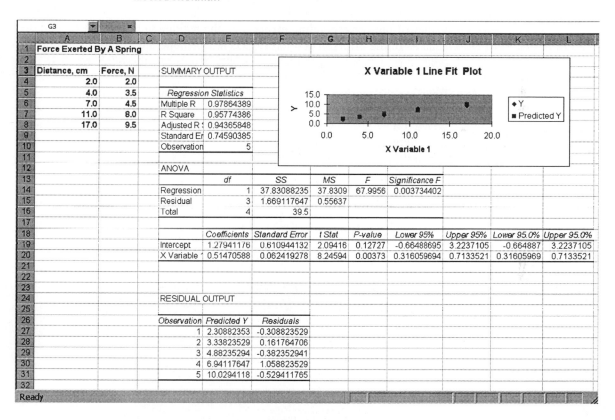

Figure 5.9

The coefficients in the regression equation are given in cells E19 and E20, under the heading *Coefficients*. Thus, the *y*-intercept (1.279412) is shown in cell E19, and the slope (0.514706) in cell E20. Also, the predicted *y*-values are shown in cells E27 through E31, and the corresponding error terms (residuals) in cells F27 through F31.

A plot of the given data points and the corresponding points generated by the regression line are shown in the graph at the top of the worksheet. The graph was dragged to this location to improve its legibility. It is still difficult to read, however, not only because of its size (which can easily be increased), but also because the resulting curve fit is not shown as an interconnected set of points on a straight line (compare with the results shown in Fig. 5.7).

Problems

5.1 Recreate the worksheet shown in Fig. 5.7, including the graph. Reposition the graph if necessary, so that columns C, D, and E are accessible. Use the trendline formula to generate a set of y-values that correspond to the x-values given in column A. Place these y-values in column C. Generate the values of the corresponding errors (i.e., the residuals) in column D, and the squares of the errors in column E. Then determine the sum of the values in column E. Compare with the values obtained in Examples 5.2 and 5.4.

5.2 Recreate the worksheet shown in Fig. 5.7. Extend the trendline across the entire graph (i.e., from $x = 0$ to $x = 20$). Then delete the trendline and create a new one that passes through the origin. Notice the r^2 value that is now obtained and compare it with the original r^2 value. What do you conclude about the relationship between the r^2 value and the quality of the curve fit?

5.3 A polymeric material contains a solvent that dissolves as a function of time. The concentration of the solvent, expressed as a percentage of the total weight of the polymer, is shown in the following table as a function of time (repeated from Exercise 3.6).

Solvent Concentration (weight percent)	Time (sec)
55.5	0
44.7	2
38.0	4
34.7	6
30.6	8
27.2	10
22.0	12
15.9	14
8.1	16
2.9	18
1.5	20

Enter the data into an Excel worksheet, plot the data, and fit a straight line through the data. Determine the equation of the line and the corresponding r^2 value.

5.4   Repeat Problem 5.3 using the following set of data:

Solvent Concentration (weight percent)	Time (sec)
30.2	0
44.7	2
22.5	4
41.3	6
28.8	8
14.0	10
26.2	12
11.0	14
23.4	16
14.5	18
4.2	20

Compare the results with those obtained for Prob. 5.3.   Which data set results in a better fit, and why?

5.5   A popular consumer magazine tabulated the following list of weight versus overall gasoline mileage for several different sizes and types of cars:

Weight (lbs)	Mileage (mpg)	Weight (lbs)	Mileage (mpg)
2775	33	3325	20
2495	27	3200	21
2405	29	3450	19
2545	28	3515	21
2270	34	3495	19
2560	24	4010	19
3050	23	4205	17
3710	24	2900	24
3085	23	2555	28
2940	21	2790	21
2395	26	2190	34

From these data, develop a straight-line correlation (i.e., an equation) for gasoline mileage as a function of weight.   Based upon your results, how well are the given data represented by the straight-line relationship?   How might a better relationship be obtained?

5.3 DERIVATION OF THE LEAST SQUARES EQUATIONS

For those of you who are more mathematically inclined, this section describes how Equations (5.2) and (5.3) were obtained. You can skip this section if you are not interested.

Our goal is to determine the values of a and b in the equation $y = ax + b$ in such a manner that the sum of the squares of the errors will be minimized. In Equation (5.1) we wrote each error term as $e_i = y_i - f(x_i)$. Now let us substitute the equation for the straight line into the general expression for $f(x)$. Hence, we can write the error term as

$$e_i = y_i - (ax_i + b) \tag{5.7}$$

If we let z represent the sum of the squares of the errors (*SSE*), we can write

$$z = e_1{}^2 + e_2{}^2 + e_3{}^2 + \cdots$$

$$z = [y_1 - (ax_1 + b)]^2 + [y_2 - (ax_2 + b)]^2 + [y_3 - (ax_3 + b)]^2 + \cdots$$

$$z = \sum_{i=1}^{n} [y_i - (ax_i + b)]^2 \tag{5.8}$$

Our goal, then, is to determine the values of a and b that will minimize z in Equation (5.8). To do so, we must set the derivatives of z with respect to a and b equal to zero. Thus, we first take the derivative of z with respect to a, holding b constant, and set the result equal to zero. We then take the derivative of z with respect to b, holding a constant, and set it equal to zero. (These are called *partial derivatives*.) We can write these derivatives as

$$\frac{\partial z}{\partial a} = -2 \sum_{i=1}^{n} x_i [y_i - (ax_i + b)] = 0$$

$$\frac{\partial z}{\partial b} = -2 \sum_{i=1}^{n} [y_i - (ax_i + b)] = 0$$

Now let us divide each equation by 2 and then write the second equation ahead of the first equation, resulting in

$$\sum_{i=1}^{n} [ax_i + b - y_i] = 0$$

$$\sum_{i=1}^{n} [ax_i^2 + bx_i - x_i y_i] = 0$$

Finally, we use the distributive rule for addition and then factor the constants a and b out of the resulting summation terms. This results in the final form of the least squares equations.

$$a\sum_{i=1}^{n} x_i + bn = \sum_{i=1}^{n} y_i \qquad (5.2)$$

$$a\sum_{i=1}^{n} x_i^2 + b\sum_{i=1}^{n} x_i = \sum_{i=1}^{n} x_i y_i \qquad (5.3)$$

Remember that Equations (5.2) and (5.3) apply only when fitting a *straight line* to a set of data. The method of least squares can also be used to fit other types of equations to a data set, though the individual equations will differ from Equations (5.2) and (5.3). The method used to obtain these equations is the same, however, as that described above. We will say more about the use of other types of equations with the method of least squares in the next section.

5.4 FITTING OTHER FUNCTIONS TO A SET OF DATA

The method of least squares can be used to fit many different types of functions through a set of data points. Thus, we need not be confined to straight-line relationships when fitting a curve through a data set. In fact, the data obtained in many engineering applications may be better represented by an exponential function, a power function, or a polynomial rather than a straight line.

Exponential Functions

We have already discussed the fact that the exponential function

$$y = ae^{bx} \qquad (5.9)$$

governs many different phenomena in engineering (see Sec. 3.4). If we take the natural log of each side of Equation (5.9), we obtain

$$\ln y = \ln a + bx \qquad (5.10)$$

If we let $u = \ln y$ and $c = \ln a$, we can rewrite Equation (5.10) as

$$u = bx + c \qquad (5.11)$$

which is the equation for a straight line (but *not* the straight line $y = ax + b$ that we considered earlier in this chapter). Thus, we can fit an exponential function to a set of data by proceeding as we did for a straight line, substituting ln y for y, b for a, and c for b in Equations (5.2) and (5.3); that is,

$$b\sum_{i=1}^{n} x_i + cn = \sum_{i=1}^{n} \ln y_i \tag{5.12}$$

$$b\sum_{i=1}^{n} x_i^2 + c\sum_{i=1}^{n} x_i = \sum_{i=1}^{n} x_i \ln y_i \tag{5.13}$$

After solving Equations (5.12) and (5.13) for b and c, the original coefficient a is then obtained by taking the antilog of c; that is,

$$a = e^c \tag{5.14}$$

The entire procedure is illustrated in the following example.

Example 5.5 Fitting an Exponential Function to a Set of Data

The transient behavior of a capacitor has been studied by measuring the voltage drop across the device as a function of time. The following data have been obtained.

Time, sec	Voltage	Time, sec	Voltage
0	10	6	0.5
1	6.1	7	0.3
2	3.7	8	0.2
3	2.2	9	0.1
4	1.4	10	0.07
5	0.8	12	0.03

With electronic devices of this type, the voltage generally varies exponentially with time. We will therefore fit an exponential function to the current set of data, using the method of least squares.

To apply the method of least squares to this data set, let us represent the time as x and the voltage as y. We then form the following table:

i	x_i	x_i^2	y_i	$\ln y_i$	$x_i \ln y_i$
1	0	0	10	2.302585	0
2	1	1	6.1	1.808289	1.808289
3	2	4	3.7	1.308333	2.616666
4	3	9	2.2	0.788457	2.365371
5	4	16	1.4	0.336472	1.345888
6	5	25	0.8	−0.223144	−1.115720
7	6	36	0.5	−0.693147	−4.158882
8	7	49	0.3	−1.203973	−8.427811
9	8	64	0.2	−1.609438	−12.875504
10	9	81	0.1	−2.302585	−20.723265
11	10	100	0.07	−2.659260	−26.592600
12	12	144	0.03	−3.506558	−42.078695
Sums:	67	529		−5.653969	−107.836263

Substituting the appropriate values into Equations (5.12) and (5.13), we obtain the following two least squares equations:

$$67b + 12c = -5.653969$$

$$529b + 67c = -107.836263$$

The solution to these equations is $b = -0.492318$ and $c = 2.277612$. Hence,

$$a = e^C = e^{2.277612} = 9.753358$$

and the desired exponential function is

$$y = ae^{bx} = 9.753358e^{-0.492318x}$$

Though the setup in the previous example enhances your understanding of the least squares process using an exponential function, it is not necessary when using Excel. In fact, Excel does all of the work for you, as was the case when we fit a straight line to a data set.

The procedure for fitting an exponential function to a set of data in Excel is the same as the procedure outlined in Sec. 5.2 for a straight line. Now, however, we select an exponential function when fitting the trendline to the data. The procedure is illustrated in the following example.

Example 5.6 Fitting an Exponential Function to a Set of Data in Excel

Fit an exponential function through the data given in Example 5.5 by plotting the data within an Excel worksheet and then fitting an exponential trendline through the graph. Display the equation of the exponential function and its corresponding r^2 value.

The solution proceeds in the same manner as described in Example 5.3, where we fit a straight line to a set of data. Thus, we enter the given data into a worksheet, plot the data as an *x-y* graph, and pass a trendline through the graph. Now, however, we select an exponential trendline rather than a linear (straight line) trendline.

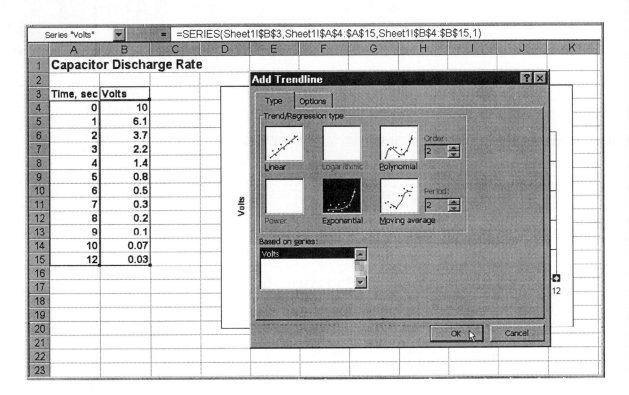

Figure 5.10

Figure 5.10 shows the worksheet containing the data and the line graph, with the Trendline dialog box superimposed over the line graph. Notice that the Exponential icon is highlighted, indicating that an exponential function will be passed through the data.

The graph containing the trendline is shown in Fig. 5.11. Note that the trendline appears to fit the data very well. The equation of the trendline is shown as

$$y = 9.7534e^{-0.4923x}$$

which agrees with the result obtained in Example 5.5. In addition, we see that the r^2 value is 0.9988, which indicates a good fit.

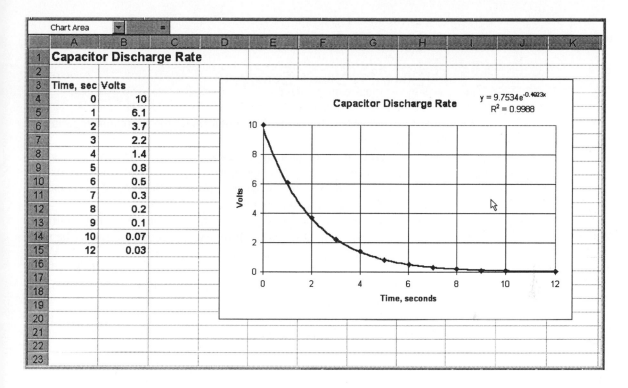

Time, sec	Volts
0	10
1	6.1
2	3.7
3	2.2
4	1.4
5	0.8
6	0.5
7	0.3
8	0.2
9	0.1
10	0.07
12	0.03

Figure 5.11

It should be mentioned that the Analysis Toolpak Regression feature, described at the end of Sec. 5.2, can also be used to fit an exponential function to a data set. When using this feature, however, the y-values are not automatically transformed into logarithmic form. Thus, the natural logarithms of the y-values must be entered into the worksheet explicitly (they can, of course, be calculated directly from the given y-values, using the LN function). The cell addresses of these ln y values are then specified as the Input Y Range within the Regression dialog box.

Logarithmic Functions

A *logarithmic function* is an equation of the form

$$y = a \ln x + b \tag{5.15}$$

This is clearly a linear function between y and $\ln x$. Hence, the applicable least squares equations in this case are

$$a \sum_{i=1}^{n} \ln x_i + bn = \sum_{i=1}^{n} y_i \tag{5.16}$$

$$a \sum_{i=1}^{n} (\ln x_i)^2 + b \sum_{i=1}^{n} \ln x_i = \sum_{i=1}^{n} (\ln x_i) y_i \tag{5.17}$$

Solving these two equations for a and b will determine the particular logarithmic function that best fits the data.

In Excel, the procedure for passing a logarithmic trendline through a given data set is identical to that for a straight line or an exponential function, except that the logarithmic function is selected when specifying the type of trendline (as shown in Fig. 5.10 for an exponential trendline).

The logarithmic function can be defined in terms of base-10 logarithms as well as natural logarithms. In either case, however, the function can only be used with data sets in which the values of the independent variable (the x-values) are positive since logarithms are undefined for zero or negative values. (In Fig. 5.10, one of the x-values is equal to zero. Hence, the logarithmic function is unavailable for that particular data set.)

Use of the logarithmic function in engineering applications is relatively uncommon. Hence, we will not discuss it further (see, however, Prob. 5.7 at the end of this section).

Power Functions

In Sec. 3.5 we considered the power function, which is an equation of the form

$$y = ax^b \tag{5.18}$$

This equation, like the exponential equation, describes many phenomena that occur in science and engineering.

To see how the method of least squares applies to a power function, we take the natural log of each side of Equation (5.18), resulting in

$$\ln y = \ln a + b \ln x \tag{5.19}$$

Now if we let $u = \ln x$, $v = \ln y$, and $c = \ln a$, we obtain

$$v = bu + c \tag{5.20}$$

which is the equation for a straight line. (Remember, however, that this straight line is different than those encountered earlier in this chapter.) Therefore, we can apply the method of least squares to the logs of the variables.

The least squares equations for a power function are obtained by substituting $\ln x$ for x, $\ln y$ for y, b for a, and c for b in Equations (5.2) and (5.3), resulting in

$$b\sum_{i=1}^{n}\ln x_i + cn = \sum_{i=1}^{n}\ln y_i \qquad (5.21)$$

$$b\sum_{i=1}^{n}(\ln x_i)^2 + c\sum_{i=1}^{n}\ln x_i = \sum_{i=1}^{n}(\ln x_i)(\ln y_i) \qquad (5.22)$$

Once these equations have been solved for b and c, the original coefficient a is obtained as

$$a = e^c \qquad (5.23)$$

The use of Equations (5.21) through (5.23) to fit a power function to a set of data parallels the use of Equations (5.12) through (5.14) to fit an exponential function to a set of data, as illustrated in Example 5.5 (see Prob. 5.9 at the end of this section).

In Excel, all of this is handled automatically, by passing a trendline through a plot of the data. The procedure is the same as that described earlier. Now, however, we specify a power function when selecting the type of trendline, as described in the next example.

Example 5.7 Fitting a Power Function to a Set of Data

A chemical engineer is studying the rate at which a reactant is consumed in a chemical reaction involving the manufacture of a polymer. The following data have been obtained, showing the reaction rate (in moles per second) as a function of the concentration of the reactant (moles per cubic foot).

Concentration, moles/cu ft	Reaction Rate, moles/sec
100	2.85
80	2.00
60	1.25
40	0.67
20	0.22
10	0.072
5	0.024
1	0.0018

With reactions of this type, the reaction rate is generally proportional to the concentration of the reactant raised to some power. Therefore, we will use Excel to determine the values of the constant of proportionality and the power to which the reactant concentration is raised by fitting a power function to the data. That is, we will determine the values for the parameters a and b in the expression

$$RR = aC^b$$

where RR is the reaction rate and C is the concentration of the reactant.

To solve this problem, we again enter the data into an Excel worksheet and plot the data as an *x-y* graph. We then pass a power trendline through the graph and determine the parameters in the resulting equation. Figure 5.12 shows the worksheet containing the data, with the Trendline dialog box superimposed over the graph. The Power icon will be activated, indicating the selection of a power function, when we click on the OK button.

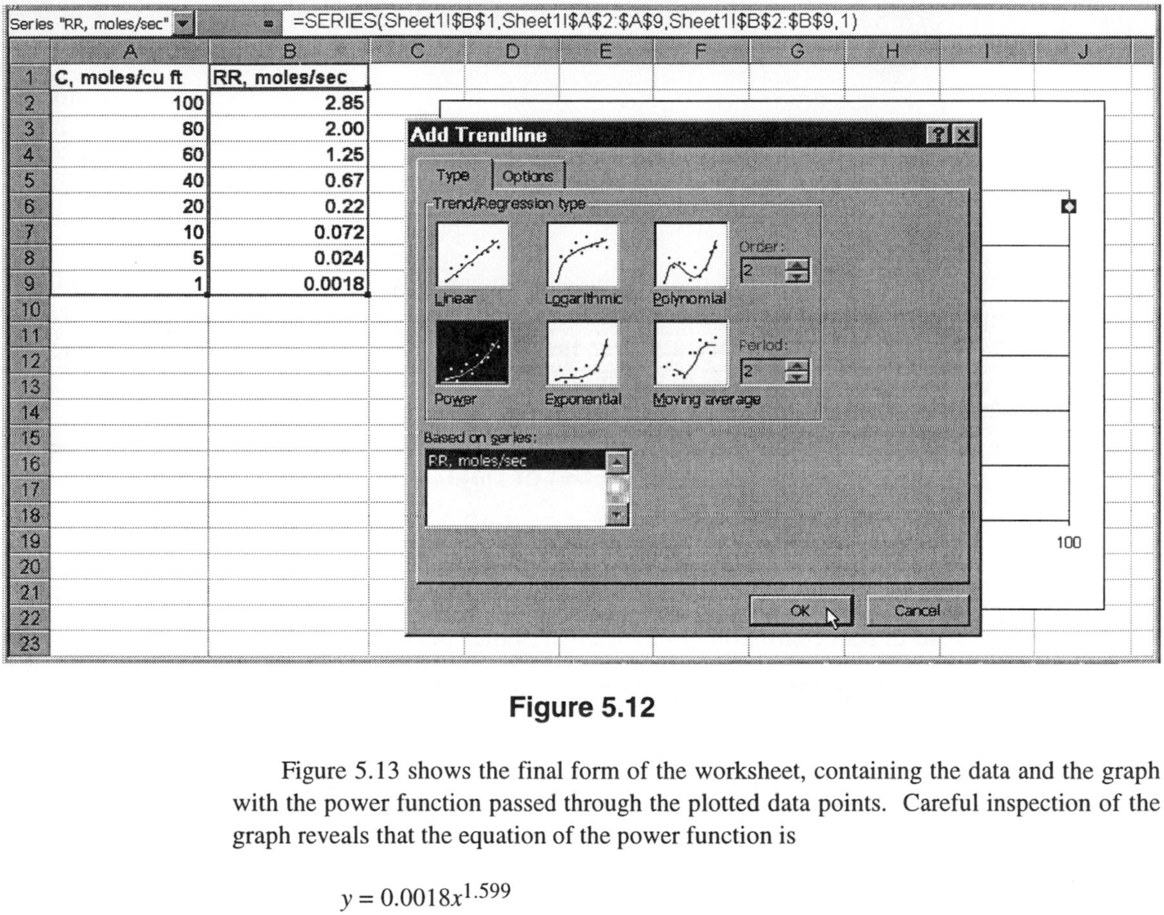

Figure 5.12

Figure 5.13 shows the final form of the worksheet, containing the data and the graph with the power function passed through the plotted data points. Careful inspection of the graph reveals that the equation of the power function is

$$y = 0.0018x^{1.599}$$

and the resulting r^2 value is 1. Thus, we conclude that the reaction rate can indeed be represented as a function of the reactant concentration using the power function

$$RR = 0.0018C^{1.599}$$

Remember that the method of least squares relates the *logarithm* of *y* to the *logarithm* of *x* when applied to a power function. Therefore all of the *x*- and *y*-values in the data set must be positive since the logarithm of a nonpositive value is undefined. If one or more data points is nonpositive, the Power icon will be inactive when selecting a trendline type (as, for example, in Fig. 5.10).

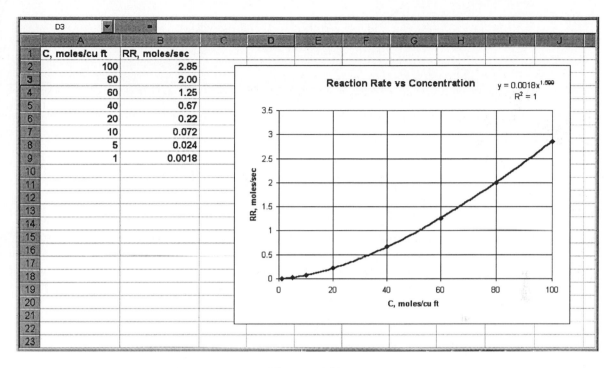

	A	B
1	C, moles/cu ft	RR, moles/sec
2	100	2.85
3	80	2.00
4	60	1.25
5	40	0.67
6	20	0.22
7	10	0.072
8	5	0.024
9	1	0.0018

Figure 5.13

The Analysis Toolpak Regression feature can also be used to fit a power function to a data set using the method of least squares. When doing so, however, remember that the logarithms of the data must first be determined manually and placed in separate columns (or separate rows) within the worksheet. Either natural logarithms or base-10 logarithms can be used, though you must be consistent within a given data set.

Polynomials

The method of least squares can also be used to fit polynomials to a data set. The good news is that logarithmic transformations are not required. The bad news, however, is that a kth-degree polynomial requires the solution of $k+1$ simultaneous equations in $k+1$ unknowns. Thus, fitting a cubic equation involves solving four equations in four unknowns, and so on.

Let us write a kth-degree polynomial as

$$y = c_1 + c_2x + c_3x^2 + \cdots + c_{k+1}\,x^k \tag{5.24}$$

If we were to fit this polynomial to a set of data using the method of least squares, the resulting simultaneous equations would be

$$nc_1 + c_2 \sum_{i=1}^{n} x_i + c_3 \sum_{i=1}^{n} x_i^2 + \cdots + c_{k+1} \sum_{i=1}^{n} x_i^k = \sum_{i=1}^{n} y_i \tag{5.25}$$

$$c_1 \sum_{i=1}^{n} x_i + c_2 \sum_{i=1}^{n} x_i^2 + c_3 \sum_{i=1}^{n} x_i^3 + \cdots + c_{k+1} \sum_{i=1}^{n} x_i^{k+1} = \sum_{i=1}^{n} x_i y_i \tag{5.26}$$

. . . . .

$$c_1 \sum_{i=1}^{n} x_i^k + c_2 \sum_{i=1}^{n} x_i^{k+1} + c_3 \sum_{i=1}^{n} x_i^{k+2} + \cdots + c_{k+1} \sum_{i=1}^{n} x_i^{2k} = \sum_{i=1}^{n} x_i^k y_i \tag{5.27}$$

Thus, we have $(k+1)$ equations in $(k+1)$ unknowns. The unknowns are $c_1, c_2, \ldots,$ c_{k+1}. The straight-line curve fit discussed in Sec. 5.2 is a special case, resulting in two equations in two unknowns. If we wanted to fit a quadratic equation to the data, we would solve three equations in three unknowns, and so on.

There are several different techniques available for solving simultaneous algebraic equations such as those expressed by Equations (5.25) through (5.27). We will not present any of these techniques at this time because Excel solves the simultaneous equations for us when fitting a polynomial to the data. Note, however, that Chap. 8 discusses techniques for solving systems of simultaneous algebraic equations in Excel (see also Prob. 5.10 at the end of this section).

Example 5.8 Fitting a Polynomial to a Set of Data

The following table presents the time for a high-performance sports car to reach various speeds. The times are given in seconds and the speeds in miles per hour. (The top speed is considered to be the independent variable in this application, and time is the *dependent* variable.) Fit an equation to the data, thus obtaining an accurate mathematical relationship between acceleration time and top speed.

Top Speed, mph	Time, sec
30	1.9
40	2.8
50	3.8
60	5.2
70	6.5
80	8.3
90	10.4
100	12.7
110	15.6
120	19.0
130	23.2
140	31.2
150	45.1

Figure 5.14 shows an Excel worksheet containing the data and a corresponding *x-y* graph. From the shape of the graph, it is obvious that a straight-line curve fit is not satisfactory. Moreover, some experimentation with the power function and the logarithmic function shows that neither of these functions provides a satisfactory fit, either. The fit resulting from the use of an exponential function is not bad, though it appears that a better fit might be obtained using a polynomial.

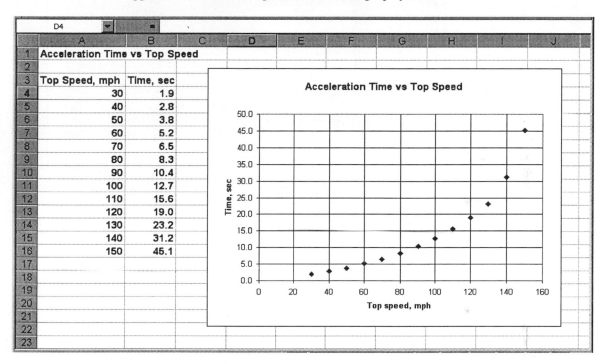

Figure 5.14

Let us pass a polynomial-type trendline through the data, using the same general approach as with the other functions discussed in this chapter. Thus, Fig. 5.15 shows the Trendline dialog box superimposed over the line graph, with a polynomial of order 5 (i.e., a fifth-degree polynomial) being selected. The decision to use this particular polynomial is entirely arbitrary. (Note that the order of the polynomial can be increased or decreased by clicking on the appropriate arrow shown next to the data entry box labeled Order.)

The resulting trendline (rearranged somewhat to show the equation more clearly) is shown in Fig. 5.16. From visual inspection, we can see that the fifth-degree polynomial

$$y = 1 \times 10^{-8} x^5 - 5 \times 10^{-6} x^4 + 8 \times 10^{-4} x^3 - 0.0525 x^2 + 1.777 x - 20.958$$

fits the data very well. (Note that y represents the time to reach top speed, in seconds, and x represents the top speed, in mph.) The corresponding value of $r^2 = 0.9998$ further supports this argument.

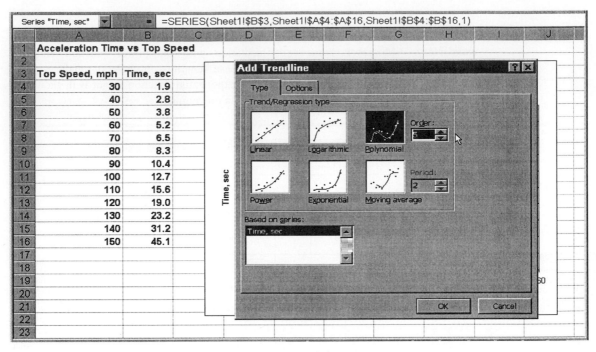

Figure 5.15

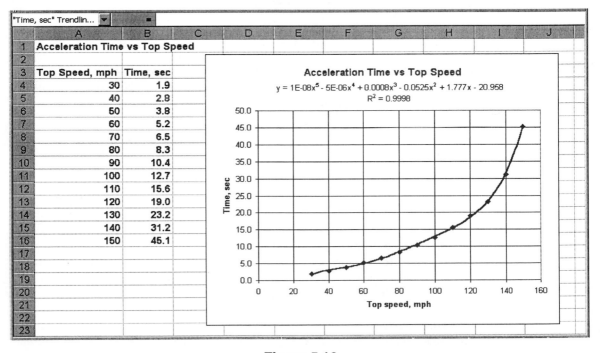

Figure 5.16

Since the first two coefficients are very small (1×10^{-8} and -5×10^{-6}, respectively), it might be argued that the fifth-degree polynomial is actually not required in order to obtain a satisfactory fit — a third-degree polynomial (i.e., a cubic equation) might be sufficient. Keep in mind, however, that these small coefficients will be multiplied by large numbers (the values of x^5 and x^4) when x takes on modestly large values. Thus, the magnitude of the first two terms in the polynomial might be significant. (We will pursue this issue further in Example 5.10, where we fit several different functions to the current set of data.)

Some warnings about higher-degree polynomials: First, the results may be inaccurate due to numerical errors in the curvefitting procedure. Second, even if the curvefitting is done accurately, use of the resulting curve fit may involve the product of very large and very small numbers, resulting in inaccurate results. And finally, the trendlines shown in Excel involve rounded coefficients, whose use may produce sizeable errors with relatively large or relatively small values of x.

Problems

5.6 For the data given in Example 5.5, carry out the following calculations:

(a) Calculate the sum of the squares of the errors using the (correct) least squares error criterion

$$e_i = \ln y_i - (\ln a + bx_i)$$

(b) Calculate the sum of the squares of the errors using the error criterion

$$e_i = y_i - ae^{bx_i}$$

(c) Use Equation (5.6) to determine a value of SST for this data set.

(d) Solve Equation (5.5) for SSE, using the value of SST obtained in part (c) and the value of $r^2 = 0.9988$ obtained in Example 5.6. Compare with the values for the sum of the squares of the errors obtained in parts (a) and (b). What do you conclude from this comparison?

5.7 The following data represent the temperature as a function of vertical depth within a chemically active settling pond. Fit the data to a logarithmic function using the method of least squares, as expressed by Equations (5.16) and (5.17).

Distance, cm	Temp, °C	Distance, cm	Temp, °C
0.1	21.2	390	45.9
0.8	27.3	710	47.7
3.6	31.8	1200	49.2
12	35.6	1800	50.5
120	42.3	2400	51.4

5.8 Enter the data given in Prob. 5.7 into an Excel worksheet and fit an appropriate trendline to the aggregate of the data. Compare the results obtained using an exponential function, a logarithmic function, a power function, and a fifth-degree polynomial.

5.9 Use Equations (5.21) through (5.23) to fit a power function to the data given in Example 5.7. Compare with the results obtained in Example 5.7, using Excel.

5.10 Verify that the curve fit obtained in Example 5.8 is correct by carrying out the following calculations:

(*a*) Write the least squares equations required for a fifth-degree polynomial, using Equations (5.25) through (5.27) as a guide. (This requires writing six equations in six unknowns.)

(*b*) Enter the data given in Example 5.8 into an Excel worksheet.

(*c*) Create additional columns within the worksheet to represent the components of the various summations appearing in the six simultaneous equations. Calculate the appropriate sum at the bottom of each column.

(*d*) Verify that the least squares equations are satisfied by substituting the summation terms into the six simultaneous equations.

5.11 Enter the data given in Example 5.7 into an Excel worksheet and carry out the following calculations:

(*a*) Verify that the results obtained in Example 5.7 are correct by fitting a power function to the data.

(*b*) Fit a straight line, an exponential function and a logarithmic function to the data. Compare with the results obtained in part (*a*).

(*c*) Fit several different polynomials of varying degrees through the data. Compare the results with each other and with the results obtained in parts (*a*) and (*b*) above.

Additional curvefitting problems are presented at the end of this chapter.

5.5 SELECTING THE BEST FUNCTION FOR A GIVEN DATA SET

Now that we know how to fit various functions to a data set, one question remains: How does one determine which function to fit to a particular set of data? There is no simple answer to this question — the issue is usually decided by trial and error. There are, however, some guidelines that may prove helpful, as outlined below:

1. Try to plot the data as a straight line. If this is successful, it will indicate the function that should be used to represent the data.

2. If a straight-line relationship cannot be obtained, try fitting different types of curves to the data. Use visual assessment to recognize a good fit, aided by the resulting values for the sum of the squares of the errors (*SSE*) and the r^2 parameter.

3. If a satisfactory curve fit cannot be obtained using the first two guidelines, try plotting the data differently (for example, try plotting y versus $1/x$, $1/y$ versus x, etc.).

4. In some cases, a better fit may be obtained by *scaling* the data so that the magnitude of the x-values is more-or-less the same as the y-values.

Each step is discussed separately below.

Plotting the Data as a Straight Line

If a data set can be plotted as a straight line on arithmetic coordinates, you should clearly fit a straight line to the data set. Recall, however, that other equations plot as straight lines with different coordinate systems, as explained in Secs. 3.3 and 5.4. The results are summarized in the following table:

Equation Type	*Equation*	*Coordinate System*
Exponential	$y = ae^{bx}$	log y versus x (semi-log)
Logarithmic	$y = a \ln x + b$	y versus log x
Power	$y = ax^b$	log y versus log x (log-log)

We have already seen that a given data set can easily be plotted with various coordinate systems in Excel by first plotting the data as an x-y graph (i.e., an XY chart) and then selecting logarithmic coordinates for one or both axes. Thus, *if the data appear as a straight line when plotted in one of these coordinate systems, we will have a clear indication of what type of trendline to fit to the data.*

Example 5.9 Obtaining a Straight-Line Plot Using Different Coordinate Systems

In Example 5.7, a power function was used to represent a set of data. The selection of the power function was based upon previous familiarity with the process from which the data had been obtained. Verify the desirability of the power function by plotting the data on arithmetic coordinates, semi-log coordinates (both log y versus x and x versus log y), and log-log coordinates. Seek a coordinate system that will result in a straight-line relationship.

We begin by plotting the data on ordinary arithmetic coordinates. This is the same plot that was obtained in Example 5.7, when fitting a trendline to the data (see Fig. 5.13). From the arithmetic graph shown in Fig. 5.17, we see that this data set does not plot as a straight line on arithmetic coordinates.

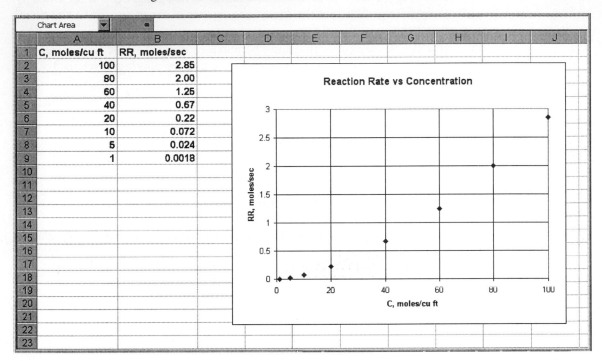

Figure 5.17

We then click on the y-axis and choose Selected Axis/Scale/Logarithmic from the Format menu, as described in Chap. 3 (see Sec. 3.4). We will then obtain the semi-log plot shown in Fig. 5.18. Clearly, the data do not plot as a straight-line on semi-logarithmic coordinates. In fact, the departure from a straight-line plot is now much worse than it was before.

We then restore the arithmetic coordinates along the y-axis, and select a logarithmic scale along the x-axis. This results in the plot shown in Fig. 5.19. The departure from a straight line is now about as bad as in Fig. 5.18, though the curvature of the data is in the opposite direction (i.e., convex rather than concave, as viewed from the upper left).

Finally, we restore the logarithmic coordinates along the y-axis, while retaining the logarithmic coordinates along the x-axis. This results in a log-log graph, as shown in Fig. 5.20. Now the data are represented very accurately as a straight line. (Remember that we are really looking at a plot of log y versus log x, not a plot of y versus x.)

We conclude that the data cannot be represented accurately by the equation for a straight line, an exponential function, or a logarithmic function. The data can, however, be represented accurately by a power function, as originally shown in Example 5.7.

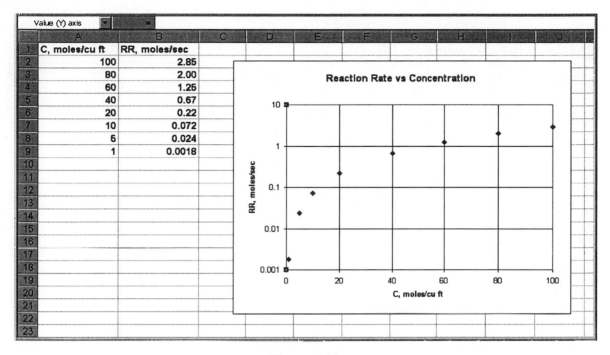

Figure 5.18

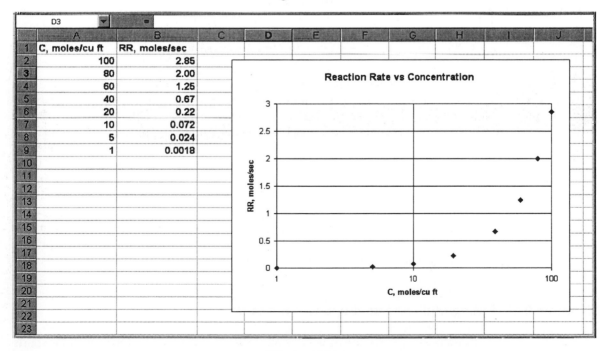

Figure 5.19

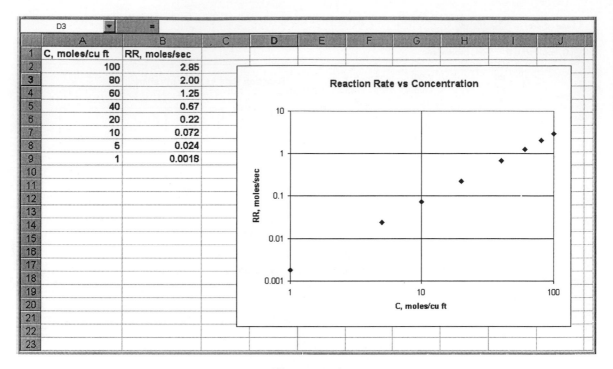

Figure 5.20

Fitting Multiple Functions to a Given Data Set

It sometimes happens that a given data set cannot be plotted as a straight line in any of the commonly used coordinate systems. We may still be able to fit an equation to the data accurately, however, using a polynomial. Moreover, we may also be able to obtain an acceptable, if not excellent, fit using an exponential function, a logarithmic function or a power function, even though we were not able to plot the data as a straight line using semi-log or log-log coordinates. Thus, we may find that several different functions will represent the data reasonably well. The question then arises as to which function best fits the data.

This question can generally be answered by visual inspection. If several different functions appear to fit the data more or less equally well, however, the sums of the squares of the errors (*SSE*) and the r^2 values may be helpful in deciding which function fits best.

Example 5.10   Fitting Multiple Functions to a Set of Data

In Example 5.8, we represented some performance data for several popular sports cars with a fifth-degree polynomial. Expand this study by fitting several other commonly used functions to the data. Determine which function best fits the data by comparing their respective *SSE* and r^2 values.

For comparative purposes, several different functions were passed through the data. The results are summarized in the following table (you will be asked to reproduce these results in Prob. 5.17):

Function	Visual Assessment	r^2	SSE
Straight line	Poor	0.8384	312.0
2nd-degree polynomial	Poor	0.965	66.237
Power	Poor	0.9715	55.03
3rd-degree polynomial	Fair	0.9909	17.57
Exponential	Good	0.9922	15.06
4th-degree polynomial	Good	0.9981	3.669
5th-degree polynomial	Excellent	0.9998	0.3862
6th-degree polynomial	Excellent	0.9999	0.1931

In each case, the value of *SSE* was obtained from Equation (5.5), using a calculated value of 1930.9 for *SST*. (Note, however, that in the case of the exponential function it is not actually *SSE* that is minimized by the method of least squares but a sum of the squares of the errors based upon the logs of y. Similarly, with a power function, the method of least squares minimizes a sum of the squares of the errors based upon the logs of y and x.)

The functions are listed in the order of increasing quality. Notice that the sums of the squares of the errors (*SSE*) show much greater variation than the r^2 values, which always remain between 0.8 and 1. Thus, the *SSE* values are a better predictor of curve-fit quality than the r^2 values. Notice also that the subjective visual assessments largely reinforce the comparisons based upon the *SSE* values.

Finally, it is interesting to compare the curve fits obtained with a third-degree and a fifth-degree polynomial, since this was a topic of some conjecture in Example 5.8. We see that the fifth-degree polynomial provides a much better fit than the third-degree polynomial. Thus, the argument that the last two terms in the fifth-degree polynomial are insignificant because of their small coefficients is invalid.

Substituting Other Variables for *y* and *x*

Sometimes we seek a linear relationship (i.e., a straight line) when plotting the data. This may be desirable because of some special significance that may be associated with the slope of the line or its intercept. In such situations, it may be helpful to replace one or both of the variables with a related function. For example, it may be helpful to substitute $1/y$ for y, $1/x$ for x, $\sqrt{x}$ for x, and so on, and then fit a straight line to the resulting plot.

Usually, some knowledge of the process being studied will suggest the type of substitution. For example, it is known that the logarithms of chemical reaction rates are proportional to the reciprocal of the absolute temperature. Thus, chemical engineers customarily plot the log of the reaction rate against $1/T$ (actually, they plot the reaction rate versus $1/T$ using semi-log coordinates) and expect to obtain a straight line. Similarly, it is known that the discharge velocity of a fluid leaving a tank is proportional to the square root of the height of the fluid above the discharge point. Thus, we would expect to obtain a straight line by plotting v versus $\sqrt{h}$ or v^2 versus h. A straight-line curve fit could be obtained in either of these situations by plotting the data with the appropriate substitutions.

If nothing is known about the expected behavior of the data based upon underlying principles, it may be helpful to try various variable substitutions using trial and error. Experienced data analysts refer to this as "playing" with the data. The results obtained can be surprisingly effective.

Example 5.11 Variable Substitution

The following data represent the rate at which an oxygenation reaction occurs within a water purification chamber as a function of temperature. The data were obtained by a civil engineer, who must now plot the data and fit an appropriate equation through the data.

Temperature, °K	Reaction Rate, moles/sec
253	0.12
258	0.17
263	0.24
268	0.34
273	0.48
278	0.66
283	0.91
288	1.22
293	1.64
298	2.17
303	2.84
308	3.70

Figure 5.21 shows a plot of reaction rate (RR) versus temperature. A cubic equation (i.e., a third-degree polynomial) has been passed through the data, resulting in the equation

$$RR = 2\times10^{-5}T^3 - 0.0171T^2 + 4.4148T - 381.8$$

where RR represents the reaction rate, in moles per second, and T represents the absolute temperature, in degrees K. The plot indicates a good fit, as verified by the corresponding

r^2 value of 0.9999. However, it is known that chemical reaction rates generally vary with absolute temperature in accordance with the formula

$$RR = ke^{-E/RT}$$

where E is the *activation energy* and R is a known physical constant. Therefore, it is preferable to plot the reaction rate against the reciprocal of the absolute temperature. Moreover, if a straight line can be obtained on a semi-logarithmic plot, we will know that the exponential form of the equation is valid, and the slope of the line can be used to determine the activation energy.

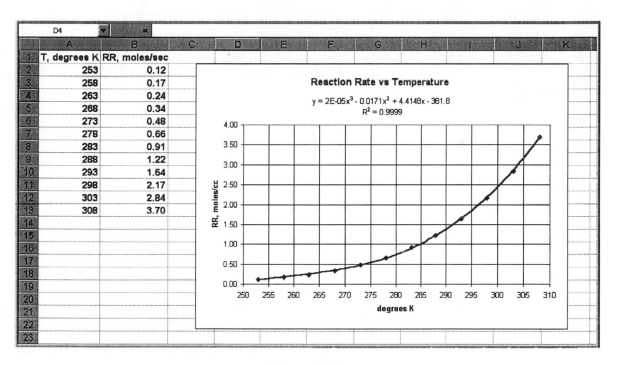

Figure 5.21

Figure 5.22 shows a plot of RR versus $1/T$ on semi-logarithmic coordinates. The data clearly plot as a straight line on semi-log coordinates. Again, we see an excellent curve fit, resulting in the equation

$$RR = 3 \times 10^7 e^{-4887.2/T}$$

with a corresponding r^2 value of 0.9999.

We can now determine the activation energy from the quotient $E/R = 4887.2$. Thus, if the universal constant R has a value of 1.987 (cal)(°K)/(mole), the activation energy for this reaction is $4887.2 \times 1.987 = 9710.9$ cal/mole.

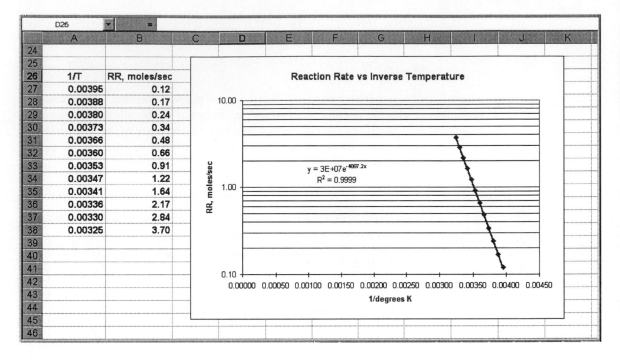

Figure 5.22

Scaling the Data

When fitting a curve to a set of data, it is sometimes helpful to *scale* the data so that the magnitudes of the *y*-values do not differ significantly from the magnitudes of the *x*-values. In such situations, the recommended procedure is to redefine one of the variables so that it is expressed in terms of some multiple of the original unit of measurement. For example, when plotting the velocity of a rocket as a function of time, it might be better to plot the velocity in *thousands of miles* per hour versus time in *seconds*, rather than in *miles* per hour (large numbers) versus time in *hours* (small numbers). This technique may result in a considerably better curve fit, particularly if the original *x*-values differ substantially (i.e., by several orders of magnitude) from the original *y*-values.

Example 5.12 Scaling a Data Set

A group of environmental engineers and scientists have made a prediction of the damage to the ozone layer in the northern hemisphere as a function of time, based upon present trends related to the use of fluorocarbon compounds. The following scenario has been suggested. (The thickness of the ozone layer is expressed in terms of an arbitrary scale ranging from 0 to 1.)

Year	Ozone Layer
1995	1.00
1996	0.97
1997	0.88
1998	0.76
1999	0.63
2000	0.50
2001	0.39
2002	0.30
2003	0.22
2004	0.17
2005	0.13
2006	0.11
2007	0.10

Fit an equation to the data, expressing the ozone layer value as a function of time.

Figures 5.23 and 5.24 show two different Excel worksheets, each containing an x-y graph of the data. The first worksheet (Fig. 5.23) contains the data as given, whereas the second worksheet (Fig. 5.24) contains redefined x-values, obtained by subtracting 1995 from each of the given values. In each graph, the data points define a distinctive S-shaped curve. A fifth-degree polynomial has been passed through each data set, since S-shaped curves are generally best represented by higher-order polynomials.

We see that both data sets can be represented accurately with a fifth-degree polynomial. In particular, the first data set results in the equation

$$y = 9 \times 10^{-6}x^5 - 0.0879x^4 + 352.14x^3 - 7.05758 \times 10^5 x^2 + 7 \times 10^8 x - 3 \times 10^{11}$$

with a value of $r^2 = 1$, whereas the second data set results in

$$y = 9 \times 10^{-6}x^5 - 0.0004x^4 + 0.007x^3 - 0.0481x^2 + 0.0117x + 1$$

with $r^2 = 1$. Each of these r^2 values indicates an excellent fit, though the quality of each fit is readily apparent simply by inspection.

Some spreadsheet programs, including earlier versions of Excel, are unable to obtain a satisfactory fit by passing a polynomial through the first data set. However, the second data set consistently results in an excellent fit. Thus, in some situations, the second data set is much easier to fit.

Problems

5.12 Try fitting a quadratic equation (i.e., a second-degree polynomial) and a power function to the data set given in Example 5.11. Compare the results with those given in Example 5.11.

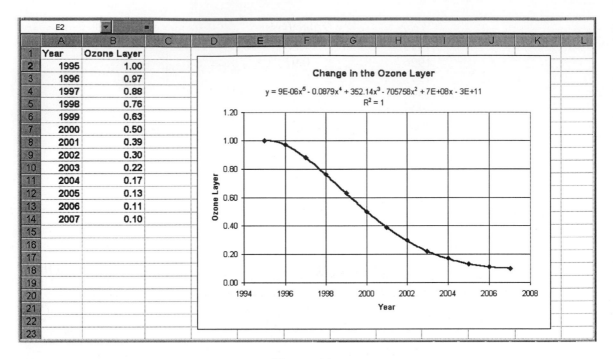

Figure 5.23

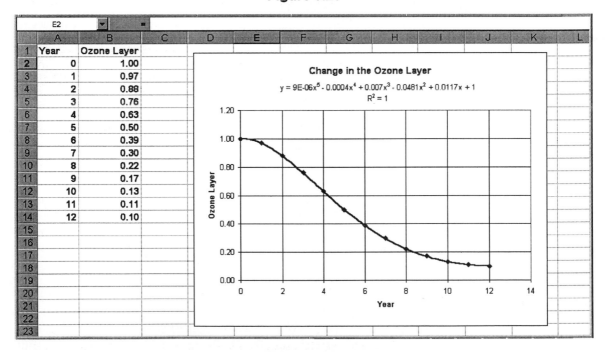

Figure 5.24

5.13 Plot the data given in Example 5.12 two different ways within an Excel worksheet:

(a) Plot y versus $x / 1000$, where y represents the ozone level and x represents the year.

(b) Plot $1000 \, y$ versus x.

Fit a fifth-degree polynomial to each data set. Compare each curve fit with the results obtained in the example. Does one method for modifying the data seems better than the other? If so, can you suggest a reason for this behavior?

5.14 Try fitting some other polynomials through the modified data given in Example 5.12 (see Fig. 5.24). What is the lowest-degree polynomial that will fit the data well?

5.15 Use Excel's **Regression** feature to fit a fifth-degree polynomial to the data given in Example 5.8. To do so, enter the x-values in one column, the x^2 values in the next column, the x^3 values in the next column, and so on, up to and including the values of x^5. Finally, enter the y-values in the last column. When you are prompted for the range of x-values within the **Regression** dialog box (see Fig. 5.8), enter a range that includes the first five columns (the values of x, x^2, x^3, etc.). Compare the outcome with the results obtained in Example 5.8.

5.16 Verify the results in Example 5.10 by fitting each type of function to the original data set. Assess each curve visually and compare your assessment with that given in the example. In addition, verify the value of *SSE* for each function, using the method described in the example.

5.17 A computer magazine has evaluated several popular desktop computers for price and performance. The cost and performance ratings (in arbitrary units) are listed below. Plot the cost versus performance in an Excel worksheet. Fit the best possible equation through the data.

Unit No.	Cost	Performance	Unit No.	Cost	Performance
1	$2480	72	6	$2570	112
2	2260	86	7	1750	94
3	2500	95	8	2000	109
4	1980	89	9	2200	113
5	2210	99	10	2240	122

Use the resulting curve fit to estimate the following:

(a) How much would you expect to pay for a computer with a performance rating of 105?

(b) What performance rating would you expect from a computer that cost $2400?

(c) What performance rating would you expect from a computer that cost $1500?

5.18 Repeat Prob. 5.17, forcing the resulting trendline through the origin. Does this make a significant difference in the answers to the three questions?

5.19 Several engineering students have built a wind-driven device that generates electricity. The following data have been obtained with the device (repeated from Prob. 3.9):

Wind velocity, mph	Power, watts
0	0
5	0.26
10	2.8
15	7.0
20	15.8
25	28.2
30	46.7
35	64.5
40	80.2
45	86.8
50	88.0
55	89.2
60	90.3

Fit an appropriate equation to the data with the intercept set to zero. Use the equation to answer the following questions:

(a) How much power will be generated when the wind velocity is 32 mph?

(b) What wind velocity will be required in order to generate 75 watts of power?

5.20 Fit an appropriate equation to the voltage versus time data presented in Prob. 3.5 (the data are repeated here for your convenience). Use the resulting equation to answer the following questions:

(a) What voltage would you expect after 1.5 seconds?

(b) What voltage would you expect after 15 seconds?

(c) How long will it take for the voltage to decline to 1.5 volts?

(d) How long will it take for the voltage to reach 0.01 volts?

Seconds	Volts
0	9.8
1	5.9
2	3.9
3	2.1
4	1.0
5	0.8
6	0.6
7	0.4
8	0.3
9	0.2
10	0.1

5.21 The following data describe the current, in milliamps, passing through an electronic device as a function of time (repeated from Prob. 3.10).

Seconds	Milliamps
0	0
1	1.06
2	1.51
3	1.63
4	1.57
5	1.43
6	1.26
7	1.08
8	0.92
9	0.77
10	0.64
12	0.44
14	0.30
16	0.20
18	0.14
20	0.091
25	0.034
30	0.012

(a) Fit the best possible equation to the entire data set.

(b) Try breaking the data set into two segments and fit an equation through each segment. Compare with the results obtained in part (a).

(c) Using the results of parts (a) and (b), estimate the current passing through the device at 0.5 seconds and at 22.8 seconds.

(d) Using the results of parts (a) and (b), estimate the times when the current passing through the device will be exactly 1 milliamp.

5.22 A recent car magazine contains fuel economy and vehicle weight data for a number of new cars. The following table contains the weight, in pounds force, and the corresponding EPA highway mileage figures for several of these vehicles.

Vehicle	Weight (lbs)	mpg
1	2684	28
2	2505	31
3	4123	19
4	3410	25
5	2647	29
6	3274	24
7	4407	14
8	3010	24
9	3055	25
10	2860	27
11	4130	18
12	3675	22
13	2375	33
14	3885	20
15	3180	26
16	3780	21
17	2778	32
18	3150	14
19	3425	23
20	2480	41

(a) Enter the data into an Excel worksheet. Then prepare an x-y graph of mpg versus weight. Identify any "outliers" (i.e., any data points that fall outside of the general cluster).

(b) Fit the best possible equation to the entire data set.

(c) Remove the outliers and then fit the best possible equation to the remaining data. Does the removal of the outliers significantly affect the outcome of the curve fit?

(d) What mileage would you expect from a car weighing 3000 pounds, based upon the results of part (b)? What mileage would you expect based upon the results of part (c)?

5.23 An engineer has determined the shear forces acting upon two beams within a structure. The following data represent the force (in newtons) as a function of distance (in cm) from the left end of each beam.

Distance, cm	Member A Force, newtons	Member B Force, newtons
0	0	0
1	3	0.03
2	6	0.20
3	8	0.57
4	9	0.79
5	11	1.15
6	12	1.29
7	14	1.36
8	15	1.60
9	16	1.62
10	18	1.68
12	20	1.93
14	22	2.10
16	24	2.08
18	26	2.17
20	28	2.28
25	33	2.25
30	38	2.39
35	42	2.42
40	46	2.50
50	54	2.47
60	61	2.54
70	68	2.57
80	75	2.62
90	82	2.60
100	88	2.65

(a) Enter the data into an Excel worksheet in tabular form, as shown above. Leave some extra space so that you can "play" with the data, as described below.

(b) Plot each data set (i.e., force versus distance) in a manner that will result in a linear (straight-line) relationship. You may need to use some ingenuity to do this. (*Hint:* Try plotting force (F) versus distance (x) on arithmetic coordinates, semi-log coordinates, and log-log coordinates. You might also try plotting $1/F$ versus x, F versus $1/x$, or $1/F$ versus $1/x$.)

(c) From the graph that yields the best straight-line relationship for each data set, determine an equation for the force as a function of distance. Use the two resulting equations to estimate the force in each member at a distance of 45 cm from the pin.

5.24 The following data describe the sequence of chemical reactions A→B→C. The concentrations of A, B, and C (in moles per liter) are tabulated as a function of time (in seconds).

Time	Conc A	Conc B	Conc C
0	5.0	0.0	0.0
1	4.5	0.46	0.02
2	4.1	0.84	0.06
3	3.7	1.2	0.13
4	3.4	1.4	0.22
5	3.0	1.6	0.33
6	2.7	1.8	0.45
7	2.5	1.9	0.58
8	2.3	2.0	0.72
9	2.0	2.1	0.87
10	1.8	2.2	1.0
12	1.5	2.2	1.3
14	1.2	2.2	1.6
16	1.0	2.1	1.9
18	0.83	2.0	2.2
20	0.68	1.85	2.5
25	0.41	1.5	3.1
30	0.25	1.2	3.5
35	0.15	0.93	3.9
40	0.09	0.71	4.2

(a) Enter the data into an Excel worksheet and determine an equation that will accurately represent concentration as a function of time for each of the substances A, B, and C. Note that a separate equation will be required for each data set.

(b) Using the equations determined in part (a), estimate the concentration of A, B, and C after 23 seconds.

(c) Using the equation obtained in part (a) for concentration of B versus time, determine when the concentration of B will be maximized.

5.25 An engineering student has carried out a series of measurements of tensile force versus elongation for a structural steel cylindrical sample with a diameter of 0.5 inches and a length of four inches. The data cover only the elastic (linear) region. From these measurements, the student has obtained the following table of stress versus strain, where the stress is defined as the force per unit area (in pounds force per square inch, or psi), and the strain is defined as the elongation per unit of original length (dimensionless). The student wishes to determine the *modulus of elasticity* (also known as *Young's modulus*) by measuring the slope of the line representing a plot of stress versus strain. (The modulus of elasticity is a property of the material being tested. It is independent of the dimensions of the cylinder or the magnitude of the forces applied, as long as the forces fall within the elastic region.)

Strain	Stress (psi)	Strain	Stress (psi)
0.1×10^{-3}	3016	0.7×10^{-3}	21200
0.2×10^{-3}	5983	0.8×10^{-3}	24228
0.3×10^{-3}	9191	0.9×10^{-3}	24261
0.4×10^{-3}	12178	1.0×10^{-3}	32205
0.5×10^{-3}	14908	1.1×10^{-3}	30966
0.6×10^{-3}	18292	1.2×10^{-3}	33392

(*a*) Enter the data into an Excel spreadsheet and generate a line graph of stress versus strain.

(*b*) Fit a straight line to the data.

(*c*) Determine the modulus of elasticity for this material from the equation of the straight line.

5.26 The following table presents stress–strain data for the same structural steel sample described in Prob. 5.25. Now, however, the data extend beyond the elastic region to the point of failure (i.e., the point at which the cylindrical sample will pull apart into two separate pieces).

Strain	Stress (psi)	Strain	Stress (psi)
0.02	29737	0.14	57046
0.04	37166	0.16	56593
0.06	44820	0.18	53448
0.08	44074	0.20	52103
0.10	49161	0.22	49185
0.12	53002	0.24	45386

(*a*) Enter the data into an Excel spreadsheet and generate an *x-y* graph of stress versus strain.

(*b*) Fit an appropriate curve to the data.

(*c*) Use the resulting equation to estimate the stress at a strain of 0.115. What is the corresponding elongation?

ADDITIONAL READING

Akai, T. J. *Applied Numerical Methods for Engineers.* New York: Wiley, 1994.

Barnes, J. W. *Statistical Analysis for Engineers and Scientists: A Computer-Based Approach.* New York: McGraw-Hill, 1994.

Chapra, S. C. and T. P. Canale. *Numerical Methods for Engineers, with Personal Computer Applications.* New York: McGraw-Hill, 1985.

Devore, J. L. *Probability and Statistics for Engineers and the Sciences.* 3d ed. Pacific Grove, California: Brooks/Cole Publishing Co., 1991.

Hogg, R. V. and J. Ledolter. *Engineering Statistics.* New York: Macmillan, 1987.

Mendenhall, W. and T. Sincich. *Statistics for the Engineering and Computer Sciences.* 2d ed. San Francisco: Dellen Publishing Co., 1988.

Milton, J. S. and J. C. Arnold. *Introduction to Probability and Statistics: Principles and Applications for Engineering and the Computing Sciences.* 2d ed. New York: McGraw-Hill, 1990.

CHAPTER 6

INTERPOLATING BETWEEN DATA POINTS

Interpolation is a procedure for determining a value of y corresponding to some specified value of x, when x falls between two known values, say x_i and x_{i+1}. The known values will be members of a set of x-y data, for example, $P_1 = (x_1, y_1)$, $P_2 = (x_2, y_2)$, $\cdots$, $P_n(x_n, y_n)$, whose values are precisely known. Our goal is to determine the interpolated value of y as accurately as possible.

One approach is to pass an equation through the *aggregate* of data points, using the methods discussed in the last chapter. Calculations of this type are useful when working with *measured* data, which usually involve some scatter in the individual data values. When working with *precise* data points, however, it may be better to pass an equation through the data points themselves. This usually provides more accurate interpolated values, particularly if the data points are equally spaced. The equation need not be passed through *all* of the data points — just the data points within the immediate area of interest. The actual number of data points included in the calculation generally depends on the nature of the curve and the level of accuracy required in the interpolated value.

The simplest way to do this is to pass a *straight line* through the two data points surrounding the point of interest (i.e., surrounding the x-value whose corresponding y-value is to be determined by interpolation). This procedure is known as *linear interpolation*. Unfortunately, this method is not very accurate if the data points exhibit very much curvature. Another approach, which is more accurate but also more complicated, is to pass a *polynomial* through *several* of the data points surrounding the point of interest. This is known as *polynomial interpolation*. The method works best when the given data points are equally spaced.

151

Unlike Excel's automated Trendline and Regression features, which are based upon the method of least squares, there is no automated procedure for carrying out linear or polynomial interpolation in Excel. Rather, the interpolating line or interpolating polynomial must be passed through the data points using the defining mathematical formulas. Fortunately, these formulas are easy to use within a spreadsheet environment. In this chapter, we will see how Excel can conveniently be used to interpolate between precise data points, using either linear or polynomial interpolation.

6.1 LINEAR INTERPOLATION

Suppose we are given two data points, (x_1, y_1) and (x_2, y_2), and we wish to determine the value of y corresponding to some specified value of x, where $x_1 < x < x_2$. One way to approach this problem is to connect the given data points with a straight line. The desired value of y can then be determined from the straight line, as illustrated in Fig. 6.1.

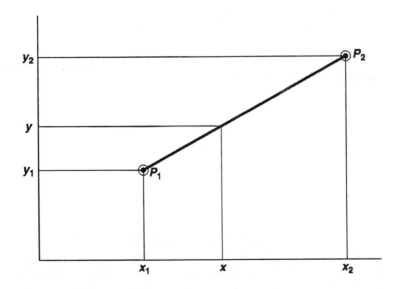

Figure 6.1

Mathematically, let us proceed as follows. From simple proportionality, we can write

$$\frac{y - y_1}{y_2 - y_1} = \frac{x - x_1}{x_2 - x_1} \tag{6.1}$$

Solving for y, we obtain

$$y = y_1 + \frac{y_2 - y_1}{x_2 - x_1}(x - x_1) \tag{6.2}$$

Equation (6.2) is the traditional form of the linear interpolation formula. It is sometimes written as

$$y = y_1 + \frac{\Delta y}{h}(x - x_1) \tag{6.3}$$

where Δy represents $y_2 - y_1$ and h represents the interval spacing, $x_2 - x_1$. This form of the interpolation equation may be easier to work with when carrying out a hand calculation since there is less of a tendency to confuse the various y- and x-values.

Example 6.1 Linear Interpolation

Suppose we are given the two data points (2.0, 5.5) and (2.1, 7.2). Determine the value of y corresponding to $x = 2.03$ using linear interpolation.

Let us use the form of the linear interpolation formula given by Equation (6.3) to solve this problem. We begin by identifying the various terms in the equation:

$$x_2 = 2.1 \qquad\qquad y_2 = 7.2$$

$$x_1 = 2.0 \qquad\qquad y_1 = 5.5$$

$$h = (2.1 - 2.0) = 0.1 \qquad \Delta y = (7.2 - 5.5) = 1.7$$

$$x = 2.03$$

We can now calculate the desired value of y as

$$y = y_1 + \frac{\Delta y}{h}(x - x_1) = 5.5 + \frac{1.7}{0.1}(2.03 - 2.0) = 6.01$$

Hence, the value of y corresponding to $x = 2.03$ is 6.01. That is, the interpolated data point is (2.03, 6.01).

In Excel, the use of Equation (6.2) may be more straightforward than Equation (6.3) since it does not involve the calculation of any intermediate parameters. The recommended procedure is to enter the tabulated x-values in one column and the corresponding y-values in another, as in earlier chapters of this book. The given x-value, corresponding to the unknown y-value, can be entered into a separate cell. It is then a simple matter to determine the interpolated y-value using an Excel formula to represent Equation (6.2). The procedure is illustrated in the following example.

Example 6.2 Linear Interpolation in Excel

Solve the linear interpolation problem described in Example 6.1 using Excel.

Figure 6.2 shows an Excel worksheet in which the tabulated x-values (2.0 and 2.1) are entered into cells A2 and A3, and the corresponding y-values (5.5 and 7.2) are entered into cells B2 and B3. Cell A5 contains the x-value (2.03) whose y-value is to be determined. Cell B5 contains the desired interpolated y-value (6.01). Note that cell B5 is a calculated value, as indicated by the formula shown in the formula bar at the top of the figure. Examination of this formula reveals that it is equivalent to Equation (6.2).

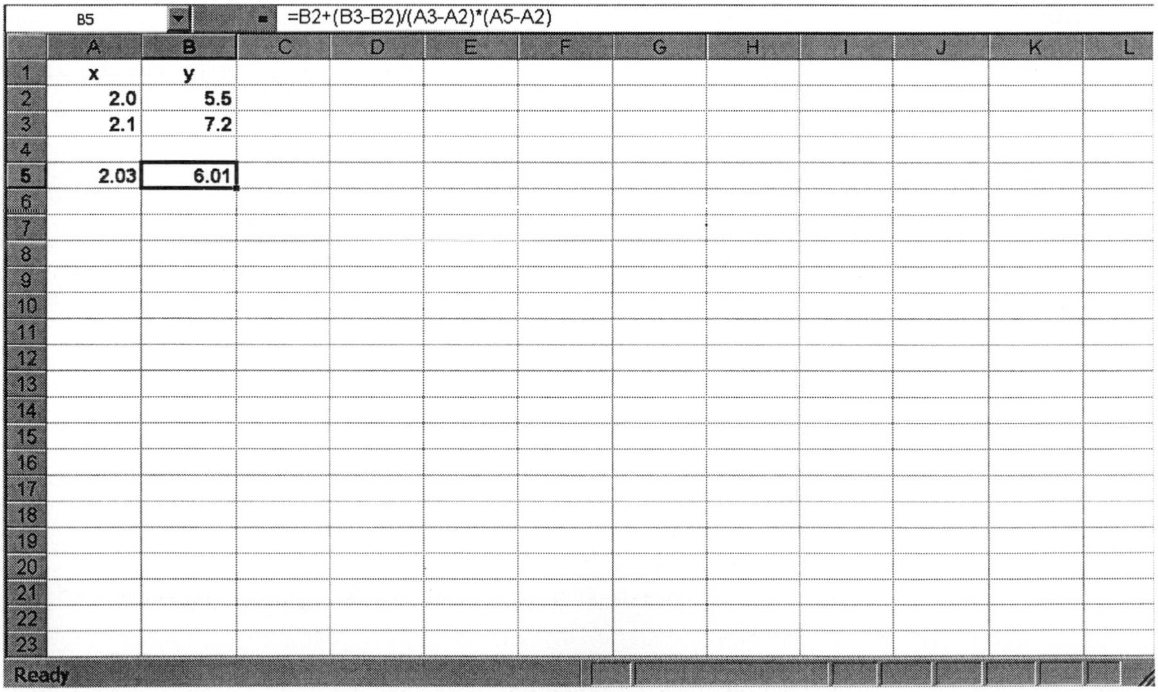

B5			=B2+(B3-B2)/(A3-A2)*(A5-A2)									
	A	B	C	D	E	F	G	H	I	J	K	L
1	x	y										
2	2.0	5.5										
3	2.1	7.2										
4												
5	2.03	6.01										

Figure 6.2

Problems

6.1   The heat capacity of iron is given below as a function of temperature.

Temperature, °C	Heat Capacity, Kcal/(kg)(°C)
0	0.1055
100	0.1168
200	0.1282
300	0.1396
400	0.1509
500	0.1623
600	0.1737
700	0.1805

Determine the heat capacity at each of the following temperatures, using linear interpolation:

(a)   80 °C   (c)   410 °C

(b)   335 °C   (d)   675 °C

6.2   The temperature of air in a "standard atmosphere" is given below as a function of altitude.

Altitude, ft	Temperature, °F
0	59.0
5,000	41.2
10,000	23.3
15,000	5.5
20,000	−12.3
25,000	−30.2
30,000	−48.0
35,000	−65.8
40,000	−67.0
50,000	−67.0

Determine the temperature at each of the following altitudes using linear interpolation:

(a)   6,530 ft   (c)   18,800 ft

(b)   12,400 ft   (d)   25,300 ft

6.3   The gravitational acceleration varies with latitude (distance from the equator) and altitude above sea level. The following table gives the gravitational acceleration at various latitudes, at sea level.

Latitude, °	Gravitational Acceleration, m/sec^2
0	0.978039
10	0.978195
20	0.978641
30	0.979329
40	0.980171
50	0.981071
60	0.981918
70	0.982608
80	0.983059
90	0.983217

The following table gives corrections as a function of altitude. These values are independent of latitude.

Altitude, m	Correction, m/sec^2
200	-0.617×10^{-4}
300	-0.926×10^{-4}
400	-1.234×10^{-4}
500	-1.543×10^{-4}
600	-1.852×10^{-4}
700	-2.160×10^{-4}
800	-2.469×10^{-4}
900	-2.777×10^{-4}

Using these two tables and linear interpolation, answer each of the following questions:

(*a*)   What is the gravitational acceleration at sea level and a latitude of 18.5°?

(*b*)   What is the correction term for an altitude of 275 m above sea level?

(*c*)   What is the gravitational acceleration at a latitude of 18.5° and an altitude of 275 m above sea level?

(*d*)   What is the gravitational acceleration in Pittsburgh, which is located at a latitude of 40.5° and an altitude of 235 m above sea level?

6.4 Compound interest factors are used to determine the increase in a sum of money if the money is allowed to accumulate for several years with interest payable at a certain rate and compounded annually. If P represents the initial sum of money and F is the future sum after n years, we can write

$$F = fP$$

where f is the compound interest factor for n years and a specified interest rate. Several values for the compound interest factor are given below as a function of n, based upon an interest rate of 10 percent per year.

n	f	n	f
0	1.000	25	12.183
5	1.649	30	20.086
10	2.718	35	33.115
15	4.482	40	54.598
20	7.389	45	90.017

(a) Suppose $1000 is deposited in a bank account that pays 10 percent interest, compounded annually. Using the tabulated values and linear interpolation, determine how much money will accumulate after eight years.

(b) How much money will accumulate after 12 years?

(c) How long will it take your money to triple if it earns interest at 10 percent per year, compounded annually?

(d) Suppose you plan to retire in 42 years. If you continue to earn interest at 10 percent per year, compounded annually, how much will have accumulated by the time you are ready to retire?

6.5 Problem 5.21 presented data for the current, in milliamps, passing through an electronic device as a function of time. A portion of the data is reproduced below, for your convenience.

Seconds	Milliamps	Seconds	Milliamps
0	0	10	0.64
1	1.06	12	0.44
2	1.51	14	0.30
3	1.63	16	0.20
4	1.57	18	0.14
5	1.43	20	0.091
6	1.26	25	0.034
8	0.92	30	0.012

Use linear interpolation to answer the following questions, based upon these tabulated values:

(*a*) What current will flow through the device after 0.7 seconds?

(*b*) What current will flow through the device after 12.75 seconds?

(*c*) When will the current reach a value of 1.00 milliamps?

(*d*) When will the current reach a value of 0.1 milliamps? When will it reach 0.01 milliamps?

6.2 POLYNOMIAL INTERPOLATION USING FORWARD DIFFERENCES

Now suppose we wish to fit an interpolating polynomial through a number of equally spaced data points. This may provide a more accurate interpolated value than that obtained by linear interpolation, particularly if the data set exhibits more or less uniform curvature throughout the range of interest, as illustrated in Fig. 5.3. One way to do this is to construct an interpolating polynomial using forward differences, based upon equally spaced data (i.e., data whose successive x-values are separated by the same amount).

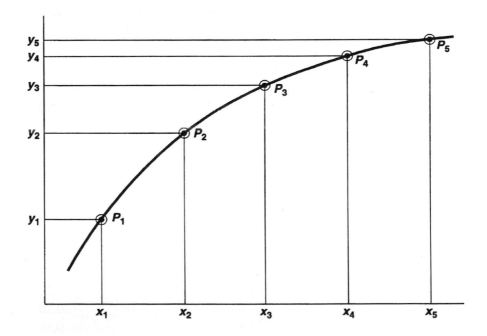

Figure 6.3

A *forward difference* is defined as

$$\Delta y_i = y_{i+1} - y_i \tag{6.4}$$

It is called a *forward* difference because it is obtained by subtracting the current point (y_i) from the forward point (y_{i+1}). We can extend this concept to higher differences by writing

$$\Delta^2 y_i = \Delta y_{i+1} - \Delta y_i \tag{6.5}$$

$$\Delta^3 y_i = \Delta^2 y_{i+1} - \Delta^2 y_i \tag{6.6}$$

. . . . .

$$\Delta^k y_i = \Delta^{k-1} y_{i+1} - \Delta^{k-1} y_i \tag{6.7}$$

and so on. Thus, we can construct the following table of forward differences for a data set consisting of six data points:

x_1	y_1	Δy_1	$\Delta^2 y_1$	$\Delta^3 y_1$	$\Delta^4 y_1$	$\Delta^5 y_1$
x_2	y_2	Δy_2	$\Delta^2 y_2$	$\Delta^3 y_2$	$\Delta^4 y_2$	
x_3	y_3	Δy_3	$\Delta^2 y_3$	$\Delta^3 y_3$		
x_4	y_4	Δy_4	$\Delta^2 y_4$			
x_5	y_5	Δy_5				
x_6	y_6					

Notice that each row has one less entry than the preceding row, resulting in an upper-triangular layout. This is a consequence of the manner in which the forward differences are defined.

Example 6.3 Constructing a Forward-Difference Table

Construct a forward-difference table for the following six data points:

x_i	y_i
1	1
2	2
3	10
4	44
5	141
6	366

The forward-difference table is shown below.

x_i	y_i	Δy_i	$\Delta^2 y_i$	$\Delta^3 y_i$	$\Delta^4 y_i$	$\Delta^5 y_i$
1	1	1	7	19	18	10
2	2	8	26	37	28	
3	10	34	63	65		
4	44	97	128			
5	141	225				
6	366					

Be sure you understand how each of the differences was obtained.

Once a difference table has been constructed, the entries are used in the *Gregory-Newton forward interpolation* formula, which is given below:

$$y = y_1 + \frac{u}{h}\Delta y_1 + \frac{u(u-h)}{2!h^2}\Delta^2 y_1 + \frac{u(u-h)(u-2h)}{3!h^3}\Delta^3 y_1 + \cdots \qquad (6.8)$$

Equation (6.8) is a polynomial in u. It is based upon the clever use of a Taylor series expansion for $y(x)$, in which the derivatives are replaced with forward differences.

In order to use Equation (6.8), you must first select a known data point, say (x_1, y_1), for which several of the forward differences are known. This data point will have a corresponding row, known as the *base row*, in the forward-difference table. Usually, the base row will be the first row in the forward-difference table since this row contains the largest number of forward differences. In principle, however, it need not necessarily be the first row — one of the lower rows in the table can also be used.

The variable u appearing in Equation (6.8) represents the distance from the unknown point x to the base row value, x_1. Thus, $u = x - x_1$. (Note that u will be negative if you interpolate behind the base row; that is, if $x < x_1$.) Also, y_1, Δy_1, $\Delta^2 y_1$, etc. represent tabulated base row values, and h represents the spacing between the given x-values (which is constant).

Example 6.4 Interpolation Using Forward Differences

Use Equation (6.8) and the forward-difference table developed in Example 6.3 to estimate the value of y when $x = 1.7$.

Let us select the first row in the forward difference table as the base row since its x-value ($x_1 = 1$) is close to the given point and it contains the largest number of forward differences. Hence, the following numerical values will be used in Equation (6.8):

$h = 1$

$x = 1.7$ $x_1 = 1$ $u = (x - x_1) = (1.7 - 1) = 0.7$

$y_1 = 1$ $\Delta y_1 = 1$ $\Delta^2 y_1 = 7$ $\Delta^3 y_1 = 19$ $\Delta^4 y_1 = 18$ $\Delta^5 y_1 = 10$

Substituting these values into Equation (6.8), we obtain

$$y = 1 + \frac{0.7}{1}(1) + \frac{0.7(0.7-1)}{2(1)^2}(7) + \frac{0.7(0.7-1)(0.7-2)}{6(1)^3}(19) +$$

$$\frac{0.7(0.7-1)(0.7-2)(0.7-3)}{24(1)^4}(18) + \frac{0.7(0.7-1)(0.7-2)(0.7-3)(0.7-4)}{120(1)^5}(10)$$

$$= 1 + 0.7 - 0.735 + 0.8645 - 0.470925 + 0.1726725$$

$$= 1.5312475$$

Therefore, we conclude that the value of y is approximately 1.531 when $x = 1.7$.

Note that this value was actually obtained by passing a fifth-degree interpolating polynomial through all six of the given data points and then determining the value of y from the interpolating polynomial.

This process lends itself very naturally to spreadsheet implementation. To do so, proceed as follows:

1. Enter the given data points into the first two columns of a worksheet.
2. Create the forward difference table from the given data points. (Note that the entire difference table can be constructed by entering only one formula. For example, to calculate the forward difference in cell C3, enter the formula =B4–B3. Then copy this formula to other parts of the worksheet, as required.)
3. Enter the value of x whose corresponding y-value is to be determined.
4. Evaluate the unknown y-value by entering a formula that represents Equation (6.8).

The entire process is illustrated in the next example.

Example 6.5 Interpolation Using Forward Differences in Excel

Reconstruct the solution to the problem stated in Example 6.4 using an Excel worksheet.

C3	▼	■	=B4-B3								
	A	B	C	D	E	F	G	H	I	J	K	L
1	x	y	Dy	D2y	D3y	D4y	D5y					
2	1	1	1	7	19	18	10					
3	2	2	8	26	37	28						
4	3	10	34	63	65							
5	4	44	97	128								
6	5	141	225									
7	6	366										

Figure 6.4

Figure 6.4 shows an Excel worksheet containing the forward-difference table. Cell C3, containing one of the forward differences, is highlighted. Note that the formula used to generate this value corresponds to the example in step 2 above. The remaining differences were obtained with similar formulas. (Actually, the formula =B3–B2 was entered into cell C2 and then copied to the other cells, as needed.)

In Fig. 6.5, we see the complete solution. Notice first that the values for u and h have been entered into cells J2 and J3, respectively. These values and their corresponding labels have been enclosed within a rectangle for easy identification. Cell A9 contains the specified x-value (1.7), and cell B9 contains the corresponding y-value (1.531248). This value agrees with the value obtained in Example 5.4, as expected.

Notice that the y-value was obtained using the lengthy two-row formula shown in the formula bar. This formula corresponds to Equation (6.8). You should examine this formula carefully and be sure that you understand the relationship between the formula and Equation (6.8). In particular, be sure you understand what is represented by each cell address.

When constructing a difference table, it sometimes happens that the values in one of the difference columns will all be the same so that the higher differences (successive columns) will all be zero. This is an indication that the given data points are in fact points on a polynomial. Therefore, an interpolating polynomial can be passed through at least some, and perhaps all, of the data points. In particular, if the nth-differences are all the same (i.e., if the values within the column are all constant), then an nth-degree polynomial will pass through any group of $(n+1)$ consecutive values of the given data set.

Furthermore, if one of the columns in the difference table contains the same constant or can be *assumed* to contain the same constant, this constant value can be used to fill in the lower portion of the table. Once the entire table has been filled in, *any* row within the table can be used as the base row with equal precision.

B9			=B2+J2*C2/J3+J2*(J2-J3)*D2/(2*J3^2)+J2*(J2-J3)*(J2-2*J3)*E2/(6*J3^3)+J2*(J2-J3)*(J2-2*J3)*
A	B		(J2-3*J3)*F2/(24*J3^4)+J2*(J2-J3)*(J2-2*J3)*(J2-3*J3)*(J2 4*J3)*G2/(120*J3^5)

	x	y	Dy	D2y	D3y	D4y	D5y			
1	x	y	Dy	D2y	D3y	D4y	D5y			
2	1	1	1	7	19	18	10		u=	0.7
3	2	2	8	26	37	28			h=	1
4	3	10	34	63	65					
5	4	44	97	128						
6	5	141	225							
7	6	366								
8										
9	1.7	1.531248								
10										

Ready

Figure 6.5

Example 6.6 Constructing an Extended Forward-Difference Table

Construct a forward-difference table for the following seven data points. (Note that the first six data points are the same as those given in Example 6.3.)

x_i	y_i
1	1
2	2
3	10
4	44
5	141
6	366
7	822

The forward-difference table is shown below.

x_i	y_i	Δy_i	$\Delta^2 y_i$	$\Delta^3 y_i$	$\Delta^4 y_i$	$\Delta^5 y_i$	$\Delta^6 y_i$
1	1	1	7	19	18	10	0
2	2	8	26	37	28	10	
3	10	34	63	65	38		
4	44	97	128	103			
5	141	225	231				
6	366	456					
7	822						

Now we see that the fifth differences (the values in the second to last column) are the same so that the sixth difference is zero. This is an indication that a fifth-degree polynomial can be passed through any consecutive six of the seven given data points.

If we *assume* that *all* of the fifth differences are equal to 10, we can fill in the lower portion of the table using Equation (6.7), rearranged in the following form:

$$\Delta^{k-1} y_{i+1} = \Delta^k y_i + \Delta^{k-1} y_i \tag{6.9}$$

We first back-calculate the remaining fourth differences, then the third differences, and so on, until the entire table has been filled in. The forward-difference table then becomes

x_i	y_i	Δy_i	$\Delta^2 y_i$	$\Delta^3 y_i$	$\Delta^4 y_i$	$\Delta^5 y_i$	$\Delta^6 y_i$
1	1	1	7	19	18	10	0
2	2	8	26	37	28	10	*0*
3	10	34	63	65	38	*10*	*0*
4	44	97	128	103	*48*	*10*	*0*
5	141	225	231	*151*	*58*	*10*	*0*
6	366	456	*382*	*209*	*68*	*10*	*0*
7	822	*838*	*591*	*277*	*78*	*10*	*0*

(The newly added values are shown in bold italic type.) When the forward-difference table is constructed in this manner, any row in the table can be selected as the base row.

Be sure you understand how these new values were obtained.

When carrying out polynomial interpolation, you need not use the entire data set—a few rows surrounding the given x-value are generally sufficient. The number of rows required to obtain an accurate interpolated result generally depends upon the nature of the data. As a rule, the more curvature in the data, the larger the number of rows (see Prob. 6.15 below).

Problems

6.6 Construct an Excel worksheet containing the forward-difference table shown in Fig. 6.4. (Generate the differences using formulas—do not simply type in the numerical values.) Then use the difference table to obtain interpolated values of y corresponding to the following x-values:

(*a*)   1.50   (*b*)   1.35   (*c*)   2.50

In each case, use the first row (corresponding to $x = 1$) as the base row.

6.7 Repeat Example 6.5, solving for the value of y corresponding to $x = 1.7$. Now, however, use the *second* row of the difference table (corresponding to $x = 2$) as the base row. Compare your answer with that obtained in Example 6.5. Why are the results different? Which answer would you assume is more accurate, and why?

6.8 Repeat Prob. 6.6 (*c*) using the *second* row of the difference table as the base row. Compare your answer with that obtained in Prob. 6.6. Which answer do you prefer, and why?

6.9 Construct an Excel worksheet containing the extended forward-difference table given in Example 6.6. (Use the appropriate formulas to determine the values of the forward differences.) Then determine the value of y corresponding to $x = 1.7$, using the second row in the table as the base row. Repeat the solution using the fifth row as the base row. Compare your answers with the results obtained in Example 6.5 and Prob. 6.7.

6.10 Construct an Excel worksheet containing the extended forward-difference table given in Example 6.6. (Use the appropriate formulas to determine the values of the forward differences.) Then use the difference table to obtain interpolated values of y corresponding to the following x-values:

(*a*)   1.35   (*b*)   4.20   (*c*)   6.70   (*d*)   2.50

In each case, use the third row (corresponding to $x = 3$) as the base row. Compare your answers with those obtained in Prob. 6.6 where appropriate.

6.11 Repeat Prob. 6.1(*a*) using forward-difference polynomial interpolation based upon the first four data points. Compare your answers with the results obtained in Prob. 6.1(*a*), using linear interpolation.

6.12 Repeat Prob. 6.1(c) using forward-difference polynomial interpolation based upon the last four data points. Compare your answers with those obtained in Prob. 6.1(c), using linear interpolation.

6.13 Repeat Prob. 6.5(a) using forward-difference polynomial interpolation based upon:

(a) The first three data points (quadratic interpolation).

(b) The first four data points (cubic interpolation).

(c) The first six data points (interpolation with a fifth-degree polynomial).

Compare your answers with those obtained using linear interpolation in Prob. 6.5(a).

6.14 Construct an Excel worksheet containing a forward-difference table for the following set of data:

x_i	y_i	x_i	y_i
0.5	0.125	3.5	42.875
1.5	3.375	4.5	91.125
2.5	15.625	5.5	166.375

Fill in the table as completely as possible. Then use the table to estimate the value of y corresponding to each of the following x-values:

(a) 3.0   (b) 1.2   (c) 1.7   (d) 5.0

For this data set, does it matter which row in the forward-difference table is used as the base row?

6.15 For the data given in Prob. 6.5, determine the current flowing through the device at 12.75 seconds using polynomial interpolation. Use the data corresponding to 12, 14, 16, and 18 seconds, with the 12-second row as the base row, when carrying out the calculations. Compare your results with those obtained in Prob. 6.5(b), using linear interpolation.

6.3 POLYNOMIAL INTERPOLATION USING BACKWARD DIFFERENCES

Forward-difference interpolation works well when the interpolation point is near the *beginning* of the data set or the data set is so large that a subset can be selected with the interpolation point near the beginning of the subset. Backward differences generally provide a more accurate result, however, when the data set is limited in size and the interpolation point falls nearer to the *end* of the data set.

A *backward difference* is defined as

$$\nabla y_i = y_i - y_{i-1} \tag{6.10}$$

We use the term *backward* difference because we are subtracting the backward point (y_{i-1}) from the current point (y_i). As with forward differences, we can extend this idea to higher differences by writing

$$\nabla^2 y_i = \nabla y_i - \nabla y_{i-1} \tag{6.11}$$

$$\nabla^3 y_i = \nabla^2 y_i - \nabla^2 y_{i-1} \tag{6.12}$$

$$\cdots\cdots$$

$$\nabla^k y_i = \nabla^{k-1} y_i - \nabla^{k-1} y_{i-1} \tag{6.13}$$

and so on. Thus, if we have a data set consistsing of six data points, we can construct the following backward-difference table:

x_1	y_1					
x_2	y_2	∇y_2				
x_3	y_3	∇y_3	$\nabla^2 y_3$			
x_4	y_4	∇y_4	$\nabla^2 y_4$	$\nabla^3 y_4$		
x_5	y_5	∇y_5	$\nabla^2 y_5$	$\nabla^3 y_5$	$\nabla^4 y_5$	
x_6	y_6	∇y_6	$\nabla^2 y_6$	$\nabla^3 y_6$	$\nabla^4 y_6$	$\nabla^5 y_6$

Now each row has one more entry than the preceding row, resulting in a lower-triangular layout. This is a result of the manner in which the backward differences are defined. (Compare with the forward-difference table shown in the previous section.)

Example 6.7 Constructing a Backward-Difference Table

Construct a backward-difference table for the following six data points. (These are the same data points used in Example 6.3.)

x_i	y_i
1	1
2	2
3	10
4	44
5	141
6	366

The backward-difference table is shown below.

x_i	y_i	∇y_i	$\nabla^2 y_i$	$\nabla^3 y_i$	$\nabla^4 y_i$	$\nabla^5 y_i$
1	1					
2	2	1				
3	10	8	7			
4	44	34	26	19		
5	141	97	63	37	18	
6	366	225	128	65	28	10

It is interesting to compare this table with the corresponding forward-difference table constructed in Example 6.3. We see that the numerical values in both tables are the same. Their arrangement, however, is different, as the values now cluster toward the lower-left rather than the upper-left portion of the table.

Be sure you understand how each of the differences was obtained.

To carry out backward-difference interpolation, we use the *Gregory-Newton backward interpolation* formula, which is given below:

$$y = y_1 + \frac{u}{h}\nabla y_1 + \frac{u(u+h)}{2!h^2}\nabla^2 y_1 + \frac{u(u+h)(u+2h)}{3!h^3}\nabla^3 y_1 + \cdots \tag{6.14}$$

Equation (6.14) is analogous to the Gregory-Newton forward difference formula given by Equation (6.8). It is a polynomial in u, obtained by expanding $y(x)$ in a Taylor series and then replacing the derivatives with backward differences.

The use of Equation (6.14) again requires the selection of a base row within the backward-difference table. The base row will typically be the last row in the backward-difference table since this row contains the largest number of backward differences. In principle, however, it need not necessarily be the last row—one of the higher rows in the table can also be used.

In Equation (6.14), as in Equation (6.8), y_1, ∇y_1, $\nabla^2 y_1$, etc. represent the tabulated base row values, and h represents the spacing between the given x-values (which is again assumed to be constant). Also, the variable u appearing in Equation (6.14) again represents the distance from the unknown point x to the base row value, x_1. Thus, $u = x - x_1$. Now, however, the base row will be *below* the interpolation point; hence, the value of u will be negative. We will see how Equation (6.14) is used to carry out backward interpolation in the next example.

Example 6.8 Interpolation Using Backward Differences

Use Equation (6.14) and the backward-difference table developed in Example 6.7 to estimate the value of y when $x = 5.7$.

Let us select the last row in the backward-difference table as the base row, since its x-value ($x_1 = 6$) is close to the given point and it contains the largest number of backward differences. Hence, the following numerical values will be used in Equation (6.14):

$h = 1$

$x = 5.7$ $x_1 = 6$ $u = (x - x_1) = (5.7 - 6) = -0.3$

$y_1 = 366$ $\nabla y_1 = 225$ $\nabla^2 y_1 = 128$ $\nabla^3 y_1 = 65$ $\nabla^4 y_1 = 28$ $\nabla^5 y_1 = 10$

Substituting these values into Equation (6.14), we obtain

$$y = 366 + \frac{-0.3}{1}(225) + \frac{-0.3(-0.3+1)}{2(1)^2}(128) + \frac{-0.3(-0.3+1)(-0.3+2)}{6(1)^3}(65) +$$

$$\frac{-0.3(-0.3+1)(-0.3+2)(-0.3+3)}{24(1)^4}(28) +$$

$$\frac{-0.3(-0.3+1)(-0.3+2)(-0.3+3)(-0.3+4)}{120(1)^5}(10)$$

$$= 366 - 67.5 - 13.44 - 3.8675 - 1.12455 - 0.2972025$$

$$= 279.7707475$$

Therefore, we conclude that the value of y is approximately 279.8 when $x = 5.7$.

It is interesting to compare this value with a value obtained by linear interpolation. In the latter case, we have

$$y = 141 + \frac{(366 - 141)}{(6 - 5)}(5.7 - 5) = 298.5$$

This value differs markedly from that obtained using backward-difference polynomial interpolation. The difference in the two values is not surprising since the table clearly shows considerable curvature near the bottom of the data set. It is generally *assumed* that the value obtained using polynomial interpolation is more accurate, though this is not always true (more about this later).

The backward-difference interpolation technique lends itself nicely to spreadsheet implementation. The procedure is entirely analogous to that used with forward differences.

Example 6.9 Interpolation Using Backward Differences in Excel

Construct an Excel worksheet to recreate the difference table given in Example 6.7 and the solution to the problem stated in Example 6.8.

B9

=B7+J6*C7/J7+J6*(J6+J7)*D7/(2*J7^2)+J6*(J6+J7)*(J6+2*J7)*E7/(6*J7^3)+J6*(J6+J7)*(J6+2*
J7)*(J6+3*J7)*F7/(24*J7^4)+J6*(J6+J7)*(J6+2*J7)*(J6+3*J7)*(J6+4*J7)*G7/(120*J7^5)

	x	y	Dy	D2y	D3y	D4y	D5y				
1	x	y	Dy	D2y	D3y	D4y	D5y				
2	1	1									
3	2	2	1								
4	3	10	8	7							
5	4	44	34	26	19						
6	5	141	97	63	37	18			u=	-0.3	
7	6	366	225	128	65	28	10		h=	1	
8											
9	5.7	279.7707									

Ready

Figure 6.6

Figure 6.6 shows the worksheet containing the backward-difference table that we built by hand in Example 6.8. The key features are directly analogous to the worksheet containing the forward-difference table shown in Fig. 6.5. The interpolated value of y, corresponding to the given value of $x = 5.7$, is shown in cell B9 to be 279.7707. This value agrees with the value obtained by hand in Example 6.8. Notice that the interpolated y-value is determined by a lengthy Excel formula. This formula corresponds to the Gregory-Newton backward interpolation formula expressed by Equation (6.14).

In the last section we described a technique for filling in the empty lower portion of a forward-difference table when all of the values in one of the columns are the same. We can use a similar technique to fill in the empty *upper* portion of a backward-difference table. To do so, we begin with the high-order differences and work up and to the left, using a rearranged form of Equation (6.13); i.e.,

$$\nabla^{k-1}y_{i-1} = \nabla^{k-1}y_i - \nabla^k y_i \tag{6.15}$$

The procedure parallels that described in Example 6.6 for a forward-difference table. The details are left as an exercise (see Prob. 6.18 at the end of the chapter).

The use of an interpolating polynomial generally produces satisfactory results when the known data points exhibit relatively uniform curvature, as shown in Fig. 6.3, and the higher differences tend to decrease in magnitude.   Forward differences generally work best when the unknown point falls near *beginning* of the data set (i.e., at the top of the forward-difference interpolation table), and backward differences work best when the unknown point falls near the *end* of the data set.   There is also a slightly more complicated technique, based upon the use of *central differences* that is best suited for an unknown point located near the *center* of the data set. (see Nakamura, 1993).

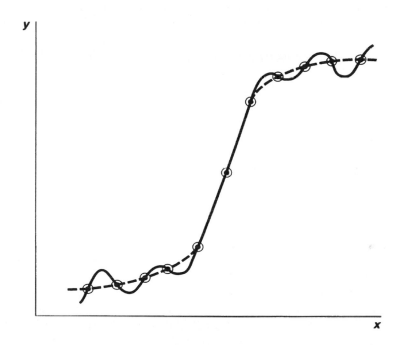

Figure 6.7

Some data sets exhibit little curvature in some areas but sharp curvature in certain localized areas, as shown in Fig. 6.7. When this happens, an interpolating polynomial may oscillate wildly between the data points, even though it passes through the data points themselves. In such situations, we may obtain more accurate results by passing a *lower-degree* polynomial (e.g., a quadratic or cubic function) through a few points within immediate region of the unknown point.

Interpolating polynomials can also produce erroneous results if they are used to extrapolate beyond the range of the data. This is particularly true of high-degree polynomials, which can oscillate wildly outside of their defined region. Therefore, polynomial-based extrapolation should generally be avoided.

6.4 OTHER INTERPOLATION TECHNIQUES

The subject of interpolation has been studied extensively, and many other techniques have been developed in addition to those described in this chapter. We mention a few in passing, for your information All of these methods can be implemented in Excel through the use of formulas, though none makes use of difference tables.

If an nth-degree polynomial is fit through $n+1$ data points, that polynomial will be unique; that is, only one nth-degree polynomial can be passed through a given set of $n+1$ data points. Therefore, you will obtain the *same* polynomial, regardless of the way the polynomial may be arranged algebraically.

The Gregory-Newton formulas offer one approach to representing an nth-degree polynomial in terms of $n+1$ data points. Another is the *Lagrangian form* of the interpolating polynomial. In the latter case, the polynomial is written as

$$y = f_1 y_1 + f_2 y_2 + \cdots + f_i y_i + \cdots + f_n y_n \tag{6.16}$$

where $f_1, f_2, \cdots, f_i, \cdots, f_n$ are polynomials defined as

$$f_1 = \left[\frac{(x-x_2)(x-x_3)\cdots(x-x_n)}{(x_1-x_2)(x_1-x_3)\cdots(x_1-x_n)} \right] \tag{6.17}$$

$$f_2 = \left[\frac{(x-x_1)(x-x_3)\cdots(x-x_n)}{(x_2-x_1)(x_2-x_3)\cdots(x_2-x_n)} \right] \tag{6.18}$$

$\cdots\cdots$

$$f_i = \left[\frac{(x-x_1)(x-x_2)\cdots(x-x_{i-1})(x-x_{i+1})\cdots(x-x_n)}{(x_i-x_1)(x_i-x_2)\cdots(x_i-x_{i-1})(x_i-x_{i+1})\cdots(x_i-x_n)} \right] \tag{6.19}$$

$\cdots\cdots$

$$f_n = \left[\frac{(x-x_1)(x-x_2)\cdots(x-x_{n-1})}{(x_n-x_1)(x_n-x_2)\cdots(x_n-x_{n-1})} \right] \tag{6.20}$$

Note that $f_i = 1$ when $x = x_i$, and $f_i = 0$ when $x = x_j$ (where x_j refers to one of the other tabulated x-values). Thus, Equation (6.16) results in $y = y_i$ when $x = x_i$.

The advantage of the Lagrangian form of the interpolating polynomial is that we can apply it to a set of *unequally* spaced data. On the other hand, the Lagrangian polynomial can oscillate wildly between tabulated points, particularly in an area in which the separation between tabulated points is relatively large.

Some other interpolation techniques involve the use of *Chebychev polynomials* or *Hermite polynomials.* The use of Chebychev polynomials is attractive because the error may be more evenly divided over the given data set. The error is minimized by carrying out the interpolation at particular, specified locations within an interval selected by the user. Interpolation involving the use of Hermite polynomials is best suited to data points whose derivatives are known.

Of particular interest is a relatively new interpolation technique involving the use of *cubic splines.* In this method, cubic polynomials are passed through pairs of data points in such a manner that each polynomial passes through a pair of data points, and the first two derivatives are continuous over the entire range of data points. The method is repeated for each pair of successive data points. The result is a smooth, well-behaved curve, even when passing through regions of sharp, local curvature, as illustrated in Fig. 6.7. The computational details are somewhat complicated since the method requires setting up and solving simultaneous equations. (We will discuss the use of spreadsheets to solve simultaneous equations in Chap. 8)

Problems

6.16 Construct an Excel worksheet containing the backward-difference table shown in Fig. 6.6. (Generate the differences using Equations (6.10) through (6.13)—do not simply type in the numerical values.) Then use the difference table to obtain interpolated values of *y* corresponding to the following *x*-values:

(*a*)   4.85   (*b*)   3.50   (*c*)   5.20

In each case, use the last row (corresponding to $x = 6$) as the base row.

6.17 Repeat Example 6.8, solving for the value of y corresponding to $x = 5.7$. Now, however, use the *second* last row of the difference table (corresponding to $x = 5$) as the base row. Compare your answer with that obtained in Example 6.8. Why are the results different? Which answer would you assume is more accurate, and why?

6.18 Repeat Prob. 6.16 using the *second* last row of the difference table as the base row. Compare your answers with those obtained in Prob. 6.16. Which answers do you prefer, and why?

6.19 Construct an Excel worksheet containing an extended backward-difference table that parallels the forward-difference table given in Example 6.6. (Use the appropriate formulas to determine the values of the backward differences.) Then determine the value of *y* corresponding to $x = 5.7$, using the sixth row (i.e., $x = 6$) in the table as the base row. Then repeat the solution using the second row ($x = 2$) as the base row. Compare your answers with the results obtained in Example 6.8 and Prob. 6.17.

6.20 Use the extended backward-difference table created in Prob. 6.19 to obtain interpolated values of y corresponding to the following x-values:

(a)   4.85   (b)   3.50   (c)   2.25   (d)   −0.88

In each case, use the fourth row (corresponding to $x = 4$) as the base row. Compare your answers with those obtained in Prob. 6.16 where appropriate.

6.21 Repeat Prob. 6.1(c) using backward-difference polynomial interpolation based upon the four center points (temperatures ranging from 200 to 500°C). Compare your answers with the results obtained in Prob. 6.1(c) and Prob. 6.12.

6.22 Repeat Prob. 6.5(b) using backward-difference polynomial interpolation based upon:

(a)   Three data points (corresponding to 10, 12, and 14 seconds).

(b)   Four data points (8, 10, 12, and 14 seconds).

(c)   Five data points (6, 8, 10, 12, and 14 seconds).

Compare your answers with those obtained in Prob. 6.5(b). What can you conclude about the use of additional data points?

6.23 Repeat Prob. 6.5(b) using *forward*-difference polynomial interpolation based upon:

(a)   Three data points (corresponding to 12, 14, and 16 seconds).

(b)   Four data points (12, 14, 16, and 18 seconds).

(c)   Five data points (12, 14, 16, 18, and 20 seconds).

Compare your answers with the corresponding results obtained in Prob. 6.22 (i.e., compare the results of part (a) with the results of Prob. 6.22(a), and so on). Why are the corresponding results different? What does this tell you about the use of polynomial interpolation?

6.24 Construct an Excel worksheet containing a backward-difference table for the following set of data:

x_i	y_i	x_i	y_i
0.5	0.125	3.5	42.875
1.5	3.375	4.5	91.125
2.5	15.625	5.5	166.375

Fill in the table as completely as possible. Then use the table to estimate the value of y corresponding to each of the following x-values:

(a)   3.0   (b)   1.2   (c)   1.7   (d)   5.0

For this data set, does it matter which row in the backward-difference table is used as the base row? Compare your answers with those obtained in Prob. 6.14.

6.25 Repeat Prob. 6.24(*a*) using

(*a*) Linear interpolation

(*b*) Lagrangian interpolation, based upon Equations (6.16) through (6.20). (*Hint*: Enter the x_i values in one column, the y_i values in the next column, and the corresponding values for f_i, calculated from Equations (6.17) through (6.20), in the third column. Then use Equation (6.16) to determine the desired *y*-values.)

Compare your answers with those obtained in Prob. 6.24(*a*).

6.26 Using the data given in Prob. 6.5, determine the current flowing through the electronic device at $t = 21$ seconds. Determine the current three ways:

(*a*) Using linear interpolation.

(*b*) Using Lagrangian interpolation based upon the given data points corresponding to $t = 18$, 20, and 25 seconds.

(*c*) Using Lagrangian interpolation based upon the last five data points.

6.27 Engineers have studied the properties of steam extensively. The following table contains data indicating the equilibrium relationship between the temperature and pressure of saturated steam. (Think of this as the boiling point of water as a function of pressure.)

T, °F	*P, psia*	*T*, °F	*P, psia*
32	0.0886	350	134.62
50	0.1780	400	247.25
100	0.9487	450	422.61
150	3.716	500	680.80
200	11.525	550	1045.6
212	14.696	600	1543.2
250	29.82	650	2208.8
300	67.01	700	3094.1

A heat exchanger is being designed to operate at a temperature of 480°F. If water is the heat-transfer fluid, what is the minimum pressure that the heat exchanger must be able to withstand so that the water does not flash to steam? Determine your answer using

(*a*) Linear interpolation.

(*b*) Four-point Lagrangian interpolation, based upon the two surrounding data points in each direction (above and below 480°F).

Compare your answers with the known value of 566.12 psia.

ADDITIONAL READING

Akai, T. J. *Applied Numerical Methods for Engineers*. New York: Wiley, 1994.

Carnahan, B., H. A. Luther and J. O. Wilkes. *Applied Numerical Methods*. New York: Wiley, 1969.

Chapra, S. C. and T. P. Canale. *Numerical Methods for Engineers, with Personal Computer Applications*. New York: McGraw-Hill, 1985.

Mayo, W. E. and M. Cwiakala. *Introduction to Computing for Engineers*. New York: McGraw-Hill, 1991.

Nakamura, S. *Applied Numerical Methods in C*. Englewood Cliffs, N. J.: Prentice-Hall, 1993.

CHAPTER 7

SOLVING
SINGLE EQUATIONS

Engineers are often required to solve complicated algebraic equations. These equations may represent cause-and-effect relationships between system variables, or they may be the result of applying a physical principle to a specific problem situation. In either case, the result generally provides a detailed understanding of the problem at hand.

For example, the relationship between pressure, volume and temperature for many real gases can be determined by Van der Waals's equation of state, which is written as

$$\left(P+\frac{a}{V^2}\right)(V-b)=RT \tag{7.1}$$

where P is the absolute pressure, V is the volume per mole, T is the absolute temperature, R is the ideal gas constant (0.082054 liter atm / mole °K), and a and b are constants that are unique to each particular gas. (Note that Equation (7.1) reduces to the well-known ideal gas equation if a and b both equal zero. Hence, the parameters a and b represent the departure from ideal gas behavior.)

Suppose we are working with a real gas for which a and b are known. If the pressure and volume are specified, it is very easy to solve Equation (7.1) for the temperature. Similarly, it is very easy to solve Equation (7.1) for the pressure if the temperature and volume are specified. It is much more difficult, however, to solve Equation (7.1) for the volume if the temperature and pressure are specified since the equation is *nonlinear* in terms of volume (because of the V^2 term in the denominator on the left-hand side). In fact, the solution involves finding a positive real root of a cubic equation. Moreover, the solution may be complicated by the presence of multiple real roots, which requires some judgment in the selection of the correct root.

In this chapter, we will see several different approaches to solving nonlinear algebraic equations. We begin by discussing the characteristics of nonlinear equations and show how approximate solutions can be obtained using graphical techniques. We will then see how real roots can be obtained using Excel's automated **Goal Seek** and **Solver** features. The chapter concludes by showing two popular mathematical techniques for obtaining detailed numerical solutions.

7.1 CHARACTERISTICS OF NONLINEAR ALGEBRAIC EQUATIONS

Let's begin by reviewing what is known about algebraic equations and their solutions. An equation is said to be *linear* if it can be rearranged so that the variable representing the unknown quantity appears only to the first power. For example, Ohm's law, written as $\Delta V = iR$, is linear with respect to all three variables (ΔV, i and R). Linear equations have one and only one root, and it is always real. Hence, linear equations are very easy to solve using the techniques of elementary algebra.

On the other hand, an equation is said to be *nonlinear* if it *cannot* be rearranged so that the variable representing the unknown quantity appears only to the first power. Thus, if x represents the unknown quantity, a nonlinear equation might include x raised to some power other than one, or it might include the log of x, the sine of x, etc. Nonlinear equations generally cannot be solved using the techniques of elementary algebra (though there are exceptions, such as the well-known formula for obtaining the roots of a quadratic equation). Hence, they must be solved either graphically or numerically. Moreover, there may be multiple real roots, or there may not be any real roots at all (they may be expressed as complex variables; i.e., $x = u + iv$, where i is an imaginary number, defined as the square root of -1). Clearly, nonlinear equations are much more difficult to solve than linear equations.

A *polynomial* equation is a special case of a nonlinear equation. A polynomial equation can include powers of x but not terms such as $\log x$, $\sin x$, e^x, etc. The highest power of x is called the *degree* of the polynomial (it is also called the *order* of the polynomial). Thus, a quadratic equation is a second-degree polynomial because the highest power of x is 2; a cubic equation is a third-degree polynomial because the highest power of x is 3, and so on.

The following information is known about polynomial equations:

1. An nth-degree polynomial can have no more than n real roots.

2. If the degree of a polynomial is *odd*, there will *always* be *at least one* real root.

3. Complex roots always exist in *pairs*, if they exist at all. Each pair of complex roots consists of *complex conjugates*; that is, $x_1 = u + iv$ and $x_2 = u - iv$.

We will confine our attention to real roots of equations within this book.

Example 7.1 Identifying the Real Roots of a Polynomial

The following equation determines the location of the maximum shear force along a beam, where x represents the distance from the left end of the beam.

$$2x^3 - \frac{5}{x^2} = 3$$

How many real roots might there be for this equation?

To answer this question, let us first rearrange the equation into the form

$$2x^5 - 3x^2 - 5 = 0$$

When written in this form, it is easy to see that the equation is a fifth-degree polynomial. Therefore, there will be at least one real root. If there is *only* one real root, there will be two pairs of complex roots. There may also be three real roots and one pair of complex roots. Or there may be five real roots.

When real roots do exist, we have no assurance that any of them will be positive. They can be negative, or they can equal zero. It is not unusual for multiple real roots to be mixed in sign (some positive, some negative, etc.) In such situations, the physical situation described by the equation generally suggests which root to choose.

Problems

7.1   What can you conclude about the number of real roots for each of the following equations?

(*a*) $3x + 10 = 0$

(*b*) $3x^2 + 10 = 0$

(*c*) $3x^3 + 10 = 0$

(*d*) $x + \cos x = 1 + \sin x$

7.2   Rearrange Van der Waals's equation of state into the form $f(x) = 0$. Once you have completed the rearrangement, what can you conclude about the number of real roots for Van der Waal's equation if the volume per mole (V) is considered the unknown quantity?

7.3 In order to determine the temperature distribution within a one-dimensional solid, engineers must often solve the equation

$$x \tan x = c$$

where c is a known positive constant. What can you conclude about the number of real roots for this equation?

7.2 GRAPHICAL SOLUTIONS

Approximate graphical solutions are easy to obtain, particularly with a spreadsheet. Basically, the procedure is to write the equation in the form $f(x) = 0$ and then plot $f(x)$ versus x. The points where $f(x)$ crosses the x-axis (i.e., the values of x that cause $f(x)$ to equal 0) are real roots of the equation. Generally, these points can be read directly from the graph. You can also interpolate between the tabulated values to find the point where $f(x) = 0$.

Example 7.2 Solving a Polynomial Equation Graphically

In Example 7.1 we established that the equation

$$f(x) = 2x^5 - 3x^2 - 5 = 0$$

has at least one real root and possibly more. Prepare an Excel worksheet in which $f(x)$ is tabulated as a function of x over the interval $-10 \le x \le 10$. Plot the data as an x-y graph and determine the real roots that fall within this interval.

The worksheet containing the tabulated data and the corresponding x-y graph is shown in Fig. 7.1. Notice the Excel formula that is used to generate the contents of cell B2; that is, =2*A2^5−3*A2^2−5. This formula (with relative cell addressing) was used to generate all of the remaining values in column B. From the tabulated values, we see that the function crosses the x-axis at a point between $x = 1$ and $x = 2$. (Using linear interpolation, we obtain the crossover point as $x = 1.113$.) Hence, the equation appears to have one real root within the specified interval, at approximately $x = 1.113$.

The x-y graph in Fig. 7.1 is not very helpful because the range of the y-values is too large. Therefore, let us retabulate and plot the data over the interval $0 \le x \le 2$ using a finer set of x-values. The results are shown in Fig. 7.2. Now we can see quite clearly, from both the tabulated data and the graph, that $f(x)$ crosses the x-axis at a point slightly beyond $x = 1.4$.

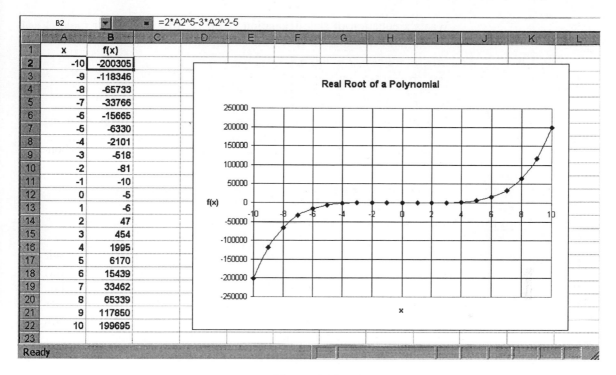

Figure 7.1

We conclude that the given equation has only one real root within the original interval, and its approximate value is $x = 1.4$. (Using linear interpolation, we obtain a value of $x = 1.403$.) The value read from the graph (or the interpolated value) may be sufficiently accurate; if so, the problem is solved. However, we will see how a more precise value can be obtained later in this chapter, using several different numerical methods.

Problems

7.4   Determine a real root for the equation given in Prob. 7.1(c) using the graphical method described above.

7.5   Determine the positive real roots for the equation given in Prob. 7.1(d) using the graphical method described above.

7.6   The Van der Waals's constants for carbon dioxide have been determined to be $a = 3.592$ liter^2 atm / mole^2 and $b = 0.04267$ liter / mole. Solve Van der

Waals's equation of state [Equation (7.1)] for the volume per mole (V) of carbon dioxide under the following conditions:

(a) $P = 1$ atm, $T = 300°K$

(b) $P = 10$ atm, $T = 400°K$

(c) $P = 0.1$ atm, $T = 300°K$

In each case, compare your solution with the volume obtained using the ideal gas law ($PV = RT$). Remember that $R = 0.082054$ liter atm / mole °K.

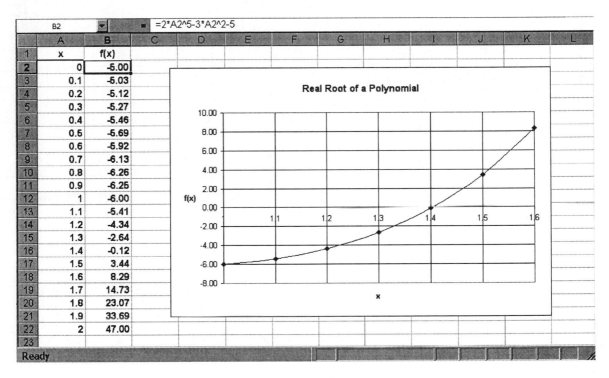

Figure 7.2

7.7   The Van der Waals's constants for an organic compound have been determined as $a = 40.0$ liter^2 atm / mole^2 and $b = 0.2$ liter / mole. Solve Van der Waals's equation of state [Equation (6.1)] for the volume per mole (V) of this compound under the following conditions:

(a) $P = 1$ atm, $T = 300°K$

(b) $P = 10$ atm, $T = 400°K$

(c) $P = 0.1$ atm, $T = 300°K$

In each case, compare your solution with the results obtained for carbon dioxide in Prob. 7.6 and the results obtained using the ideal gas law.

7.8   For the equation given in Prob. 7.3, determine

(*a*)   The three smallest positive roots.

(*b*)   The two largest negative roots (i.e., the two negative roots closest to the origin).

Assume a value of $c = 2$ in all calculations.

7.3 SOLVING EQUATIONS IN EXCEL USING GOAL SEEK

The material in the last two sections was intended to provide you with a general understanding of the characteristics of single algebraic equations and a simple method for determining approximate roots.   Now let us see how accurate solutions can be obtained quickly and easily using Excel's Goal Seek feature. This feature permits rapid solutions of algebraic equations using *iterative* (i.e., trial-and-error) root-finding techniques, based upon a series of successive refinements derived from an initial guess (see Sec. 7.6).

To solve a single algebraic equation with Goal Seek, proceed as follows:

1. Enter an initial guess in one of the cells on the worksheet.

2. Enter a formula for the equation, in the form $f(x) = 0$, in another cell. Within this formula, express the unknown quantity x as the cell address containing the initial guess.

3. Select Goal Seek from the Tools menu.

4. When the Goal Seek dialog box appears, enter the following information:

(*a*)   The address of the cell containing the formula in the Set cell entry location.

(*b*)   The value 0 in the To value entry location.

(*c*)   The address of the cell containing the initial value in the entry location labeled By changing cell. Then select OK.

A new dialog box labeled Goal Seek Status will then appear, telling you whether or not Excel has been able to solve the problem (i.e., whether convergence has been obtained). If a solution has been obtained, the value of the root will appear in the cell originally containing the initial guess. The value in the cell containing the formula will show a value that is close to zero (but usually not exactly zero). This last value will also appear within the Goal Seek Status dialog box.

The likelihood of obtaining a converged solution will be enhanced if the initial guess is as close as possible to the desired root. Thus, we would like to use an approximate value for the root as the initial guess. This value can often be established using the graphical or tabular techniques described in the last section.

A warning message will be generated if the computation does not converge or if an inappropriate value (such as the square root of a negative number) is about to be generated during the course of the computation.

Example 7.3 Solving a Polynomial Equation in Excel Using Goal Seek

In Example 7.2, we found that the equation

$$f(x) = 2x^5 - 3x^2 - 5 = 0$$

has a real root at approximately $x = 1.4$. Let us now make use of Excel's Goal Seek feature to obtain a more accurate solution. We will select the value $x = 1.4$ as a first guess since we already know that this is the approximate value of the root.

Figure 7.3 shows an Excel worksheet with the initial guess (1.4) entered into cell B3, and the formula for $f(x)$ entered into cell B5. (Notice the Excel formula for $f(x)$ shown in the formula bar.) The numerical value -0.12352 appearing in cell B5 is the value of $f(x)$ evaluated at $x = 1.4$.

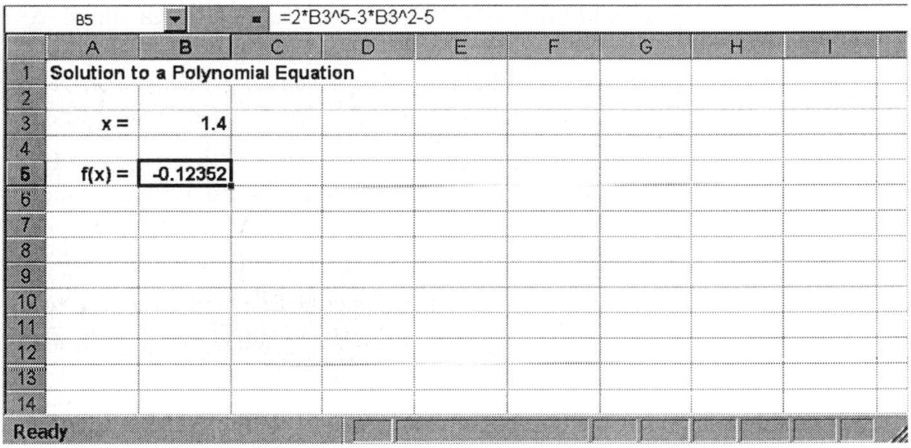

Figure 7.3

Our goal, of course, is to determine the value of x that causes $f(x)$ to equal zero. To do so, we select Goal Seek from the Tools menu. Figure 7.4 shows the resulting Goal Seek dialog box. The information that has been entered into the dialog box indicates that the value of $f(x)$ in cell B5 will be driven to zero (or approximately zero) by changing the value of x in cell B3.

The resulting solution is shown in Fig. 7.5. The Goal Seek Status dialog box, which has replaced the previous dialog box, indicates that a solution has been found that drives the value of the formula to -3.5625×10^{-5} (approximately zero). The corresponding value of x, which is the desired solution, now appears in cell B3. Thus, Excel has found a solution of $x = 1.404085$. This value is consistent with the approximate solution obtained earlier. The corresponding value of $f(x)$, $-3.6E-5$, appears in cell B5. This is simply a duplication (rounded) of the value shown in the Goal Seek Status dialog box.

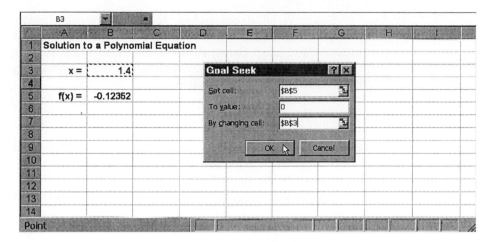

Figure 7.4

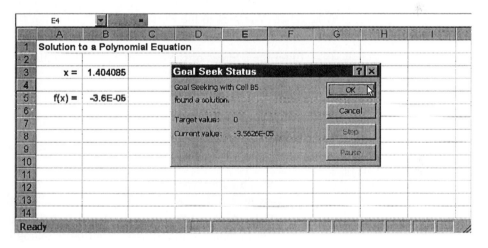

Figure 7.5

7.4 SOLVING EQUATIONS IN EXCEL USING SOLVER

Single algebraic equations can also be solved using Excel's Solver feature, though Solver is intended primarily for solving constrained optimization problems (see Chap. 11). The advantage to using Solver is that you can specify constraints (i.e., auxiliary conditions) on the independent variable. For example, you can restrict the independent variable to nonnegative values or require that the independent variable falls within prescribed bounds. Otherwise, the procedure for using Solver is very similar to that used with Goal Seek.

Solver is an Excel add-in, like the Analysis Toolpak used to generate histograms (see Chap. 4). To install the Solver, choose Add-Ins from the Tools menu. Then select Solver Add-In from the resulting Add-Ins dialog box. Once the Solver feature has been installed, it will remain installed unless it is removed, by reversing the above procedure.

To solve a single equation with Solver, proceed as follows:

1. Enter an initial guess in one of the cells on the worksheet.

2. Enter an equation for $f(x)$ in another cell, expressed as an Excel formula. Within this formula, express the unknown quantity x as the address of the cell containing the initial guess.

3. Select Solver from the Tools menu.

4. When the Solver Parameters dialog box appears, enter the following information:

    (*a*) The address of the cell containing the formula for $f(x)$ in the Set Target Cell location.

    (*b*) Select Value of in the Equal to line. Then enter 0 within the associated data area. (In other words, set the target cell equal to a value of zero.)

    (*c*) Enter the address of the cell containing the initial value for x in the area labeled By Changing Cells.

    (*d*) If you wish to restrict the range of x, click on the Add button under the heading Subject to the Constraints. Then provide the following information within the Add Constraint dialog box:

        (*i*) The address of the cell containing the initial value for x in the Cell Reference location.

        (*ii*) Specify the type of constraint (i.e., $\leq$ or $\geq$) from the pull-down menu.

        (*iii*) Enter the bounding value in the Constraint data area. Then select OK to return to the Solver Parameters dialog box.

    Note that you can change or delete any constraint after it has been added.

    (*e*) When all of the required information has been entered correctly, select Solve. This will initiate the actual solution procedure.

A new dialog box labeled Solver Results will then appear, telling you whether or not Solver has been able to solve the problem (i.e., whether convergence has been obtained). If a solution has been obtained, the value of the root will appear in the cell originally containing the initial guess. The cell containing the formula will show a value that is zero or very close to zero.

You should understand that Solver and Goal Seek use different mathematical solution procedures. Hence, the values obtained with Solver may not be identical to those obtained using Goal Seek, though they should be very close. Remember that both features arrive at their respective solutions through a series of refined approximations rather than providing exact answers.

Example 7.4 Solving a Polynomial Equation in Excel Using Solver

In the previous examples we solved the polynomial equation

$$f(x) = 2x^5 - 3x^2 - 5 = 0$$

and found a positive real root in the vicinity of $x = 1.4$. Let us now use Excel's Solver feature to find a root of this equation, with the added restriction that $x \geq 0$.

Figure 7.6 shows an Excel worksheet containing an initial guess in cell B3 and a formula for the the given equation in cell B5, as before. We also see the Solver Parameters dialog box, which was obtained by selecting Solver from the Tools menu. The cell address of the equation (B5) has been entered into the Set Target Cell area. Notice the selection requiring that the value of the target cell be set equal to zero. The cell address containing the initial guess (B3) has also been entered into the By Changing Cells area of the dialog box. However, the nonnegative restriction on x has not yet been entered.

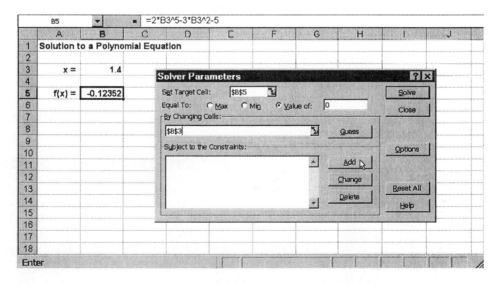

Figure 7.6

To add the nonnegativity restriction, we select Add near the bottom of the Solver Parameters dialog box, resulting in the Add Constraint dialog box shown in Fig. 7.7. We then enter the cell address of the independent variable (B3) in the left data area, we select the type of constraint ($\geq$) in the center data area, and we provide the bounding value (0) in the right data area. Once the constraint has been entered correctly, we select OK at the bottom of the dialog box, which returns us to the Solver Parameters dialog box, as shown in Fig. 7.8. Now the entire problem, including the constraint specification, appears at the bottom of the Solver Parameters dialog box.

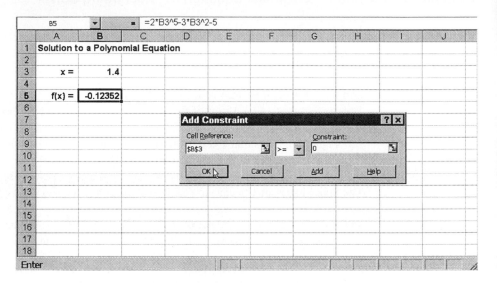

Figure 7.7

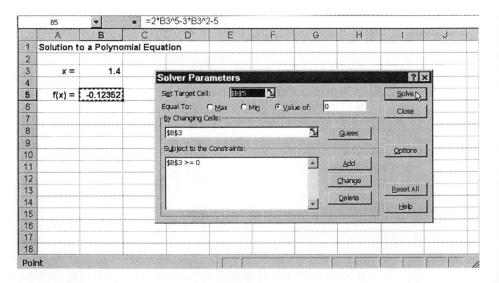

Figure 7.8

Selecting Solve from the Solver Parameters dialog box initiates the solution procedure. Once a solution has been obtained, the resulting values appear in cells B3 and B5, as shown in Fig. 7.9. Thus, we see that the desired solution is $x = 1.404086$, resulting in a value of $f(x) = 3.59 \times 10^{-8}$. These values are very close to (though slightly different) than those obtained in Example 7.3.

In addition, we see that the Solver Results dialog box replaces the Solver Parameters dialog box. Our response to the Solver Results dialog box is Keep Solver Solution followed by OK. Neither the Reports nor the Save Scenario feature should be selected since we already have the information being sought.

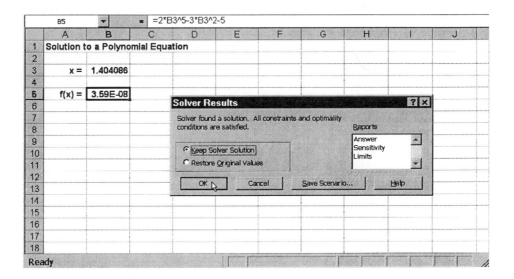

Figure 7.9

Solver includes a number of additional features that are not needed when solving single algebraic equations. We will discuss several of these features in Chap. 11, where we explain how Solver is used to solve a more complicated class of problems.

Problems

Solve each of the following problems in Excel using either Goal Seek or Solver.

7.9 Determine a real root for the equation given in Prob. 7.1(c):

$$3x^3 + 10 = 0$$

7.10 Determine the two smallest positive real roots for the equation given in Prob. 7.1(*d*):

$$x + \cos x = 1 + \sin x$$

7.11 Solve Van der Waals's equation [Equation (7.1)] for the volume per mole (V) of an organic compound at 10 atm pressure and 400°K. The Van der Waals's constants for this particular compound are a = 40.0 liter^2 atm/mole^2 and b = 0.2 liter/mole. Compare your answer with the results obtained in Prob. 7.7(b).

7.12 Determine the smallest positive root and the largest negative root (the negative root closest to the origin) for the equation originally given in Prob. 7.3, with $c = 2$:

$$x \tan x = 2$$

Compare your answers with the results obtained in Prob. 7.8.

7.13 An angular support member within a bridge is subject to a coaxial compressive force. If the horizontal force component is 120 lbs$_f$ and the vertical force component is 175 lb$_f$, what angle does the member make with the horizontal?

7.14 Many consumers buy expensive items, such as cars and houses, by borrowing the purchase cost from a bank and then repaying the loan on a constant monthly basis. If P is the total amount of money that is borrowed initially, the amount of the monthly payment, A, can be determined from the formula

$$A = P\left[\frac{i(1+i)^n}{(1+i)^n - 1}\right]$$

where i is the monthly interest rate, expressed as a *fraction* (not a percentage), and n is the total number of payments.
Suppose you borrow $10,000 to buy a car.

(*a*) If the nominal interest rate is 8 percent API (which is equivalent to a fractional monthly interest rate of 0.08/12 = 0.006667) and you borrow the money for 36 months, how much money will you have to repay each month?

(*b*) If you are required to repay $350 each month for 36 months, what is the corresponding monthly interest rate?

(c) If you choose to repay $350 each month at the 0.006667 monthly interest rate, how many payments (i.e., how many months) will be required to repay the loan?

(d) Do all of the above questions require the use of the Goal Seek feature in Excel?

7.15 An equation that sometimes arises in the design of a structure is

$$\tanh x = \tan x + 1$$

where x represents the angle (in radians, measured from the horizontal) at which an external force must be applied in order to distribute the force properly. Determine the first positive root of this equation; that is, the value of x within the interval $0 < x < \pi$, that satisfies this equation. Use Excel's TAN and TANH functions to obtain your solution.

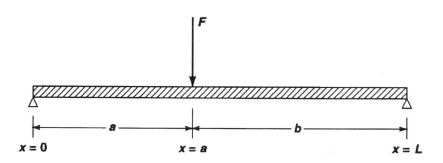

Figure 7.10

7.16 A horizontal beam of length L is supported at each end ($x = 0$ and $x = L$), as shown in Fig. 7.10. Suppose an external vertical force F is applied to the beam at location a, where a is measured from the left end of the beam (from Fig. 7.10, note that $a + b = L$). Then the vertical deflection of the beam at some point x, $x \le a$, is determined by the expression

$$y = \frac{Fbx}{6EIL}(L^2 - x^2 - b^2), \qquad 0 \le x \le a$$

where y is the vertical deflection, E is the *modulus of elasticity* (also known as *Young's modulus*), and I is the *moment of intertia* of the cross-sectional area of the beam. (Note that E depends only on the beam material, and I depends only on the beam geometry.)

Suppose a 10-ft beam is made of steel ($E = 30{\times}10^6$ psi), and the moment of inertia is $I = 5$ in^4. If a vertical force of 15,000 lb$_f$ is applied at $a = 4$ ft, determine

(a)   The maximum deflection (which occurs at $x = \sqrt{(L^2 - b^2)/3}$)

(b)   The location to the left of the vertical force (solve for $x < a$) at which the deflection is 0.75 times the maximum deflection.

Remember to use consistent units (convert ft to inches).

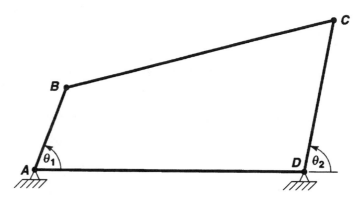

Figure 7.11

7.17 Four-bar linkages are commonly used to transmit mechanical movement in various ways. A typical four-bar linkage is shown in Fig. 7.11. Member AD is anchored to its supporting base and does not move. Member AB rotates about point A, resulting in a reciprocating movement of members BC and CD. Therefore, the angle θ_2 changes in response to the change in θ_1.

$$AD = AB\cos\theta_1 - CD\cos\theta_2 + \sqrt{(BC)^2 - (CD\sin\theta_2 - AB\sin\theta_1)^2}$$

Suppose $AB = 1.5$ in, $BC = 6$ in, $CD = 3.5$ in, and $AD = 5$ in.

(a)   Determine the value of θ_2 corresponding to $\theta_1 = 45°$.

(b)   By plotting several additional values of θ_2 against θ_1, determine the maximum value of θ_2 for this configuration.

7.18 Chemical and mechanical engineers must often determine the power required to pump a fluid through a series of pipes. To do so, we must first determine a *friction factor*. If the fluid flow is turbulent and the pipes are smooth, the friction factor can be determined as

$$1.1513 / \sqrt{f} = \ln (Re \ \sqrt{f} / 2.51)$$

In this equation f represents the friction factor, a positive value that does not exceed 0.1, and Re represents the *Reynolds number*, which is a function of the fluid properties, the pipe diameter and the nominal fluid velocity. The Reynolds number increases with increasing fluid velocity. The flow becomes turbulent at higher velocities, when the Reynolds number exceeds approximately 3000. Hence the above equation is valid only when Re exceeds 3000.

Determine the values of the friction factor corresponding to Reynolds numbers of 5000, 12,000 and 20,000. Does the friction factor increase or decrease as the Reynolds number increases?

7.19 Mechanical engineers often study the movement of damped oscillating objects. Such studies play an important role in the design of automobile suspensions, among other things.

The horizontal displacement of an object is given as a function of time by the equation

$$x = x_0 e^{-\beta t} \left[\cos (\omega t) + (\beta / \omega) \sin (\omega t) \right]$$

The parameters β and ω depend upon the mass of the object and the dynamic characteristics of the system (i.e., the spring constant and the damping constant).

For a particular system, suppose $x_0 = 8$ in, $\beta = 0.1$ sec^{-1} and $\omega = 0.5$ sec^{-1}. Carry out the following calculations for this system:

(a) Plot x versus t over the interval $0 \le t \le 30$ sec.

(b) Determine, as accurately as possible, the point where x first crosses the t-axis (i.e., determine the value of t corresponding to $x = 0$).

(c) Determine, as accurately as possible, the point where x crosses the t-axis for the second time.

(d) If x_0 and β retain their original values, what value of ω will cause x to cross the t-axis for the first time at $t = 2.5$ sec?

7.20 Electrical engineers often study the transient behavior of RLC circuits (containing resistors, inductors and capacitors). Such studies play an important role in the design of various electronic components.

Suppose an open RLC circuit has an initial charge of q_0 stored within a capacitor. When the circuit is closed (by throwing a switch), the charge flows out of the capacitor and through the circuit in accordance with the equation

$$q = q_0 e^{-Rt/(2L)} \cos\left[\sqrt{\frac{1}{LC} - \left(\frac{R}{2L}\right)^2}\; t\right]$$

In this equation, R is the resistive value of the circuit, L is the value of the inductance, and C is the value of the capacitance.

For a particular circuit, suppose $q_0 = 10^3$ coulombs, $R = 0.5 \times 10^3$ ohms, $L = 10$ henries, and $C = 10^{-4}$ farads. Carry out the following calculations for this system:

(a) Plot q versus t over the interval $0 \le t \le 0.5$ sec.

(b) Determine, as accurately as possible, the point where q first crosses the t-axis (i.e., determine the value of t corresponding to $q = 0$).

(c) Determine, as accurately as possible, the point where q crosses the t-axis for the second time.

(d) If q_0, L and C retain their original values, what value of R will cause q to cross the t-axis for the first time at $t = 0.06$ sec?

Notice the similarity between this problem and the vibrating mass described in Prob. 7.19.

7.5 INTERVAL-REDUCTION TECHNIQUES

We now turn our attention to a class of simple mathematical methods that are often used to solve nonlinear algebraic equations. Methods such as these form the basis for the procedures used in Goal Seek. Thus, these methods provide insight into how Excel goes about solving algebraic equations.

When using an interval-reduction technique, the basic idea is to determine the approximate location of a root within a specified interval and then eliminate a part of the interval, retaining the portion of the original interval containing the root. This process is repeated several times, until the interval containing the root is very small. The desired root will lie somewhere within this final small interval. (The exact location of the root will not be required if the final interval is sufficiently small.)

One of the most widely used interval reduction techniques is the *method of bisection* (also called the *method of interval halving*). To use this method, *we must begin with an interval containing one and only one real root*. (Thus, a plot of the function $f(x)$ versus x will cross the x-axis once and only once.) An appropriate interval can usually be determined by graphical or tabular analysis, as described in Sec. 7.2. The interval is then divided in half, and the half-interval containing the root is retained. The process is then repeated with the remaining half-interval. The detailed procedure is outlined below.

1. Begin with an interval $a \leq x \leq b$, which is known to contain one real root. Since $f(x)$ will cross the x-axis once within this interval, the value of $f(x)$ at each of the end points will have a different sign; that is, either $f(a) < 0$ and $f(b) > 0$, or $f(a) > 0$ and $f(b) < 0$.

2. Determine the interval midpoint, x_m, as

$$x_m = \frac{a+b}{2} \tag{7.2}$$

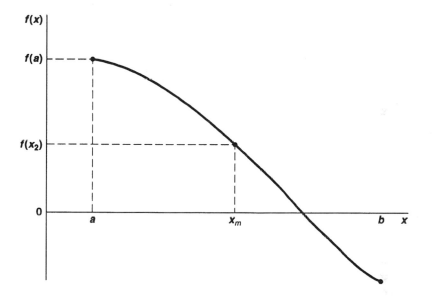

Figure 7.12

3. Evaluate $f(x_m)$ and compare the sign with that of $f(a)$.

    (a) If $f(a)$ and $f(x_m)$ both have the same sign, the root must lie within the right half-interval, as illustrated in Fig. 7.12. Then assign the value of x_m to a and repeat steps 2 and 3 using the new interval $a \leq x \leq b$.

    (b) If $f(a)$ and $f(x_m)$ differ in sign, the root must lie within the left half-interval, as illustrated in Fig. 7.13. Then assign the value of x_m to b and repeat steps 2 and 3 using the new interval $a \leq x \leq b$.

4. Repeat steps 2 and 3 until a and b are very close together. As a rule, the computation is continued until $(b - a) / (b_0 - a_0) \leq \varepsilon$, where $(b_0 - a_0)$ represents the original interval and ε represents some specified small number.

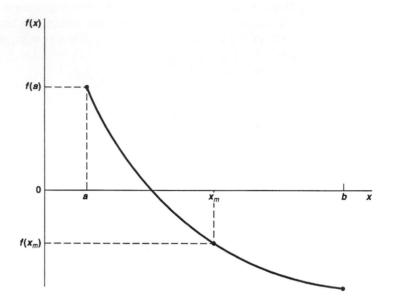

Figure 7.13

5. If ε is sufficiently small, the exact location of the root within the final interval is not critical. The root is often assumed to be located at the midpoint. Or, if a more precise result is desired, linear interpolation can be used to determine where the line connecting $f(a)$ and $f(b)$ crosses the x-axis.

Interval-reduction techniques always result in a valid solution, provided the initial interval is chosen properly. Their disadvantages are the relatively inaccurate final answer and the relatively great amount of computational effort required to obtain a solution.

Example 7.5 Solving a Polynomial Equation Using the Method of Bisection

In Example 7.2, we determined that the equation

$$f(x) = 2x^5 - 3x^2 - 5 = 0$$

has a real root within the interval $1.3 \le x \le 1.5$. Let us now determine a more accurate value for the root, using the method of bisection. We will choose a value of $\varepsilon = 0.01$ as a maximum value for the size of the final interval.

We begin with $a_0 = 1.3$, $b_0 = 1.5$ and $x_m = 1.4$. The corresponding values for $f(x)$ are $f(a_0) = -2.644141$, $f(x_m) = -0.123521$, and $f(b_0) = 3.4375$. Since $f(a_0)$ and $f(x_m)$ are both negative but $f(b_0)$ is positive, we conclude that the root lies in the right half-interval. Therefore, we set $a = 1.4$ and repeat the process within the interval $1.4 \leq x \leq 1.5$.

The series of eliminations is summarized below. At the end of each calculation, we test the stopping condition; that is, we determine whether or not the ratio of the final to the initial interval, $(b - a) / (b_0 - a_0)$, exceeds the value of $\varepsilon = 0.01$. If so, another elimination is required. The half-interval to be retained for the next calculation is indicated in parentheses.

Interval No.				$(b-a)/(b_0 - a_0)$
1	$a = 1.3$ $f(a) = -2.644141$	$x_m = 1.4$ $f(x_m) = -0.123521$	$b = 1.5$ $f(b) = 3.4375$	1 (*Retain R*)
2	$a = 1.4$ $f(a) = -0.123521$	$x_m = 1.45$ $f(x_m) = 1.51197$	$b = 1.5$ $f(b) = 3.4375$	0.5 (*Retain L*)
3	$a = 1.4$ $f(a) = -0.123521$	$x_m = 1.425$ $f(x_m) = 0.659921$	$b = 1.45$ $f(b) = 1.51197$	0.25 (*Retain L*)
4	$a = 1.4$ $f(a) = -0.123521$	$x_m = 1.4125$ $f(x_m) = 0.25986$	$b = 1.425$ $f(b) = 0.659921$	0.125 (*Retain L*)
5	$a = 1.4$ $f(a) = -0.123521$	$x_m = 1.40625$ $f(x_m) = 0.066116$	$b = 1.4125$ $f(b) = 0.25986$	0.0625 (*Retain L*)
6	$a = 1.4$ $f(a) = -0.123521$	$x_m = 1.403125$ $f(x_m) = -0.029211$	$b = 1.40625$ $f(b) = 0.066116$	0.03125 (*Retain R*)
7	$a = 1.403125$ $f(a) = -0.029211$	$x_m = 1.4046875$ $f(x_m) = 0.018325$	$b = 1.40625$ $f(b) = 0.066116$	0.015625 (*Retain L*)
8	$a = 1.403125$ $f(a) = -0.029211$	$x_m = 1.4039062$ $f(x_m) = -0.005473$	$b = 1.4046875$ $f(b) = 0.018325$	0.0078125

Within the last interval, $1.403125 \leq x \leq 1.4046875$, we see that the interval ratio has been reduced to 0.0078125, which is less than the value of $\varepsilon = 0.01$. Hence, the elimination procedure stops, and we conclude that the desired root falls within this last interval. If we interpolate linearly across the interval, we determine that the desired root is $x = 1.404084$. This value is close to the more accurate answers (1.404085 and 1.404086) obtained in Examples 7.3 and 7.4.

7.6 ITERATIVE TECHNIQUES

Iterative techniques are another widely used class of methods for solving nonlinear equations. These methods are based upon a trial-and-error strategy that begins with an initial guess and then uses this value to calculate a more refined (i.e., a more accurate) value. The newly calculated value is then used as a guess to obtain another value that is still more refined, and so on. The process continues until the last calculated value is very close to its corresponding guess.

The most widely used iterative technique is the *Newton-Raphson method*, which is based upon the formula

$$x_{i+1} = x_i - \frac{f(x_i)}{f'(x_i)} \tag{7.3}$$

In this expression, x_i represents the assumed value (i.e., the guess) for the ith iteration, x_{i+1} represents the calculated value, $f(x_i)$ represents the given equation evaluated at $x = x_i$, and $f'(x_i)$ represents the first derivative of the equation with respect to x, evaluated at x_i. The process continues until the magnitude of the ratio $(x_i - x_{i-1}) / x_i$ does not exceed some specified value; that is, until

$$\left| (x_i - x_{i-1}) / x_i \right| \le \varepsilon \tag{7.4}$$

This last condition is known as the *stopping condition* or the *convergence criterion*.

To understand the basis for the Newton-Raphson method, let x^* represent the desired root of the equation and let x represent some other point, so that $f(x^*) = 0$ but $f(x) \ne 0$ when $x \ne x^*$. Now let us express $f(x^*)$ in terms of $f(x)$ using the first two terms of a Taylor series:

$$f(x^*) = f(x) + f'(x) \, (x^* - x) \tag{7.5}$$

We now solve for x^*, making use of the fact that $f(x^*) = 0$. This results in

$$x^* = x - \frac{f(x)}{f'(x)} \tag{7.6}$$

The value of x^* obtained from Equation (7.6) will be only approximate because we have retained only the first two terms of the Taylor series expansion. If we assume, however, that the value of x^* obtained from Equation (7.6) is closer to the desired root than the original value of x, then Equation (7.6) suggests the following iterative procedure:

$$x_{i+1} = x_i - \frac{f(x_i)}{f'(x_i)} \tag{7.3}$$

This is the Newton-Raphson equation, originally shown on the previous page.

Example 7.6 Solving a Polynomial Equation Using the Newton-Raphson Method

Use the Newton-Raphson method to solve the equation

$$f(x) = 2x^5 - 3x^2 - 5 = 0$$

for a real root within the interval $1.3 \leq x \leq 1.5$. Use the value $x = 1.4$ as a first guess and $\varepsilon = 0.00001$ for the convergence criterion. Compare your answer with the results obtained in Examples 7.3 through 7.5.

In order to use the Newton-Raphson method, we must first determine $f'(x)$, the first derivative of the given equation. Applying the rules of calculus, we obtain

$$f'(x) = 10x^4 - 6x$$

Hence, we can write the Newton-Raphson equation as

$$x_{i+1} = x_i - \frac{f(x_i)}{f'(x_i)} = x_i - \frac{2x_i^5 - 3x_i^2 - 5}{10x_i^4 - 6x_i} = \frac{8x_i^5 - 3x_i^2 + 5}{10x_i^4 - 6x_i}$$

Let us refer to the initial guess, $x = 1.4$, as x_0. Then we can write

$$x_1 = \frac{8(1.4)^5 - 3(1.4)^2 + 5}{10(1.4)^4 - 6(1.4)} = 1.404115$$

$$|(x_1 - x_0) / x_1| = |(1.404115 - 1.4) / 1.404115| = 0.002931$$

$$x_2 = \frac{8(1.404115)^5 - 3(1.404115)^2 + 5}{10(1.404115)^4 - 6(1.404115)} = 1.404086$$

$$|(x_2 - x_1) / x_2| = |(1.404115 - 1.404086) / 1.404086| = 0.000021$$

$$x_3 = \frac{8(1.404086)^5 - 3(1.404086)^2 + 5}{10(1.404086) - 6(1.404086)} = 1.404086$$

$$|(x_3 - x_2) / x_3| = |(1.404086 - 1.404086) / 1.404086| = 0$$

Thus, we conclude that the desired root of this equation (rounded to seven significant figures) is $x = 1.404086$. This value agrees very closely with those obtained in Examples 7.3 and 7.4, using Excel. It is also very close to the final interpolated value $x = 1.404084$, obtained in Example 7.5 using the method of bisection. Notice, however, that the current solution converged after only three iterations, though seven elimination steps were required to obtain a similar result with the method of bisection.

Although the Newton-Raphson method converged very quickly in the last example, this is not always the case. In fact, in some situations the Newton-Raphson method (like all iterative methods) may actually *diverge*; that is, the successive values of x may move *further away* from the desired root. (Notice what happens to Equation (7.3) when $f'(x_i) = 0$.) In fact, it can be shown that the Newton-Raphson method will converge only if

$$\left| \frac{f(x)f''(x)}{f'(x)^2} \right| < 1 \tag{7.7}$$

in the vicinity of the root. As a rule, the method is much more likely to converge if the initial guess is close to the actual root.

It is interesting to compare the characteristics of iterative techniques with those of interval reduction techniques. Iterative techniques frequently converge to a solution rapidly, though convergence is not guaranteed. Interval-reduction techniques are generally less accurate, though they will always produce a solution provided the initial interval is selected properly (spanning one and only one root).

The Newton-Raphson method and the method of bisection are both implemented on computers with great frequency. In fact, both methods are often used as exercises in programming courses. As a practical matter, however, it is much easier to use Excel's Goal Seek or Solver features than to program either method from scratch. In fact, using Excel's Goal Seek feature is often preferable to a dedicated, special-purpose equation solver that someone else has already written.

Problems

7.21 For the polynomial equation solved in Example 7.6, evaluate Equation (7.7) at the root ($x = 1.404086$). Is the magnitude less than 1? Might this be related to the rapid convergence experienced in Example 7.6?

7.22 Determine a real root for the equation given in Prob. 7.1(c) using

  (*a*) The method of bisection.

  (*b*) The Newton-Raphson method.

  Compare the results and the computational effort required to obtain each solution. Compare your answers with those obtained in Probs. 7.4 and 7.9.

7.23 Determine the two smallest positive real roots for the equation given in Prob. 7.1(d) using

  (*a*) The method of bisection.

  (*b*) The Newton-Raphson method.

Compare the results and the computational effort required to obtain each solution. Compare your answers with those obtained in Probs. 7.5 and 7.10.

7.24 Solve Van der Waals's equation [Equation (7.1)] for the volume per mole (V) of an organic compound at 10 atm pressure and 400°K. The Van der Waals's constants for this particular compound are $a = 40.0$ liter^2 atm/mole^2 and $b = 0.2$ liter/mole. Solve the equation two different ways:

(a) Using the method of bisection.

(b) Using the Newton-Raphson method.

Compare the answers obtained using the two methods and compare the answers with those obtained in Probs. 7.7(b) and 7.11.

7.25 Determine the smallest positive root and the largest negative root (the negative root closest to the origin) for the following equation, originally given in Prob. 7.3, with $c = 2$ (as specified in Probs. 7.8 and 7.12):

$$x \tan x = 2$$

Solve the equation two different ways:

(a) Using the method of bisection.

(b) Using the Newton-Raphson method.

Compare the answers obtained using the two methods and compare the answers with those obtained in Prob. 7.12.

ADDITIONAL READING

Akai, T. J. *Applied Numerical Methods for Engineers.* New York: Wiley, 1994.

Carnahan, B., H. A. Luther, and J. O. Wilkes. *Applied Numerical Methods.* New York: Wiley, 1969.

Chapra, S. C. and T. P. Canale. *Numerical Methods for Engineers, with Personal Computer Applications.* New York: McGraw-Hill, 1985.

Mayo, W. E. and M. Cwiakala. *Introduction to Computing for Engineers.* New York: McGraw-Hill, 1991.

Nakamura, S. *Applied Numerical Methods in C.* Englewood Cliffs, N. J.: Prentice-Hall, 1993.

CHAPTER 8

SOLVING
SIMULTANEOUS EQUATIONS

Simultaneous algebraic equations arise in many different engineering applications. We have already seen one such application, when fitting a curve to a set of data using the method of least squares (see Sec. 5.3). There are many other applications in such diverse fields as heat conduction, solid mechanics, electrical circuits, molecular diffusion, and fluid mechanics. In fact, the solution of simultaneous algebraic equations is one of the most commonly used mathematical solution techniques.

The systems of simultaneous equations that arise in engineering can be either *linear* or *nonlinear*. In a system of *linear* equations, none of the unknown quantities are raised to a power or appear as arguments within a trigonometric function, a logarithmic function, a square root, and so on; thus, the effect of each unknown quantity is one of direct proportionality. Therefore, linear equations are easier to solve than nonlinear equations. Moreover, the techniques used to solve systems of linear equations are different than the techniques used for nonlinear equations.

We will examine both linear and nonlinear systems of equations within this chapter. We begin by describing matrix notation and its application to simultaneous linear equations. We then explain how these concepts can be used to solve systems of linear algebraic equations in Excel. Finally, we will see another approach to the solution of simultaneous algebraic equations, based upon the use of Excel's Solver feature, that can be used with either linear or nonlinear systems of equations.

Before launching into matrix notation and its associated solution techniques, however, let us see how a system of simultaneous equations can be written in general terms using traditional algebraic notation. Suppose we have a system of n simultaneous linear algebraic equations in n unknowns. Using traditional notation, we would typically write the equations in the following manner:

$$a_{11} x_1 + a_{12} x_2 + a_{13} x_3 + \cdots + a_{1n} x_n = b_1$$
$$a_{21} x_1 + a_{22} x_2 + a_{23} x_3 + \cdots + a_{2n} x_n = b_2 \qquad (8.1)$$
$$\cdots \cdots$$
$$a_{n1} x_1 + a_{n2} x_2 + a_{n3} x_3 + \cdots + a_{nn} x_n = b_n$$

Within these equations, the a_{ij} and b_i terms represent known values, whereas the x_js represent the unknown quantities. Notice that the coefficients a_{ij} have double subscripts. The first subscript (i) represents the equation number (i.e., the row number), whereas the second subscript (j) represents the number of the unknown quantity (i.e., the column number).

8.1 MATRIX NOTATION

A *matrix* is simply a two-dimensional array of numbers. The elements of a matrix are characterized by a row number and a column number. A matrix that contains only one column is called a *vector*. Thus, the system of equations given by (8.1) becomes, in matrix notation,

$$AX = B \qquad (8.2)$$

where the matrix A and the vectors X and B are written as

$$
A = \begin{bmatrix} a_{11} & a_{12} & \cdots & a_{1n} \\ a_{21} & a_{22} & \cdots & a_{2n} \\ \cdots & \cdots & \cdots & \cdots \\ a_{n1} & a_{n2} & \cdots & a_{nn} \end{bmatrix}
\qquad
X = \begin{bmatrix} x_1 \\ x_2 \\ \cdots \\ x_n \end{bmatrix}
\qquad
B = \begin{bmatrix} b_1 \\ b_2 \\ \cdots \\ b_n \end{bmatrix}
\qquad (8.3)
$$

The advantage to matrix notation is that entire matrices or vectors can be manipulated collectively using a single symbolic notation. We shall see that this simplifies the way we view the process for solving a system of simultaneous equations.

Example 8.1 Writing a System of Simultaneous Equations in Matrix Form

Write the following system of equations in matrix form:

$$2x_1 + 3x_2 = 8$$
$$4x_1 - 3x_2 = -2$$

From Equation (8.3), we can write the above equations in matrix form as

$$\begin{bmatrix} 2 & 3 \\ 4 & -3 \end{bmatrix} \begin{bmatrix} x_1 \\ x_2 \end{bmatrix} = \begin{bmatrix} 8 \\ -2 \end{bmatrix}$$

or simply

$$AX = B$$

where

$$A = \begin{bmatrix} 2 & 3 \\ 4 & -3 \end{bmatrix}, \qquad X = \begin{bmatrix} x_1 \\ x_2 \end{bmatrix}, \qquad B = \begin{bmatrix} 8 \\ -2 \end{bmatrix}$$

Matrix Multiplication

Two matrices can be multiplied by one another, resulting in a product matrix having the same number of rows as the first matrix multiplicand and the same number of columns as the second. The first multiplicand must have the same number of columns as the second multiplicand has rows.

Suppose A is an $m \times n$ matrix (m rows and n columns) and B is an $n \times p$ matrix. Note that A has n columns and B has n rows, as required. Then the matrix product $AB = C$ is defined on an element-by-element basis in the following manner:

$$\sum_{j=1}^{n} a_{ij} b_{jk} = c_{ik}, \qquad i = 1, 2, \ldots, m, \qquad k = 1, 2, \ldots, p \qquad (8.4)$$

The resulting C matrix will contain m rows and p columns.

Matrix multiplication is really not as complicated as it appears. The rule is work *across* (by columns) in A and *down* (by rows) in X.

Example 8.2 Matrix Multiplication

Suppose A is a 2×3 matrix and B is a 3×2 matrix, where each matrix is defined as

$$A = \begin{bmatrix} 1 & 2 & 3 \\ 4 & 5 & 6 \end{bmatrix}, \qquad B = \begin{bmatrix} 7 & 8 \\ 9 & 10 \\ 11 & 12 \end{bmatrix}$$

Determine the matrix product $C = AB$.

We first observe that A has the same number of columns as B has rows (namely, three). Hence, the desired matrix multiplication can be carried out. The resulting product matrix, C, will have two rows (because A has two rows) and two columns (because B has two columns).

From Equation (8.4), we can write

$$c_{11} = a_{11} b_{11} + a_{12} b_{21} + a_{13} b_{31} = (1)(7) + (2)(9) + (3)(11) = 58$$

Similarly,

$$c_{12} = a_{11} b_{12} + a_{12} b_{22} + a_{13} b_{32} = (1)(8) + (2)(10) + (3)(12) = 64$$
$$c_{21} = a_{21} b_{11} + a_{22} b_{21} + a_{23} b_{31} = (4)(7) + (5)(9) + (6)(11) = 139$$
$$c_{22} = a_{21} b_{12} + a_{22} b_{22} + a_{23} b_{32} = (4)(8) + (5)(10) + (6)(12) = 154$$

Hence, we can write the desired product as

$$C = \begin{bmatrix} 58 & 64 \\ 139 & 154 \end{bmatrix}$$

Equation (8.4) undergoes some simplification if the second multiplicand is a vector. Thus, if A is an $m \times n$ matrix and X is a vector having n elements (that is, n rows), the matrix product $AX = B$ can be written on an element-by-element basis in the following manner:

$$\sum_{j=1}^{n} a_{ij} x_j = b_i, \qquad i = 1, 2, \ldots, m \tag{8.5}$$

The resulting B vector will contain m elements (m rows) since the A matrix has m rows.

Example 8.3 Reconstructing a System of Simultaneous Equations

Beginning with the matrix form of the equations given in Example 8.1 and using Equation (8.5), reconstruct the original two equations.

From Equation (8.5), we can write the first equation as

$$\sum_{j=1}^{2} a_{1j} x_j = b_1$$

Substituting the values of the coefficients, we obtain

$$2x_1 + 3x_2 = 8$$

which is the first of the original two equations.

Similarly, we can write the second equation as

$$\sum_{j=1}^{2} a_{2j} x_j = b_2$$

or

$$4x_1 - 3x_2 = -2$$

This is the second of the original two equations.

Matrix multiplication is not commutative; that is, the product BA is generally not the same as the product AB. In fact, *the product BA will not even be defined unless the number of columns of B is the same as the number of rows of A*. If A and B are both *square $n \times n$* matrices, however, we can form either product $C1 = AB$ or $C2 = BA$, though $C1$ will generally differ from $C2$.

Other Matrix Operations

For completeness, three other matrix operations should be discussed, though they are not required to solve simultaneous equations. These operations are *matrix addition*, *matrix subtraction*, and *scalar multiplication*.

Two matrices can be added if they both contain the same number of rows and the same number of columns. Thus, if A and B are each $m \times n$ matrices, the matrix sum is written as $A + B = C$. On an element-by-element basis, the matrix sum is determined simply by adding the elements of A to the corresponding elements of B. In other words,

$$a_{ij} + b_{ij} = c_{ij} \tag{8.6}$$

Note that C will also be an $m \times n$ matrix.

Matrix subtraction is directly analogous to matrix addition. Thus, if A and B are each $m \times n$ matrices, the matrix difference is written as $A - B = C$, where C is also an $m \times n$ matrix. On an element-by-element basis, the matrix difference is determined by subtracting the elements of B from the corresponding elements of A. That is,

$$a_{ij} - b_{ij} = c_{ij} \tag{8.7}$$

Scalar multiplication involves multiplying each element of an $m \times n$ matrix by a constant k. Thus, we can write scalar multiplication as $kA = C$. On an element-by-element basis, we can write

$$k\, a_{ij} = c_{ij} \tag{8.8}$$

Again, note that C will be an $m \times n$ matrix.

These three operations are less complicated than matrix multiplication. We mention them only to provide a balanced overview of elementary matrix operations.

Example 8.4 Matrix Addition, Matrix Subtraction, and Scalar Multiplication

Suppose we are given the matrices

$$A = \begin{bmatrix} 4 & 7 \\ -2 & 5 \end{bmatrix} \qquad B = \begin{bmatrix} 1 & -8 \\ 4 & 4 \end{bmatrix}$$

Determine the matrix sum $A + B$, the matrix difference $A - B$, and the product $3A$.

Using Equations (8.6), (8.7) and (8.8), we obtain

$$A + B = \begin{bmatrix} (4+1) & (7-8) \\ (-2+4) & (5+4) \end{bmatrix} = \begin{bmatrix} 5 & -1 \\ 2 & 9 \end{bmatrix}$$

$$A - B = \begin{bmatrix} (4-1) & (7+8) \\ (-2-4) & (5-4) \end{bmatrix} = \begin{bmatrix} 3 & 15 \\ -6 & 1 \end{bmatrix}$$

$$3A = \begin{bmatrix} 3(4) & 3(7) \\ 3(-2) & 3(5) \end{bmatrix} = \begin{bmatrix} 12 & 21 \\ -6 & 15 \end{bmatrix}$$

Special Matrices

Before proceeding further, we must discuss two special matrices, the *identity matrix*, I, and the *inverse matrix*, A^{-1}.

The *identity matrix* is analogous to the constant 1 in ordinary algebra. It is a square matrix, containing n rows and n columns; for example,

$$I = \begin{bmatrix} 1 & 0 & 0 & 0 & 0 \\ 0 & 1 & 0 & 0 & 0 \\ 0 & 0 & 1 & 0 & 0 \\ 0 & 0 & 0 & 1 & 0 \\ 0 & 0 & 0 & 0 & 1 \end{bmatrix} \tag{8.9}$$

The identity matrix has the important property that

$$IA = AI = A \tag{8.10}$$

provided A is also a square matrix having n rows and n columns. Note that this is analogous to writing

$$1 \times a = a \times 1 = a \tag{8.11}$$

in ordinary algebra.

The *inverse matrix* is analogous to a reciprocal in ordinary algebra. It is a square matrix, containing n rows and n columns, which has the important property that

$$A^{-1}A = AA^{-1} = I \tag{8.12}$$

Note that this is analogous to writing

$$a^{-1}a = aa^{-1} = 1 \tag{8.13}$$

in ordinary algebra. You should understand, however, that the elements of A^{-1} are *not* the reciprocals of the elements of A. In fact, calculating A^{-1} is quite complicated, as we will see in Sec. 8.3. For now, we will let Excel compute the inverse matrix for us when it is needed.

Example 8.5 Properties of the Inverse Matrix

The inverse of the A matrix (i.e., the coefficient matrix) given in Example 8.1 is

$$A^{-1} = \begin{bmatrix} 1/6 & 1/6 \\ 2/9 & -1/9 \end{bmatrix}$$

How this inverse is obtained is relatively complicated and beyond the scope of this book. Note, however, that *the elements of the inverse matrix are not the reciprocals of the corresponding elements of A.*

Let us use the rules of matrix multiplication to show that $AA^{-1} = I$.

Writing out the components of AA^{-1}, we obtain

$$AA^{-1} = \begin{bmatrix} 2 & 3 \\ 4 & -3 \end{bmatrix} \begin{bmatrix} 1/6 & 1/6 \\ 2/9 & -1/9 \end{bmatrix}$$

$$AA^{-1} = \begin{bmatrix} (2)(1/6)+(3)(2/9) & (2)(1/6)+(3)(-1/9) \\ (4)(1/6)+(-3)(2/9) & (4)(1/6)+(-3)(-1/9) \end{bmatrix}$$

$$AA^{-1} = \begin{bmatrix} (1/3+2/3) & (1/3-1/3) \\ (2/3-2/3) & (2/3+1/3) \end{bmatrix} = \begin{bmatrix} 1 & 0 \\ 0 & 1 \end{bmatrix} = I$$

as expected. You may wish to work out the product $A^{-1}A$ and show that it is also equal to the identity matrix I (see Prob. 8.6).

Simultaneous Linear Equations

Now we will make use of the identity matrix and the inverse matrix to solve a system of simultaneous linear equations. Let us begin by writing the system of simultaneous equations as

$$AX = B \tag{8.14}$$

If we premultiply by A^{-1}, we obtain

$$A^{-1}AX = A^{-1}B \tag{8.15}$$

But $A^{-1}AX$ can be written as $IX = X$. Hence, we can write

$$X = A^{-1}B \tag{8.16}$$

Therefore, we conclude that *a system of simultaneous linear algebraic equations* $AX = B$ *can be solved by determining the inverse,* A^{-1}, *of the coefficient matrix A and then forming the product* $A^{-1}B$. The resulting product will be a vector, X, whose components are the desired unknown quantities.

Example 8.6 Solving Simultaneous Linear Equations by Matrix Inversion

Solve the system of equations given in Example 8.1 using the method described above. For your convenience, the given system of equations is repeated below.

$$2x_1 + 3x_2 = 8$$
$$4x_1 - 3x_2 = -2$$

In matrix form, we can write the system of equations as

$$AX = B$$

where

$$A = \begin{bmatrix} 2 & 3 \\ 4 & -3 \end{bmatrix} \qquad X = \begin{bmatrix} x_1 \\ x_2 \end{bmatrix} \qquad B = \begin{bmatrix} 8 \\ -2 \end{bmatrix}$$

Furthermore, in the last example we established that

$$A^{-1} = \begin{bmatrix} 1/6 & 1/6 \\ 2/9 & -1/9 \end{bmatrix}$$

Hence, the solution can be obtained from Equation (8.16) as

$$X = A^{-1}B = \begin{bmatrix} 1/6 & 1/6 \\ 2/9 & -1/9 \end{bmatrix}\begin{bmatrix} 8 \\ -2 \end{bmatrix} = \begin{bmatrix} 8/6 - 2/6 \\ 16/9 + 2/9 \end{bmatrix} = \begin{bmatrix} 1 \\ 2 \end{bmatrix}$$

Therefore, the desired solution is $x_1 = 1$, $x_2 = 2$.

Problems

8.1 Write each of the following systems of simultaneous equations in matrix form:

(a)
$$x_1 - 2x_2 + 3x_3 = 17$$
$$3x_1 + x_2 - 2x_3 = 0$$
$$2x_1 + 3x_2 + x_3 = 7$$

(b)
$$0.1x_1 - 0.5x_2 + x_4 = 2.7$$
$$0.5x_1 - 2.5x_2 + x_3 - 0.4x_4 = -4.7$$
$$x_1 + 0.2x_2 - 0.1x_3 + 0.4x_4 = 3.6$$
$$0.2x_1 + 0.4x_2 - 0.2x_3 = 1.2$$

(c) $\quad 11x_1 + 3x_2 \quad\quad\quad + \quad x_4 + 2x_5 = 51$

$\quad\quad\quad 4x_2 + 2x_3 \quad\quad\quad + \quad x_5 = 15$

$\quad\quad 3x_1 + 2x_2 + 7x_3 + \quad x_4 \quad\quad\quad = 15$

$\quad\quad 4x_1 \quad\quad + 4x_3 + 10x_4 + \quad x_5 = 20$

$\quad\quad 2x_1 + 5x_2 + \quad x_3 + \quad 3x_4 + 13x_5 = 92$

8.2   Determine the matrix product $AB = C$ for each of the following pairs of matrices:

(a) $\quad A = \begin{bmatrix} 1 & 2 \\ 3 & 4 \\ 5 & 6 \end{bmatrix}$ $\quad\quad B = \begin{bmatrix} 7 & 8 & 9 & 10 \\ 11 & 12 & 13 & 14 \end{bmatrix}$

(b) $\quad A = \begin{bmatrix} 1 & 2 & 3 \end{bmatrix}$ $\quad\quad B = \begin{bmatrix} 4 \\ 5 \\ 6 \end{bmatrix}$

(c) $\quad A = \begin{bmatrix} 1 \\ 2 \\ 3 \end{bmatrix}$ $\quad\quad\quad B = \begin{bmatrix} 4 & 5 & 6 \end{bmatrix}$

(d) $\quad A = \begin{bmatrix} 1 & 2 & 3 \\ 4 & 5 & 6 \\ 7 & 8 & 9 \end{bmatrix}$ $\quad\quad B = \begin{bmatrix} 5 \\ -3 \\ 2 \end{bmatrix}$

8.3   Reconstruct a system of simultaneous algebraic equations from the following matrix equation:

$$\begin{bmatrix} 4 & -2 & 1 \\ 1 & 3 & -7 \\ -2 & 0 & -5 \end{bmatrix} \begin{bmatrix} x_1 \\ x_2 \\ x_3 \end{bmatrix} = \begin{bmatrix} 6 \\ 34 \\ 14 \end{bmatrix}$$

8.4   Suppose we are given the following square matrices:

$$A = \begin{bmatrix} 6 & 0 & -3 \\ -1 & 4 & 9 \\ 8 & -5 & 2 \end{bmatrix} \qquad B = \begin{bmatrix} 5 & -5 & -7 \\ 0 & 9 & 2 \\ 4 & -1 & 0 \end{bmatrix}$$

Using these two matrices, carry out each of the following matrix operations:

(a) $A + B = C$ (d) $2A = C$

(b) $B + A = C$ (e) $AB = C$

(c) $A - B = C$ (f) $BA = C$

Are the matrix sums (a) and (b) equivalent? Are the matrix products (e) and (f) equivalent? What do you conclude from these two comparisons?

8.5 Suppose I is a 3×3 identity matrix and A is the following square matrix:

$$A = \begin{bmatrix} 6 & 0 & -3 \\ -1 & 4 & 9 \\ 8 & -5 & 2 \end{bmatrix}$$

(a) Determine the matrix product IA

(b) Determine the matrix product AI

Are (a) and (b) equivalent?

8.6 Determine the matrix product $A^{-1}A$ using the matrices given in Examples 8.1 and 8.5. Compare with the results obtained in Example 8.5. Is $A^{-1}A$ equivalent to AA^{-1}?

8.7 Determine whether or not the following matrix is the inverse of the A matrix given in Prob. 8.5.

$$B = \begin{bmatrix} 0.132832 & 0.037594 & 0.030075 \\ 0.185464 & 0.090226 & -0.127820 \\ -0.067670 & 0.075188 & 0.060150 \end{bmatrix}$$

8.8 Consider the system of three equations in three unknowns, written in matrix form as $AX = B$, where

$$A = \begin{bmatrix} 4 & -2 & 1 \\ 1 & 3 & -7 \\ -2 & 0 & -5 \end{bmatrix} \qquad X = \begin{bmatrix} x_1 \\ x_2 \\ x_3 \end{bmatrix} \qquad B = \begin{bmatrix} 6 \\ 34 \\ 14 \end{bmatrix}$$

Suppose the inverse of A is known to be

$$A^{-1} = \begin{bmatrix} 0.163043 & 0.108696 & -0.119570 \\ -0.206520 & -0.195652 & -0.315220 \\ -0.065220 & -0.043480 & -0.152170 \end{bmatrix}$$

Using this information, solve for the unknowns x_1, x_2 and x_3.

8.2 MATRIX OPERATIONS IN EXCEL

Matrix operations can easily be carried out in Excel. To do so, we need to make use of *arrays* since a matrix is represented as an array in Excel.

An array is a block of cells that is referenced collectively, in the same manner as a single cell. Any operation that is carried out on an array will produce the same outcome for all of the cells within the array. In addition, an array can be specified as a single argument to a function. This is equivalent to specifying the individual cells as multiple arguments.

An array is represented as a block of cells enclosed in braces ({}). For example, {A1:C3} refers to an array that is composed of the block of cells extending from cell A1 to cell C3. The braces are added automatically by Excel as a part of the array specification. Thus, you should *not* attempt to include the braces when writing an array.

To specify an array operation (i.e., a matrix operation) within an Excel worksheet, proceed as follows.

1. Select the block of cells that make up the array. Then move the cell marker to the upper left cell within the block.

2. Enter an array formula, as required. You may include cell ranges within the formula. However, after you have entered the formula, *do not* enclose the formula in braces.

3. Press **Ctrl-Shift-Enter** (on an Intel-type computer) or **Command-Return** (on a Macintosh) simultaneously. This will cause the formula to appear within a pair of braces, thus identifying the block of cells as an array. The indicated operations will then be carried out.

Example 8.7 Matrix Addition, Matrix Subtraction, and Scalar Multiplication in Excel

Repeat the operations shown in Example 8.4 within an Excel worksheet. Thus, we wish to determine the matrix sum $A + B$, the matrix difference $A - B$, and the product $3A$ from the following two matrices:

$$A = \begin{bmatrix} 4 & 7 \\ -2 & 5 \end{bmatrix} \qquad B = \begin{bmatrix} 1 & -8 \\ 4 & 4 \end{bmatrix}$$

The Excel worksheet is shown in Fig. 8.1. Within this worksheet, the elements of the A and B matrices were entered in the normal manner. (The vertical lines forming the matrix borders were added by selecting Cells/Border from the Format menu.)

In order to determine the matrix sum, we first select the block of cells extending from B6 to C7. The cell marker will appear over cell B6. We then enter the formula =B3:C4+F3:G4 (without enclosing braces) and press Ctrl-Shift-Enter (or Command-Return) simultaneously. The resulting matrix sum will then appear within cells B6 through C7, as shown in Fig. 8.1.

Notice the array formula shown in the formula bar. This formula was entered while the cell marker was highlighting cell B6, though the formula applies to the entire selected block of cells. The braces were added automatically, after pressing Ctrl-Shift-Enter (or Command-Return).

Similarly, the matrix difference was determined by selecting the block of cells extending from F6 to G7. The formula =B3:C4−F3:G4 was then entered, without the enclosing braces. The braces were added automatically, after pressing Ctrl-Shift-Enter (or Command-Return). The result of the matrix subtraction operation was then generated in cells F6 through G7.

Finally, the scalar product was determined by selecting the block of cells extending from B9 through C10, entering the formula =3*B3:C4, and then pressing Ctrl-Shift-Enter (or Command-Return).

	B6		{=B3:C4+F3:G4}					
	A	B	C	D	E	F	G	H
1	Matrix Addition, Subtraction and Scalar Multiplication							
2								
3	A =	4	7		B =	1	-8	
4		-2	5			4	4	
5								
6	A + B =	5	-1		A - B =	3	15	
7		2	9			-6	1	
8								
9	3A =	12	21					
10		-6	15					
11								
12								
Ready								

Figure 8.1

8.3 SOLVING SIMULTANEOUS EQUATIONS IN EXCEL USING MATRIX INVERSION

One way to solve a system of simultaneous linear equations in Excel is to use the matrix inversion method described in Sec. 8.1. Excel includes a library function, MINVERSE, for determining the inverse of a matrix, and another, MMULT, for carrying out matrix multiplication. Hence, it is quite simple to solve the system of equations $AX = B$ using these two functions. The general procedure is to first determine A^{-1} using the MINVERSE function and then obtain the matrix product $A^{-1}B$ using the MMULT function. The resulting product will be the vector X, which contains the desired unknown values.

The detailed procedure is as follows:

1. Enter the elements of the coefficient matrix A into an $n \times n$ block of cells.

2. Enter the right-hand side B within another block (a single column) of n cells.

3. Identify where you want the inverse A^{-1} to appear. (You must set aside another $n \times n$ block of cells.) Then highlight this block of cells, place the cell marker in the upper left cell, and enter the formula =MINVERSE( ). Place the range of the coefficient matrix inside the parentheses; for example, =MINVERSE(A1:B2). Then press Ctrl-Shift-Enter (or Command-Return) simultaneously. The elements of A^{-1} will appear within the block.

4. Now identify where you want the solution X to appear. (You must set aside a block of n cells within a single column.) Then highlight this block of cells, place the cell marker in the top cell, and enter the formula =MMULT(,). Place the range of the inverse matrix inside the parentheses, followed by a comma, followed by the range of the right-hand side; for example, =MMULT(F1:G2,D1:D2). Then press Ctrl-Shift-Enter (or Command-Return) simultaneously. The elements of X (the unknown values) will then appear within the block.

5. You may wish to verify your solution by calculating the matrix product AX and then comparing the resulting product with the original right-hand side vector, B.

Example 8.8 Solving Simultaneous Equations in Excel by Matrix Inversion

In Example 8.6, we outlined a procedure for solving the following system of simultaneous equations:

$$2x_1 + 3x_2 = 8$$
$$4x_1 - 3x_2 = -2$$

The solution procedure assumed that we knew the inverse of the coefficient matrix. Let us now use Excel to determine the inverse and then solve the system of equations.

Figure 8.2 shows an Excel worksheet containing the coefficient matrix (i.e., the A matrix) in cells A4:B5, and the right-hand side (i.e., the B vector) in cells G4:G5. Notice that the worksheet also contains space reserved for the inverse of the coefficient matrix (A^{-1}), the solution vector (i.e., the X vector), and the matrix product AX. These quantities will be determined from the given information. Each matrix or vector has been labeled and placed within a border in order to make the worksheet more legible.

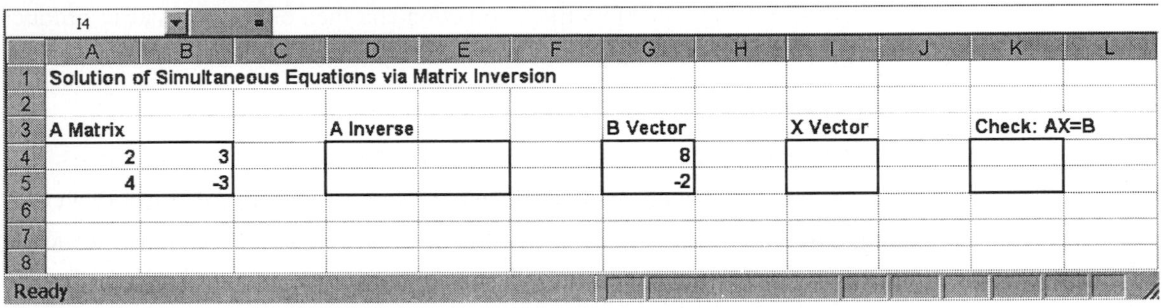

Figure 8.2

Figure 8.3 shows the elements of the inverse matrix in cells D4:E5. These values were obtained using the MINVERSE function, as shown in the formula bar.

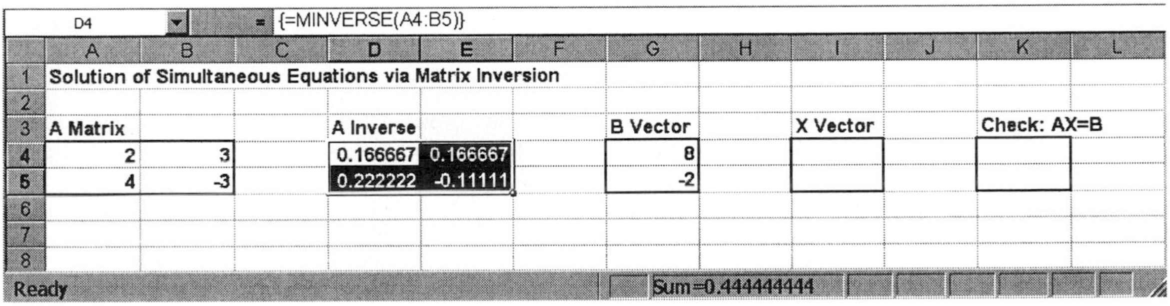

Figure 8.3

In Fig. 8.4 we see the desired solution vector in cells I4:I5. The solution vector was obtained by multiplying the inverse of the coefficient matrix by the right-hand side; that is, by forming the matrix product $A^{-1}B$. The matrix multiplication was carried out using the MMULT function, as shown in the formula bar. Thus, we can now see that the desired solution is $x_1 = 1$, $x_2 = 2$, which is consistent with the results obtained in Example 8.6.

Finally, we can check our solution by forming the product AX and comparing it to the original right-hand side vector, B. Figure 8.5 shows the resulting product in cells K4:K5. These values were obtained using the MMULT function, as can be seen in the formula bar. Clearly, the calculated value agrees with the original B vector, thus assuring us that the calculated solution is correct.

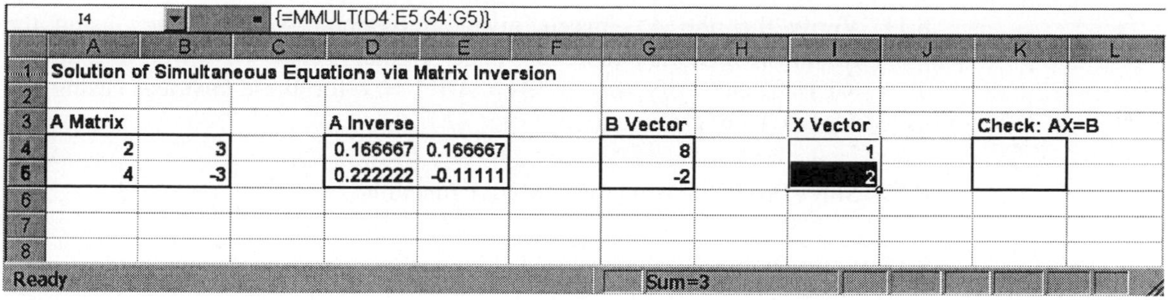

I4   {=MMULT(D4:E5,G4:G5)}

	A	B	C	D	E	F	G	H	I	J	K	L
1	Solution of Simultaneous Equations via Matrix Inversion											
2												
3	A Matrix			A Inverse			B Vector		X Vector		Check: AX=B	
4	2	3		0.166667	0.166667		8		1			
5	4	-3		0.222222	-0.11111		-2		2			
6												
7												
8												

Ready   Sum=3

Figure 8.4

K4   {=MMULT(A4:B5,I4:I5)}

	A	B	C	D	E	F	G	H	I	J	K	L
1	Solution of Simultaneous Equations via Matrix Inversion											
2												
3	A Matrix			A Inverse			B Vector		X Vector		Check: AX=B	
4	2	3		0.166667	0.166667		8		1		8	
5	4	-3		0.222222	-0.11111		-2		2		-2	
6												
7												
8												

Ready   Sum=6

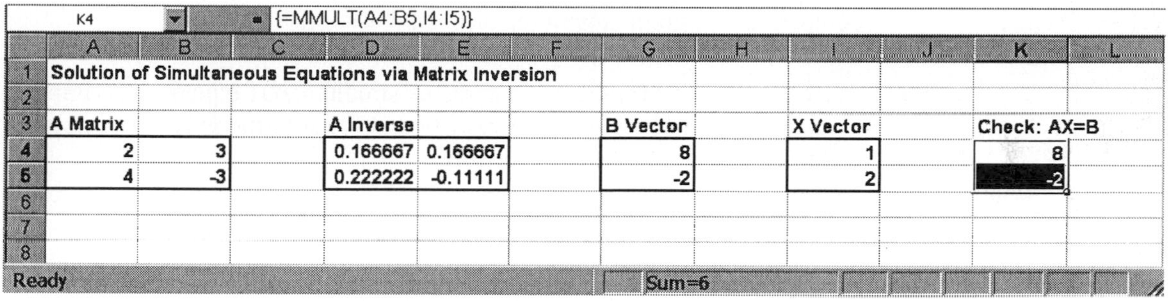

Figure 8.5

Problems

Use Excel to solve each of the following problems.

8.9 Suppose we are given the following square matrices (these matrices were originally given in Prob. 8.4).

$$A = \begin{bmatrix} 6 & 0 & -3 \\ -1 & 4 & 9 \\ 8 & -5 & 2 \end{bmatrix} \quad B = \begin{bmatrix} 5 & -5 & -7 \\ 0 & 9 & 2 \\ 4 & -1 & 0 \end{bmatrix}$$

Representing each matrix as an array, carry out each of the following matrix operations.

(a) $A + B = C$ (c) $A - B = C$

(b) $B + A = C$ (d) $2A = C$

8.10 Carry out the matrix multiplication operations given in Prob. 8.2. Use the MMULT library function to carry out the matrix multiplication.

8.11 Verify that the A^{-1} matrix given in Example 8.5 is correct using the MINVERSE library function. (The original A matrix is given in Example 8.1.) Then verify that $A^{-1}A = AA^{-1} = I$ for these matrices, using the MMULT library function.

8.12 Solve Prob. 8.5 using the MMULT function.

8.13 Solve Prob. 8.7 using the MMULT function.

8.14 Solve the system of simultaneous equations given in Prob. 8.8 using Excel. Verify that the solution is correct by forming the product AX and comparing this product with the original B vector.

8.15 Solve each of the following systems of simultaneous equations, originally given in Prob. 8.1. In each case, verify that the solution is correct using the method described in the last problem.

(a)
$$x_1 - 2x_2 + 3x_3 = 17$$
$$3x_1 + x_2 - 2x_3 = 0$$
$$2x_1 + 3x_2 + x_3 = 7$$

(b)
$$0.1x_1 - 0.5x_2 + x_4 = 2.7$$
$$0.5x_1 - 2.5x_2 + x_3 - 0.4x_4 = -4.7$$
$$x_1 + 0.2x_2 - 0.1x_3 + 0.4x_4 = 3.6$$
$$0.2x_1 + 0.4x_2 - 0.2x_3 = 1.2$$

(c)
$$11x_1 + 3x_2 + x_4 + 2x_5 = 51$$
$$4x_2 + 2x_3 + x_5 = 15$$
$$3x_1 + 2x_2 + 7x_3 + x_4 = 15$$
$$4x_1 + 4x_3 + 10x_4 + x_5 = 20$$
$$2x_1 + 5x_2 + x_3 + 3x_4 + 13x_5 = 92$$

8.16 Solve each of the following systems of simultaneous equations. In each case, verify that the solution is correct using the method described in Prob. 8.14.

(a)
$$15x_1 + 20x_2 = 25$$
$$5x_1 + 10x_2 = 12$$

(b)
$$4x_1 - 2x_2 + x_3 = 6$$
$$x_1 + 3x_2 - 7x_3 = 34$$
$$-2x_1 - 5x_3 = 14$$

(c)
$$x_1 + \quad x_2 + 5x_3 - 12x_4 = 29$$
$$x_2 \qquad\quad + 16x_4 = 21$$
$$7x_1 + 2x_2 \qquad\quad - 12x_4 = \quad 5$$
$$x_3 - \quad x_4 = \quad 6$$

8.4 SOLVING SIMULTANEOUS EQUATIONS IN EXCEL USING SOLVER

Excel's Solver feature provides an altogether different approach to the solution of simultaneous equations. This method can be used with nonlinear as well as linear systems of equations.

Remember that Solver is an Excel add-in, like the Analysis Toolpak used to generate histograms (see Chap. 4). To install the Solver, choose Add-Ins from the Tools menu. Then select Solver Add-In from the resulting Add-Ins dialog box. Once the Solver feature has been installed, it will remain installed unless it is removed by reversing the above procedure.

Now let us see how Solver can be used to solve a system of simultaneous equations. The equations need not necessarily be linear. Suppose the equations are represented as

$$f_1(x_1, x_2, \ldots, x_n) = 0 \tag{8.17}$$

$$f_2(x_1, x_2, \ldots, x_n) = 0 \tag{8.18}$$

$$\ldots\ldots$$

$$f_n(x_1, x_2, \ldots, x_n) = 0 \tag{8.19}$$

Thus, we have a system of n equations in n unknowns. We wish to find the values of $x_1, x_2, \ldots, x_n$ that cause each of the equations to equal zero.

One way to do this is to force the function

$$y = f_1{}^2 + f_2{}^2 + \ldots + f_n{}^2 \tag{8.20}$$

to zero; that is, to find the values of $x_1, x_2, \ldots, x_n$ that cause Equation (8.20) to equal zero. Since all of the terms on the right-hand side of Equation (8.20) are squares, they will all be greater than or equal to zero. Hence, the only way that y can equal zero is for each of the individual fs to also equal zero. Therefore, the values of $x_1, x_2, \ldots, x_n$ that cause y to equal zero will be the solution to the given system of equations. The general approach used with Solver is to define a *target function* consisting of the squares of the individual equations, as indicated by Equation (8.20), and to then determine the values of $x_1, x_2, \ldots, x_n$ that cause the target function to equal zero.

To solve a system of simultaneous equations with Solver, proceed as follows:

1. Enter an initial guess for each independent variable $(x_1, x_2, \ldots, x_n)$ in a separate cell on the worksheet.

2. Enter the equations for $f_1, f_2, \ldots, f_n$ and y in separate cells, expressed as Excel formulas. Within these formulas, express the unknown quantities x_1, $x_2, \ldots, x_n$ as the addresses of the cells containing the initial guess.

3. Select Solver from the Tools menu.

4. When the Solver Parameters dialog box appears, enter the following information:

    (*a*)   The address of the cell containing the formula for y in the Set Target Cell location.

    (*b*)   Select Value of in the Equal to line. Then enter 0 within the associated data area. (In other words, determine the values of $x_1, x_2, \ldots, x_n$ that will drive the target function to zero.)

    (*c*)   Enter the range of cell addresses containing the initial values of x_1, x_2, $\ldots, x_n$ in the area labeled By Changing Cells.

    (*d*)   If you wish to restrict the range of the independent variables, click on the Add button under the heading Subject to the Constraints. Then provide the following information within the Add Constraint dialog box for each of the independent variables:

    (*i*)   The cell address containing the initial value of the independent variable in the Cell Reference location.

    (*ii*)   The type of constraint (i.e., $\leq$, or $\geq$) from the pull-down menu.

    (*iii*) The limiting value in the Constraint data area.

    (*iv*)   Select OK to return to the Solver Parameters dialog box or select Add to add another constraint.

    Note that you can always change a constraint or delete a constraint after it has been added.

    (*e*)   When all of the required information has been entered correctly, select Solve. This will initiate the actual solution procedure.

A new dialog box labeled Solver Results will then appear, telling you whether or not Solver has been able to solve the problem. If a solution has been obtained, the desired values of the independent variables will appear in the cells that originally contained the initial values. The cell containing the target function will show a value that is zero or nearly zero.

Example 8.9 Solving Simultaneous Linear Equations in Excel Using Solver

Determine the solution to the following system of equations using Solver.

$$3\,x_1 + 2\,x_2 - \quad x_3 = \quad 4$$
$$2\,x_1 - \quad x_2 + \quad x_3 = \quad 3$$
$$x_1 + \quad x_2 - 2\,x_3 = -3$$

For clarity, let us refer to the first equation as $f(x_1, x_2, x_3)$, the second as $g(x_1, x_2, x_3)$, and the third as $h(x_1, x_2, x_3)$. Thus, we can write the given system of equations as

$$f = 3\,x_1 + 2\,x_2 - \quad x_3 - 4 = 0$$
$$g = 2\,x_1 - \quad x_2 + \quad x_3 - 3 = 0$$
$$h = \quad x_1 + \quad x_2 - 2\,x_3 + 3 = 0$$

We wish to find the values of x_1, x_2, and x_3 that will cause $f(x_1, x_2, x_3)$, $g(x_1, x_2, x_3)$, and $h(x_1, x_2, x_3)$ to equal zero.

Now let us form the sum

$$y = f^2 + g^2 + h^2$$

If $f = 0$, $g = 0$, and $h = 0$, then y will also equal zero. For any other values of f, g, and h (whether positive or negative), however, y will be greater than zero. Hence, we can solve the given system of equations by finding the values of x_1, x_2, and x_3 that cause y to equal zero. We will use Excel's **Solver** feature to determine the values of x_1, x_2, and x_3 that drive the target function, y, to zero.

	B11	▼	■	=B7^2+B8^2+B9^2							
	A	B	C	D	E	F	G	H	I	J	K
1	Simultaneous Linear Equations										
2											
3	x1 =	1									
4	x2 =	1									
5	x3 =	1									
6											
7	f(x1, x2, x3) =	0									
8	g(x1, x2, x3) =	-1									
9	h(x1, x2, x3) =	3									
10											
11	y =	10									
12											
13											
14											
15											
16											
17											
18											
Ready											

Figure 8.6

An Excel worksheet containing the initial setup is shown in Fig. 8.6, and the corresponding cell formulas are shown in Fig. 8.7. Note that cells B3, B4, and B5 contain the initial assumed values for x_1, x_2, and x_3. Cells B7, B8, and B9 contain the formulas for $f(x_1, x_2, x_3)$, $g(x_1, x_2, x_3)$, and $h(x_1, x_2, x_3)$; and cell B11 contains the formula for y. The values within these cells in Fig. 8.6 are based upon the initial values for x_1, x_2, and x_3.

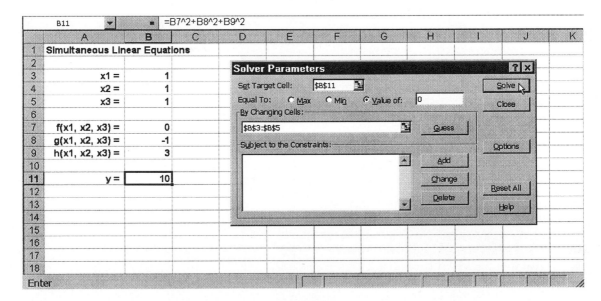

Figure 8.7

Figure 8.8

Next, we select **Solver** from the **Tools** menu, and we enter the address of the target function (B11) and the range of addresses of the independent variables (B3:B5) in the Solver Parameters dialog box, as shown in Fig. 8.8. Notice the specification that the target function be set equal to zero. We then select **Solve** to initiate the solution procedure.

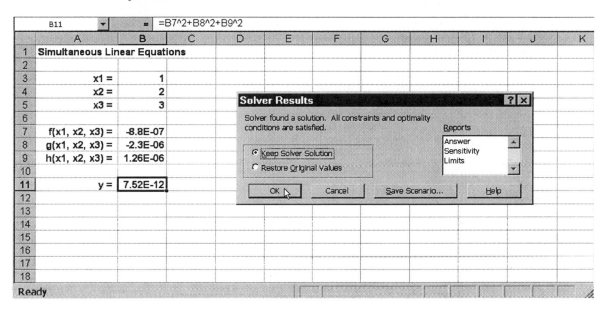

Figure 8.9

Once a solution has been found, the resulting values are shown in their respective cells within the worksheet, and the **Solver Results** dialog box appears, as shown in Fig. 8.9. Thus, we see that the solution to the three equations is $x_1 = 1$, $x_2 = 2$, and $x_3 = 3$. The corresponding values for f, g, h, and y are each approximately zero, as expected.

The equations given in the preceding example were all linear. Solver can also obtain the solution to *nonlinear* problems by following the same procedure as is used with linear problems. Remember, however, that nonlinear problems may have more than one solution, and the solutions need not necessarily be real. Hence, it may be desirable to restrict the range of the independent variables, as illustrated in the following example.

Example 8.10 Solving Simultaneous Nonlinear Equations in Excel Using Solver

Determine the solution to the following two simultaneous nonlinear equations using **Solver**. Restrict the range of the independent variables to nonnegative values.

$$x_1^2 + 2\,x_2^2 - 5\,x_1 + 7\,x_2 = 40$$

$$3\,x_1^2 - \ \ x_2^2 + 4\,x_1 + 2\,x_2 = 28$$

Let us refer to the first equation as $f(x_1, x_2)$ and the second as $g(x_1, x_2)$. Using this notation, the given equations can be written as

$$f(x_1, x_2) = \ \ x_1^2 + 2\,x_2^2 - 5\,x_1 + 7\,x_2 - 40 = 0$$

$$g(x_1, x_2) = 3\,x_1^2 - \ \ x_2^2 + 4\,x_1 + 2\,x_2 - 28 = 0$$

Thus, we seek the values of x_1 and x_2 that cause $f(x_1, x_2)$ and $g(x_1, x_2)$ to equal zero. The procedure will be to form the target function

$$y(x_1, x_2) = f^2 + g^2$$

We will then use Solver to determine the values of x_1 and x_2 that force the target function, and hence, f and g, to zero. These values will represent a solution to the given equations.

Figure 8.10 shows an Excel worksheet containing the initial problem statement. Notice that x_1 and x_2 have each been assigned an initial value of 1, as shown in cells B3 and B4. The corresponding values for f, g, and y are shown in cells B6, B7, and B9, respectively. These values were obtained by evaluating the formulas shown in Fig. 8.11, based upon the initial values for x_1 and x_2.

	B9	▼	■	=B6^2+B7^2							
	A	B	C	D	E	F	G	H	I	J	K
1	Simultaneous Nonlinear Equations										
2											
3	x1 =	1									
4	x2 =	1									
5											
6	f(x1, x2) =	-35									
7	g(x1, x2) =	-20									
8											
9	y =	1625									
10											
11											
12											
13											
14											
15											
16											
17											
18											
Ready											

Figure 8.10

We now select Solver from the Tools menu, resulting in the Solver Parameters dialog box shown in Fig. 8.12. The address of the target function (B9) is entered near the top, followed by a specification that the target function take on a value of zero. The range of independent variables (B3:B4) is then entered in the area labeled By Changing Cells.

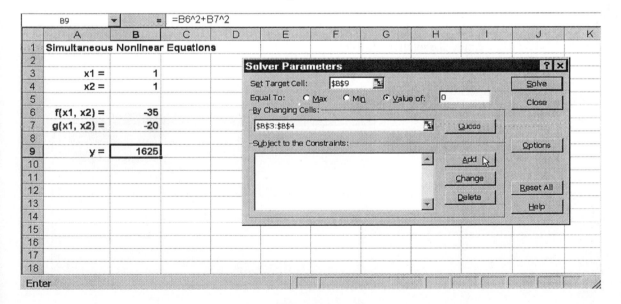

Figure 8.11

Figure 8.12

To include the nonnegativity restrictions, we select Add near the bottom of the Solver Parameters dialog box. The Add Constraint dialog box then appears, as shown in Fig.

8.13. Within this dialog box, we enter the cell address, the type of constraint ($\leq$, =, or $\geq$), and the constraining value (in this case, 0) for each independent variable.

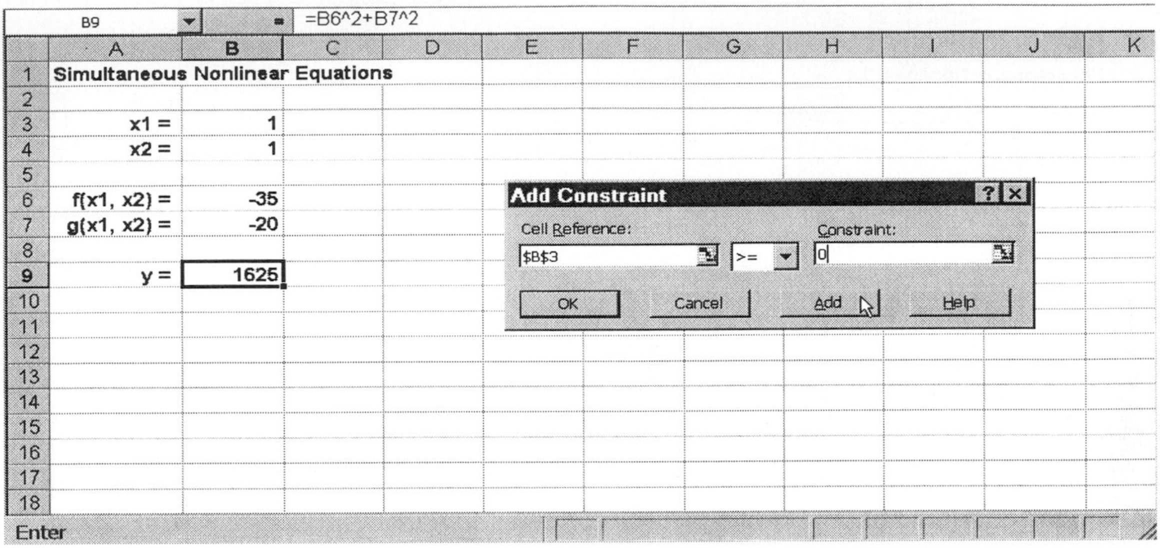

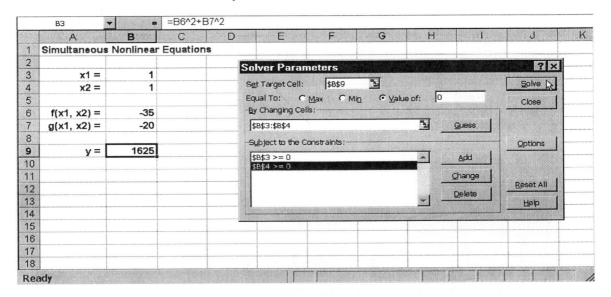

Figure 8.13

After both restrictions have been added, the Solver Parameters dialog box will reappear, with the constraints included, as shown in Fig. 8.14. We then select Solve to initiate the solution procedure.

Figure 8.14

Once a solution has been found, the resulting values are shown in cells B3 and B4 within the worksheet, and the Solver Results dialog box appears, as shown in Fig. 8.15. Thus, we see that the solution to the given equations is approximately $x_1 = 2.70$ and $x_2 = 3.37$. In addition, we see that the corresponding values for the given equations f and g, and the target function y are each very close to zero.

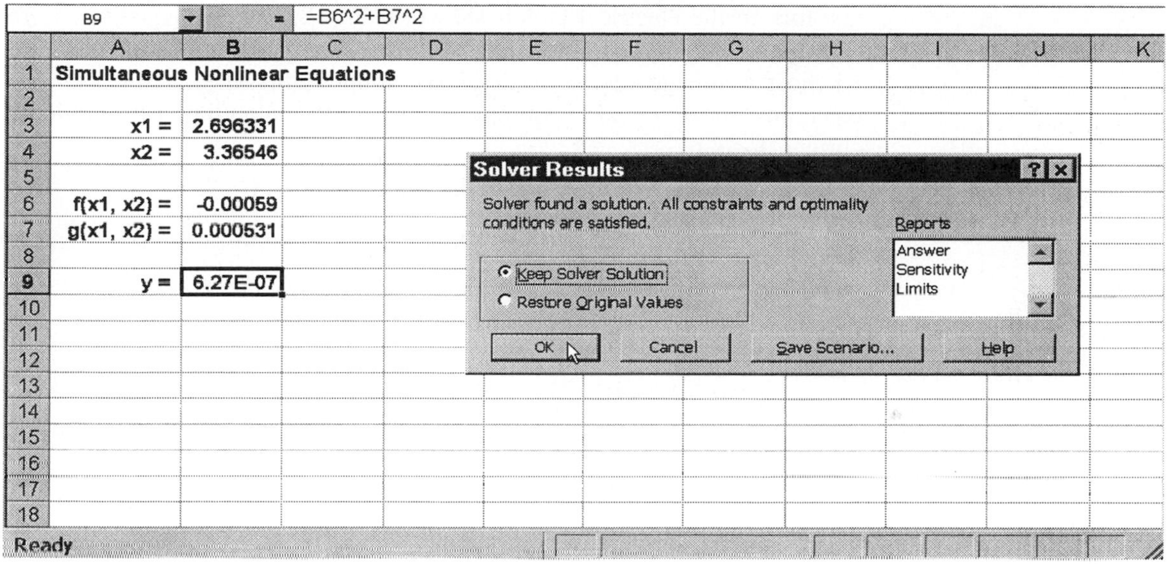

Figure 8.15

Problems

8.17 Solve Problem 8.16 (*a*) in Excel using Solver. Compare the solution with that obtained previously.

8.18 Solve Problems 8.16 (*b*) and (*c*) in Excel using Solver. Compare the effort involved in obtaining these solutions with the effort involved using the matrix-inversion method.

8.19 Solve the following systems of nonlinear algebraic equations in Excel using Solver.

(*a*) $$4x_1^3 - \sqrt{3x_2} = 20$$
$$x_1 x_2^2 + 2 / x_1 = 50$$

(*b*) $$\sin 2x_1 + x_1 \ln x_2 = -0.3$$
$$\ln (x_1^2) - \cos 3x_2 = 2$$

(c) $\sin x_1 + \cos x_2 - \ln x_3 = 0$
$\cos x_1 + 2\ln x_2 + \sin x_3 = 3$
$3\ln x_1 - \sin x_2 + \cos x_3 = 2$

8.20 The following equations describe the current flowing across each of the resistors in the electrical circuit shown in Fig. 8.16. The equations were obtained by applying Kirchoff's current law to each outer node, and Kirchoff's voltage law to each closed loop within the circuit. (These equations make certain assumptions about the direction of individual current flows.)

$i_1 + i_2 + i_3 = 2$ $10i_1 - 40i_2 + 10i_4 = 0$

$i_1 - i_4 + i_6 = 0$ $10i_4 + 2i_6 - 50i_7 = 0$

$i_6 + i_7 + i_{10} = 3$ $50i_7 - 10i_8 - 5i_{10} = 0$

$i_8 - i_9 - i_{10} = 0$ $20i_5 - 10i_8 - 10i_9 = 0$

$i_3 - i_5 - i_9 = 0$ $40i_2 - 5i_3 - 20i_5 = 0$

(a)   Rearrange these equations into a more traditional form as 10 equations in 10 unknowns.

(b)   Solve for the unknown current flows in Excel, using either of the methods described in this chapter.

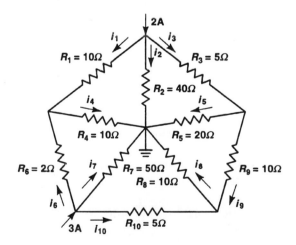

Figure 8.16

8.21 The following equations describe the electrical circuit shown in Fig. 8.17. These equations were obtained by applying Kirchoff's voltage law to each closed loop.

$$V_1 - i_1(R_1 + R_2) + i_2 R_2 = 0$$

$$V_2 - i_2(R_2 + R_3 + R_4) + i_1 R_2 + i_6 R_3 + i_3 R_4 = 0$$

$$V_3 - i_3(R_4 + R_9) + i_2 R_4 + i_6 R_9 = 0$$

$$-V_3 - i_4(R_5 + R_{10}) = 0$$

$$V_4 - i_6(R_3 + R_7 + R_8 + R_9) + i_2 R_3 + i_3 R_9 + i_5 R_7 = 0$$

$$-V_1 - i_5(R_6 + R_7) + i_6 R_7 = 0$$

Typically, the voltage sources (V_1 through V_4) and the resistances (R_1 through R_{10}) are known, and the resulting current flows within each loop (i_1 through i_6) must be determined.

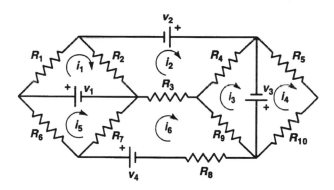

Figure 8.17

(*a*) Rearrange these equations into a more traditional form as six equations in six unknowns.

(*b*) Solve for the unknown current flows in Excel, assuming the following values for the voltages and resistances. Use either of the methods described in this chapter to obtain a solution.

$V_1 = 12$ volts	$R_1 = 100\ \Omega$	$R_6 = 150\ \Omega$
$V_2 = 5$ volts	$R_2 = 200\ \Omega$	$R_7 = 50\ \Omega$
$V_3 = 1.5$ volts	$R_3 = 100\ \Omega$	$R_8 = 100\ \Omega$
$V_4 = 5$ volts	$R_4 = 150\ \Omega$	$R_9 = 50\ \Omega$
	$R_5 = 200\ \Omega$	$R_{10} = 100\ \Omega$

8.22 The following equations describe the truss shown in Fig. 8.18.   These equations were obtained by setting the sum of the horizontal forces and the sum of the vertical forces equal to zero at each pin.

$$R_1 + T_1 \cos 60° + T_2 = 0 \qquad \text{(lower left pin)}$$

$$R_2 + T_1 \sin 60° = 0 \qquad \text{(lower left pin)}$$

$$-T_2 - T_3 \cos 60° + T_4 \cos 60° + T_5 = 0 \qquad \text{(lower middle pin)}$$

$$T_3 \sin 60° + T_4 \sin 60° = 0 \qquad \text{(lower middle pin)}$$

$$-T_5 - T_6 \cos 60° = 0 \qquad \text{(lower right pin)}$$

$$T_6 \sin 60° + R_3 = 0 \qquad \text{(lower right pin)}$$

$$-T_1 \cos 60° + T_3 \cos 60° + T_7 - F_1 \cos \theta_1 = 0 \qquad \text{(upper left pin)}$$

$$-T_1 \sin 60° - T_3 \sin 60° - F_1 \sin \theta_1 = 0 \qquad \text{(upper left pin)}$$

$$-T_4 \cos 60° - T_7 + T_6 \cos 60° - F_2 \cos \theta_2 = 0 \qquad \text{(upper right pin)}$$

$$-T_4 \sin 60° - T_6 \sin 60° - F_2 \sin \theta_2 = 0 \qquad \text{(upper right pin)}$$

The objective is to determine the internal tensile forces (T_1 through T_7) and the reactive forces (R_1, R_2, and R_3) when the external forces (F_1 and F_2) and their angles (θ_1 and θ_2) are specified.

(a)   Rearrange these equations into a more traditional form as 10 equations in 10 unknowns.

(b)   Use Excel to solve for the unknown forces, assuming that $F_1 = 10{,}000$ lb$_f$, $F_2 = 7{,}000$ lb$_f$, $\theta_1 = 75°$, and $\theta_2 = 45°$.   (Remember to convert the angles from degrees to radians when using the SIN and COS functions in Excel.)   Use either of the methods described in this chapter to obtain a solution.

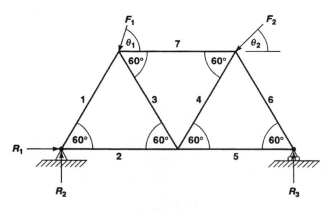

Figure 8.18

8.23 A furnace wall is made up of three separate materials, as shown in Fig. 8.19. Let k_i represent the thermal conductivity of material i, let Δx_i represent the corresponding thickness ($i = 1, 2, 3$) and let h_1 and h_2 represent the convective heat transfer coefficients at the outer surfaces. If $T_a > T_0 > T_1 > T_2 > T_3 > T_b$, then the following steady-state heat transfer equations apply:

$$h_1(T_a - T_0) = \frac{k_1}{\Delta x_1}(T_0 - T_1)$$

$$\frac{k_1}{\Delta x_1}(T_0 - T_1) = \frac{k_2}{\Delta x_2}(T_1 - T_2)$$

$$\frac{k_2}{\Delta x_2}(T_1 - T_2) = \frac{k_3}{\Delta x_3}(T_2 - T_3)$$

$$\frac{k_3}{\Delta x_3}(T_2 - T_3) = h_2(T_3 - T_b)$$

Typically, the outer temperatures (T_a and T_b), the thermal conductivities, the material thicknesses, and the convective heat transfer coefficients are known. The objective is to determine the temperatures at the surfaces (T_0 and T_3) and the temperatures at the interfaces (T_1 and T_2).

(a)   Rearrange these equations into a more traditional form as four equations in four unknowns.

(b)   Using Excel, solve for the unknown temperatures, based upon the following given information (all units are consistent):

$\Delta x_1 = 0.5$ cm	$k_1 = 0.01$ cal/(cm)(sec)(°C)
$\Delta x_2 = 0.3$ cm	$k_2 = 0.15$ cal/(cm)(sec)(°C)
$\Delta x_3 = 0.2$ cm	$k_3 = 0.03$ cal/(cm)(sec)(°C)
$T_a = 200$°C	$h_1 = 1.0$ cal/(cm^2)(sec)(°C)
$T_b = 20$°C	$h_2 = 0.8$ cal/(cm^2)(sec)(°C)

(c)   Use the results of part (b) to determine the heat flux, in cal/(cm^2)(sec). (The heat flux is represented by either the left-hand side or the right-hand side of any of the above four equations.)

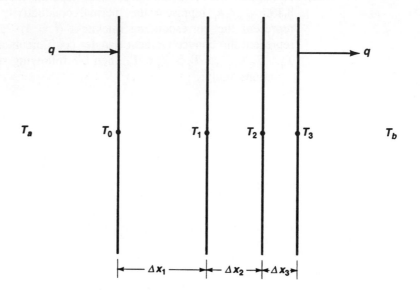

Figure 8.19

8.24 A steel company manufactures four different types of steel alloys, called A1, A2, A3, and A4. Each alloy contains small amounts of chromium (Cr), molybdenum (Mo), titanium (Ti), and nickel (Ni). The required composition of each alloy is given below.

Alloy	Cr	Mo	Ti	Ni
A1	1.6%	0.7%	1.2%	0.3%
A2	0.6	0.3	1.0	0.8
A3	0.3	0.7	1.1	1.5
A4	1.4	0.9	0.7	2.2

Suppose the alloying materials are available in the following amounts:

Material	Availability
Cr	1200 kg/day
Mo	800
Ti	1000
Ni	1500

Using Excel, determine the daily production rate for each alloy in metric tons per day (1 metric ton = 1000 kg).

8.25 The *catenary* is a classical problem concerning the shape of a uniform cable hanging under its own weight. Figure 8.20 shows a drawing of the cable and its associated coordinate system. Notice that the horizontal distance x is measured from the lowest point (so that x_a is negative and x_b is positive), and that the lowest point is a distance v above the vertical datum.

Based upon this coordinate system, the length of the cable is calculated as

$$L = v \left[\sinh(x_b / v) - \sinh(x_a / v) \right]$$

and the difference in height between the end points is calculated as

$$h = v \left[\cosh(x_b / v) - \cosh(x_a / v) \right]$$

Also, the width spanned by the cable is given by

$$w = x_b - x_a$$

Suppose a 100-ft cable is suspended in such a manner that $x_a = -30$ ft and $h = 20$ ft. Solve for the corresponding values of x_b, w, and v in Excel using **Solver**. (Note that this problem requires solving a system of *nonlinear* algebraic equations.)

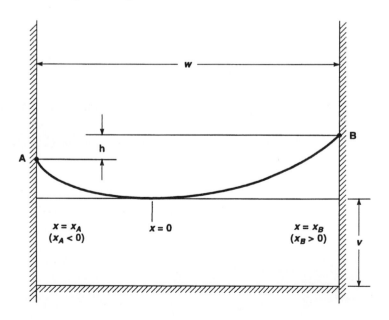

Figure 8.20

8.26 In Chap. 7 we learned that the horizontal displacement of a damped oscillating object as a function of time is given by the equation

$$x = x_0 e^{-\beta t}\left[\cos(\omega t) + (\beta / \omega)\sin(\omega t)\right]$$

(see Prob. 7.19). The parameters β and ω depend upon the mass of the object and the dynamic characteristics of the system (i.e., the spring constant and the damping constant).

Determine the values of x_0, β, and ω for an object that experiences the following displacement as a function of time:

t	x
1.0 sec	7.822 in
2.0	1.533
3.0	−2.979

Obtain your solution with Excel, using **Solver**.

ADDITIONAL READING

Akai, T. J. *Applied Numerical Methods for Engineers.* New York: Wiley, 1994.

Carnahan, B., H. A. Luther, and J. O. Wilkes. *Applied Numerical Methods.* New York: Wiley, 1969.

Chapra, S. C. and T. P. Canale. *Numerical Methods for Engineers, with Personal Computer Applications.* New York: McGraw-Hill, 1985.

Mayo, W. E. and M. Cwiakala. *Introduction to Computing for Engineers.* New York: McGraw-Hill, 1991.

Nakamura, S. *Applied Numerical Methods in C.* Englewood Cliffs, NJ: Prentice-Hall, 1993.

CHAPTER 9

EVALUATING INTEGRALS

Many engineering problems involve the evaluation of integrals. Sometimes an integral is required in order to determine an area or the effect of some accumulation with respect to time. Or an integral may be required to evaluate a scientific formula. In any event, the need to evaluate integrals occurs frequently in many different areas of engineering.

Suppose, for the example, that a gas is enclosed within a volume V. If the temperature of the gas varies with time and the gas is well mixed (so that there are no spatial variations), then the average temperature (with respect to time) is given by the following expression:

$$\overline{T} = \frac{\int_0^{t_{max}} T(t)dt}{\int_0^{t_{max}} dt} = \frac{1}{t_{max}} \int_0^{t_{max}} T(t)dt \tag{9.1}$$

where $T(t)$ represents the temperature of the gas at any given time and t represents time, where $0 \leq t \leq t_{max}$.

If the pressure-volume-temperature relationship of the gas is governed by Van der Waals's equation of state, that is,

$$\left(P+\frac{a}{V^2}\right)(V-b)=RT \tag{9.2}$$

then the average pressure (with respect to time) is given by the expression

$$\overline{P}=\frac{1}{t_{max}}\int_0^{t_{max}}P(t)dt \tag{9.3}$$

Substituting Equation (9.2) into Equation (9.3), the average pressure is given by

$$\overline{P}=\frac{R\int_0^{t_{max}}T(t)dt}{(V-b)t_{max}}-\frac{a}{V^2} \tag{9.4}$$

Thus, we must evaluate an integral in order to obtain an explicit value for either $\overline{T}$ or $\overline{P}$. In order to do so, we will need some information about the exact variation of the temperature with time. That is, we will have to replace the general expression $T(t)$ with an explicit formula for temperature as a function of time.

If $T(t)$ can be represented by some simple formula, it may be possible to evaluate the integral in Equations (9.1) and (9.3) using the rules learned in calculus. If $T(t)$ is represented by some relatively complex formula, however, or by a graph or a tabulated set of data, it will be necessary to evaluate the integral numerically.

In this chapter, we will discuss some simple numerical integration techniques and see how they can be implemented within a spreadsheet such as Excel.

9.1 THE TRAPEZOIDAL RULE

Suppose we are given a continuously varying function $y=f(x)$, defined over the interval $a\le x\le b$, as illustrated in Fig. 9.1. Then the integral

$$I=\int_a^b f(x)\,dx=\int_a^b y\,dx \tag{9.5}$$

can be interpreted as the area under the curve, as indicated in Fig. 9.1. The function $y=f(x)$ is referred to as the *integrand.*

Now suppose we approximate this irregularly shaped area with a large number of adjacent, narrow, rectangular intervals, as illustrated in Fig. 9.2. Then we can think of the integral as the sum of the areas of these rectangular intervals.

The area of each interval can be determined as the product of the height, y, and the width, Δx. Thus,

$$I = \sum_{i=1}^{n} A_i \qquad (9.6)$$

where

$$A_i = y_i \, \Delta x_i \qquad (9.7)$$

and n is the total number of intervals.

If we combine Equations (9.6) and (9.7), we obtain

$$I = \sum_{i=1}^{n} y_i \, \Delta x_i \qquad (9.8)$$

This is the basis for the *trapezoidal rule*, which is the simplest approach to carrying out a numerical integration. The larger the number of intervals and the smaller the width of each interval, the better the approximation.

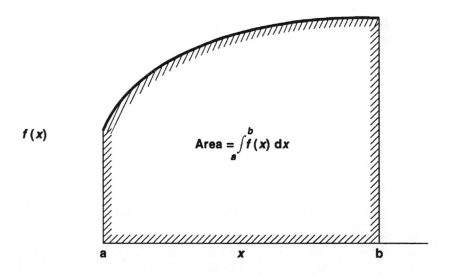

Figure 9.1

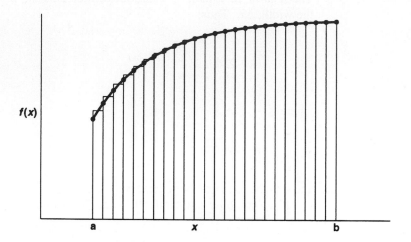

Figure 9.2

Suppose we are given a set of n data points: (x_1, y_1), (x_2, y_2), . . ., (x_n, y_n), where $x_1 = a$ and $x_n = b$. These data points define $n-1$ rectangular intervals, where the width of the ith interval is given by

$$\Delta x_i = x_{i+1} - x_i \tag{9.9}$$

The height of the rectangle associated with the ith interval can be expressed as the average of the y-values at the interval boundaries; that is,

$$\overline{y}_i = \frac{y_i + y_{i+1}}{2} \tag{9.10}$$

Hence the area of the rectangular interval is

$$A_i = \overline{y}_i \, \Delta x_i = \frac{(y_i + y_{i+1})(x_{i+1} - x_i)}{2} \tag{9.11}$$

and the entire integral can be approximated by

$$I = \int_a^b y \, dx = \sum_{i=1}^{n-1} A_i = \frac{1}{2} \sum_{i=1}^{n-1} (y_i + y_{i+1})(x_{i+1} - x_i) \tag{9.12}$$

Equation (9.12) represents the trapezoidal rule for *unequally spaced data*.

Example 9.1 The Trapezoidal Rule for Unequally Spaced Data

The current passing through an electrical inductor can be determined from the equation

$$i = \frac{1}{L}\int_0^t v\, dt$$

where

i = current (amperes)

L = inductance (henries)

v = voltage

t = time (seconds)

Suppose a current of 2.15 amperes is induced over a period of 500 milliseconds. The variation in voltage with time during this period is given below.

t (msec)	v (volts)	t	v	t	v	t	v
0	0	40	45	90	45	180	27
5	12	50	49	100	42	230	21
10	19	60	50	120	36	280	16
20	30	70	49	140	33	380	9
30	38	80	47	160	30	500	4

Determine the value of the inductance, using the above equation.

In order to determine the inductance, we must first evaluate the integral

$$I = \int_0^{0.5} v\, dt$$

If we use the trapezoidal rule, we will have 19 intervals since there are 20 data points. The calculations using the trapezoidal rule are summarized in the table below.

Note that the times have been converted to seconds within this table (column 2). The values in the fourth column of the table were obtained using Equation (9.9), and the values in the fifth column were obtained from Equation (9.10).   The last column contains the products of the values in the fourth and fifth columns (the interval areas).

Point No.	t (sec)	v (volts)	Width, Δt (sec)	Height, $\bar{v}$ (volts)	Area, $\bar{v}\,\Delta t$ (volt) (sec)
1	0	0	0.005	6.0	0.030
2	0.005	12	0.005	15.5	0.0775
3	0.01	19	0.01	24.5	0.245
4	0.02	30	0.01	34.0	0.340
5	0.03	38	0.01	41.5	0.415
6	0.04	45	0.01	47.0	0.470
7	0.05	49	0.01	49.5	0.495
8	0.06	50	0.01	49.5	0.495
9	0.07	49	0.01	48.0	0.480
10	0.08	47	0.01	46.0	0.460
11	0.09	45	0.01	43.5	0.435
12	0.10	42	0.02	39.0	0.780
13	0.12	36	0.02	34.5	0.690
14	0.14	33	0.02	31.5	0.630
15	0.16	30	0.02	28.5	0.570
16	0.18	27	0.05	24.0	1.200
17	0.23	21	0.05	18.5	0.925
18	0.28	16	0.10	12.5	1.250
19	0.38	9	0.12	6.5	0.780
20	0.50	4			

Total: 10.7675

At the bottom of the table, beneath the last column, we see that the sum of the individual areas is 10.7675 (volt) (sec). Hence, we conclude that the area under the curve (the value of the integral) is 10.7675 and the value of the inductance is

$$L = \frac{1}{i}\int_0^{0.5} v\,dt = \frac{10.7675}{2.15} = 5.0081 \text{ henries}$$

We conclude that the inductance is approximately 5 henries.

Excel does not include any special features for evaluating integrals. Since the calculations are carried out in a tabular format, however, they lend themselves very naturally to a spreadsheet solution. Thus, it is easy to evaluate an integral in Excel by following the layout given in the above example, as illustrated in the next example.

Example 9.2 The Trapezoidal Rule for Unequally Spaced Data in Excel

Repeat the problem given in Example 9.1 using Excel to carry out the numerical integration.

The procedure is the same as that used in Example 9.1. Now, however, we will determine the individual areas and the sum of the areas within an Excel worksheet, as shown in Fig. 9.3.

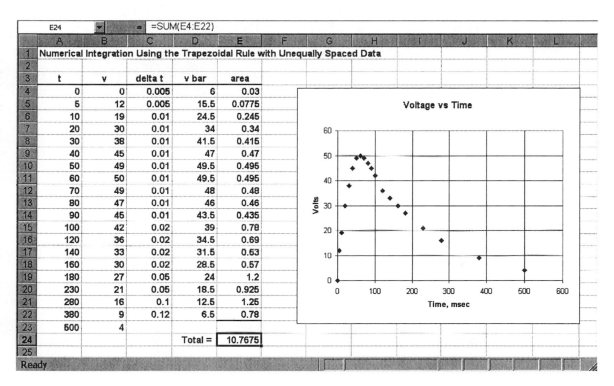

Figure 9.3

Within this worksheet, the first two columns contain the given data. The third column contains the calculated interval widths, in units of seconds rather than milliseconds. Thus, the value in cell C4 is obtained from the formula =0.001*(A5−A4), the value in cell C5 is obtained from =0.001*(A6−A5), and so on.

The fourth column contains the calculated height of each interval. Therefore, the value in cell D4 is obtained from the formula =0.5*(B5+B4), the value in cell D5 is obtained from =0.5*(B6+B5), and so on. Similarly, column five contains the area of each interval. Hence, the value in cell E4 is obtained as =C4*D4, the value in cell E5 is obtained as =C5*D5, and so on.

The sum of the areas is calculated in cell E24 from the formula =SUM(E4:E22). We see that this value is 10.7675. Therefore, we conclude that the value of the desired integral is 10.7675 (volt) (sec). This value agrees with the value obtained in the previous example. Hence, we determine that the unknown inductance is 5.0081 henries, as in the last example.

Notice that the worksheet also contains a graph of the given data, simply to illustrate the manner in which the voltage varies with time. It is the area under this curve that is represented by the desired integral.

Now suppose we have n equally spaced data points, where the successive x-values are separated by a distance Δx. Then Equation (9.12) simplifies to

$$I = \int_a^b y\, dx = \frac{1}{2}(y_1 + 2y_2 + 2y_3 + \cdots + 2y_{n-2} + 2y_{n-1} + y_n)\, \Delta x \qquad (9.13)$$

or

$$I = \left(\frac{y_1 + y_n}{2} + \sum_{i=2}^{n-1} y_i \right) \Delta x \qquad (9.14)$$

Equation (9.14) represents the trapezoidal rule for *equally spaced data*.

Example 9.3 The Trapezoidal Rule for Equally Spaced Data

Earlier in this chapter, we were given an expression for the average pressure of a Van der Waals gas when the temperature of the gas varies with time over the interval $0 \leq t \leq t_{max}$; that is,

$$\overline{P} = \frac{R \int_0^{t_{max}} T(t)dt}{(V - b)t_{max}} - \frac{a}{V^2} \qquad (9.4)$$

In this expression, P is the absolute pressure, V is the volume per mole, T is the absolute temperatue, R is the ideal gas constant (0.082054 liter atm/mole °K), and the symbols a and b are Van der Waals constants, which are unique to each particular gas.

Suppose one mole of carbon dioxide is enclosed within a one-liter container. Determine the average pressure if the temperature increases with time in accordance with the expression

$$T = 300 + 12t$$

over the interval $0 \leq t \leq 100$ sec. The Van der Waals constants for carbon dioxide are $a = 3.592$ liter^2 atm/mole^2 and $b = 0.04267$ liter/mole.

In order to determine the average pressure, we must evaluate the expression

$$\bar{P} = \frac{R \int_0^{100} (300 + 12t)\, dt}{100\,(V - b)} - \frac{a}{V^2}$$

Substituting the appropriate numerical values, this equation becomes

$$\bar{P} = \frac{0.082054 \int_0^{100} (300 + 12t)\, dt}{100\,(1 - 0.04267)} - \frac{3.592}{1^2}$$

or

$$\bar{P} = 8.571 \times 10^{-4} \int_0^{100} (300 + 12t)\, dt - 3.592$$

We must evaluate the integral in order to determine the average pressure. This particular integral can easily be evaluated using the classical rules of calculus, resulting in a value of $I = 9 \times 10^4$. Suppose, however, that we evaluate the integral using the trapezoidal rule. If we select 21 equally spaced points, we can generate the following table:

Point No	t (sec)	T (°K)	Point No	t (sec)	T (°K)
1	0	300	12	55	960
2	5	360	13	60	1020
3	10	420	14	65	1080
4	15	480	15	70	1140
5	20	540	16	75	1200
6	25	600	17	80	1260
7	30	660	18	85	1320
8	35	720	19	90	1380
9	40	780	20	95	1440
10	45	840	21	100	1500
11	50	900			

Substituting these values into Equation (9.14), we obtain

$$I = \left[\frac{300 + 1500}{2} + (360 + 420 + 480 + \cdots + 1380 + 1440) \right] 5 = 9 \times 10^4$$

In this case, the result of the trapezoidal rule integration agrees exactly with the result obtained using calculus. This is because the integrand is a straight line, so the area determined for each interval is exact. If the integrand were a more complicated function, however, we would see that the trapezoidal rule offers only an approximation to the correct answer, with the accuracy of the approximation increasing as the number of intervals (the number of points) increases.

We can now determine the average pressure of the gas, as originally requested. The pressure can be determined as

$$\overline{P} = (8.571 \times 10^{-4})(9 \times 10^4) - 3.592 = 73.547$$

Hence, the average pressure is 73.547 atmospheres.

Example 9.4 The Trapezoidal Rule for Equally Spaced Data in Excel

Use Excel to determine the average pressure of one mole of carbon dioxide in a one-liter container when the temperature increases with time, as described in Example 9.3.

In Example 9.3, we established that the average pressure is given by the expression

$$\overline{P} = 8.571 \times 10^{-4} \int_{0}^{100} (300 + 12t) \, dt - 3.592$$

Hence, this problem requires that we evaluate the given integral using Excel. To do so, we create the Excel worksheet shown in Fig. 9.4. The first column contains the times, ranging from 0 to 100 seconds. All of these values, other than the first, are generated using cell formulas. Thus, the value shown in cell A5, for example, is determined by the cell formula =A4+5. This formula is then copied to the remaining cells in column A. Fig. 9.5 shows the cell formulas used to generate the values given in Fig. 9.4.

Similarly, all of the temperatures shown in column B, other than the first, are generated with cell formulas. For example, the value shown in cell B5 is determined by the formula =B4+12*(A5−A4). This formula is then copied to the remaining cells in column B, as shown in Fig. 9.5.

F25	▼	●	=((B4+B24)/2+D25)*5										
A	B	C	D	E	F	G	H	I	J	K	L	M	N

	A	B	C	D	E	F	G	H	I	J	K	L	M	N
1	Numerical Integration Using the Trapezoidal Rule with Equally Spaced Data													
2														
3	Time	Temp												
4	0	300												
5	5	360		360										
6	10	420		420										
7	15	480		480										
8	20	540		540										
9	25	600		600										
10	30	660		660										
11	35	720		720										
12	40	780		780										
13	45	840		840										
14	50	900		900										
15	55	960		960										
16	60	1020		1020										
17	65	1080		1080										
18	70	1140		1140										
19	75	1200		1200										
20	80	1260		1260										
21	85	1320		1320										
22	90	1380		1380										
23	95	1440		1440										
24	100	1500												
25			Sum =	17100	I =	90000								
26														

Ready

Figure 9.4

Column D contains all of the temperature values, other than the first and last. These values were obtained by copying the corresponding values from column B (using Copy/Paste Special/Values). The sum of these values is shown at the bottom of column D, in cell D25. This sum represents the general summation term given in Equation (9.14).

Finally, the value of the integral (I = 90,000) is shown in cell F25. This value was determined using Equation (9.14). The general strategy was to add the average of the first and last temperatures to the value in cell D25 and then multiply this sum by the interval width (5). The highlighted cell formula shown in cell F25 of Fig. 9.5 illustrates how this strategy was carried out.

Once the value of the integral is known, we can substitute it into the equation for the average pressure; that is,

$$\bar{P} = (8.571 \times 10^{-4})\,(9 \times 10^{4}) - 3.592 = 73.547$$

Thus, we obtain the value of 73.547 atmospheres for the average pressure, as in the previous example.

	F25	▼	=	=((B4+B24)/2+D25)*5				
	A	B	C	D	E	F	G
1	Numerical Integratio						
2							
3	Time	Temp					
4	0	300					
5	=A4+5	=B4+12*(A5-A4)		360			
6	=A5+5	=B5+12*(A6-A5)		420			
7	=A6+5	=B6+12*(A7-A6)		480			
8	=A7+5	=B7+12*(A8-A7)		540			
9	=A8+5	=B8+12*(A9-A8)		600			
10	=A9+5	=B9+12*(A10-A9)		660			
11	=A10+5	=B10+12*(A11-A10)		720			
12	=A11+5	=B11+12*(A12-A11)		780			
13	=A12+5	=B12+12*(A13-A12)		840			
14	=A13+5	=B13+12*(A14-A13)		900			
15	=A14+5	=B14+12*(A15-A14)		960			
16	=A15+5	=B15+12*(A16-A15)		1020			
17	=A16+5	=B16+12*(A17-A16)		1080			
18	=A17+5	=B17+12*(A18-A17)		1140			
19	=A18+5	=B18+12*(A19-A18)		1200			
20	=A19+5	=B19+12*(A20-A19)		1260			
21	=A20+5	=B20+12*(A21-A20)		1320			
22	=A21+5	=B21+12*(A22-A21)		1380			
23	=A22+5	=B22+12*(A23-A22)		1440			
24	=A23+5	=B23+12*(A24-A23)					
25			Sum =	=SUM(D5:D23)		I =	=((B4+B24)/2+D
26							
Ready							

Figure 9.5

Problems

9.1   Evaluate the integral

$$I = \int_1^5 x^3 dx$$

using each of the following methods:

(a)   Classical calculus.

(b)   The trapezoidal rule within an Excel worksheet, using eight equally spaced intervals.

(c)   The trapezoidal rule within an Excel worksheet, using 20 equally spaced intervals.

Compare the answers obtained in each case.

9.2   For the integral given in Prob. 9.1, how many equally spaced intervals would be required using the trapezoidal rule in order to obtain an answer that is within 2 percent of the exact answer obtained with classical calculus? Use Excel to carry out the integration using the trapezoidal rule.

9.3   Evaluate the integral

$$I = \int_0^\pi \sin x \, dx$$

using each of the following methods:

(a)   Classical calculus.

(b)   The trapezoidal rule within an Excel worksheet, using 10 equally spaced intervals.

(c)   The trapezoidal rule in an Excel worksheet, using 24 equally spaced intervals.

Compare the answers obtained in each case.

9.4   The following data, representing temperature as a function of vertical depth, are taken from Prob. 5.7.

Distance, cm	Temperature, °C
0.1	21.29
0.8	27.3
3.6	31.8
12	35.6
120	42.3
390	45.9
710	47.7
1200	49.2
1800	50.5
2400	51.4

Use the trapezoidal rule within an Excel worksheet to determine the average temperature over the entire 2400 cm distance. Compare this value with a simple arithmetic average of the temperatures.

9.5   Using Excel and the methods described in Chap. 5, fit a polynomial to the voltage versus time data given in Example 9.1.   Then integrate the polynomial over the interval $0 \leq t \leq 0.5$ sec. using classical calculus. Compare the ease of solution and the accuracy of the solution with the results obtained in Examples 9.1 and 9.2.

9.6   Using Excel, integrate the polynomial obtained in Prob. 9.5 (not the actual data) using the trapezoidal rule. Carry out the integration with

(*a*)   Ten equally spaced intervals.

(*b*)   Thirty equally spaced intervals.

(*c*)   One hundred equally spaced intervals.

Compare your solution with the results obtained in Prob. 9.5. What can you conclude about the relationship between accuracy and the number of intervals?

9.7   The following data, representing reaction rate as a function of temperature, are taken from Example 5.11.

Temperature, °K	Reaction Rate, moles/sec
253	0.12
258	0.17
263	0.24
268	0.34
273	0.48
278	0.66
283	0.91
288	1.22
293	1.64
298	2.17
303	2.84
308	3.70

Use the trapezoidal rule for equally spaced intervals to determine the average reaction rate over the entire temperature range. Carry out the integration within an Excel worksheet. Compare this value with a simple arithmetic average of the reaction-rate data.

9.8   The following data, taken from Prob. 5.19, represent the power produced by a wind-driven generator built by a group of students.

Wind Velocity, mph	Power, watts
0	0
5	0.26
10	2.8
15	7.0
20	15.8
25	28.2
30	46.7

(*Data set continues on next page.*)

Wind Velocity, mph	Power, watts
35	64.5
40	80.2
45	86.8
50	88.0
55	89.2
60	90.3

Use the trapezoidal rule for equally spaced intervals to determine the average power generation over the given range of wind velocities. Evaluate the integral within an Excel worksheet.

9.9   The following stress-strain data, taken from Prob. 5.26, were obtained from a structural steel cylindrical sample with a diameter of 0.5 inches and a length of four inches.

Strain	Stress (psi)	Strain	Stress (psi)
0.02	29737	0.14	57046
0.04	37166	0.16	56593
0.06	44820	0.18	53448
0.08	44074	0.20	52103
0.10	49161	0.22	49185
0.12	53002	0.24	45386

Using the trapezoidal rule within an Excel worksheet, determine the average stress for the strain values ranging from 0 to 0.24. (Assume the data set begins at the origin; i.e., the stress is zero when the strain is zero.)

9.2 SIMPSON'S RULE

Simpson's rule is a widely used numerical integration technique that combines simplicity and accuracy. It is similar to the trapezoidal rule in the sense that it approximates an irregular area with a number of geometrically simple subintervals. Rather than construct single-interval rectangles, however, we now pass a second-order polynomial (i.e., a parabola) through successive groups of three adjacent, equally spaced data points, as illustrated in Fig. 9.6. The area under each polynomial can then be obtained by direct integration.

If the number of subintervals is even (i.e., if the number of data points is odd), then the repeated use of this parabolic approximation results in the following simple expression:

$$I = \int_a^b y\,dx = \frac{1}{3}(y_1 + 4y_2 + 2y_3 + 4y_4 + 2y_5 + \cdots + 2y_{n-2} + 4y_{n-1} + y_n)\Delta x \quad (9.15)$$

where n is the number of data points.   Notice that the interior y-values are alternately multiplied by 4 and 2.

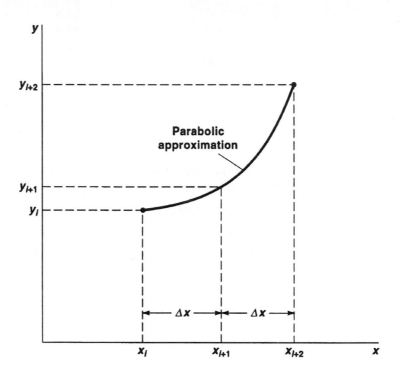

Figure 9.6

Example 9.5 Comparing Simpson's Rule with the Trapezoidal Rule

Evaluate the integral

$$I = \int_0^1 e^{-x^2}\, dx$$

using both the trapezoidal rule and Simpson's rule, with 10 equally spaced intervals in each case. Compare the results obtained with each method with the tabulated answer, which is 0.7468. (This particular integral *cannot* be evaluated using the methods of classical calculus.   The integral has, however, been evaluated numerically to a high degree of precision, and tabulated values are available in a number of reference books.)

Since there will be 10 subintervals, we will require 11 data points and Δx will have a value of 0.10. Hence we can construct the following table:

Point No	x_i	y_i	Point No	x_i	y_i
1	0	1.000	7	0.6	0.698
2	0.1	0.990	8	0.7	0.613
3	0.2	0.961	9	0.8	0.527
4	0.3	0.914	10	0.9	0.445
5	0.4	0.852	11	1.0	0.368
6	0.5	0.779			

Substituting these values into Equation (9.14) for the trapezoidal rule, we obtain

$$I = \left[\frac{1.000 + 0.368}{2} + (0.990 + 0.961 + \cdots + 0.527 + 0.445) \right](0.1) = 0.7463$$

Similarly, if we substitute these same values into Equation (9.15) for Simpson's rule, we obtain

$$I = \frac{1}{3}\left[1.000 + 4(0.990) + 2(0.961) + 4(0.914) + \cdots + \right.$$

$$\left. 2(0.527) + 4(0.445) + 0.368\right](0.1) = 0.7469$$

Summarizing the results, we have

Tabulated answer: 0.7468

Trapezoidal rule: 0.7463

Simpson's rule: 0.7469

Thus, we see that Simpson's rule results in a more accurate answer with very little additional computational effort.

Simpson's rule is easy to implement in Excel. One way to do this is to enter the x-values in one column, the y-values in the next column, followed by an additional column containing the y-values multiplied by their appropriate constants (i.e., multiplied by 4 or 2, except for the end points). The value of the integral is then obtained by summing this last column and multiplying the sum by $\Delta x/3$. The details are illustrated in the following example.

Example 9.6 Simpson's Rule in Excel

Evaluate the integral

$$I = \int_0^1 e^{-x^2} \, dx$$

using Simpson's rule with 10 equally spaced intervals within an Excel worksheet. (Recall that we evaluated this integral with Simpson's rule in the last example.)

	F16			=	=D16*(0.1/3)						
	A	B	C	D	E	F	G	H	I	J	K	L
1	Numerical Integration Using Simpson's Rule										
2											
3	x	y									
4	0.0	1.0000		1.0000							
5	0.1	0.9900		3.9602							
6	0.2	0.9608		1.9216							
7	0.3	0.9139		3.6557							
8	0.4	0.8521		1.7043							
9	0.5	0.7788		3.1152							
10	0.6	0.6977		1.3954							
11	0.7	0.6126		2.4505							
12	0.8	0.5273		1.0546							
13	0.9	0.4449		1.7794							
14	1.0	0.3679		0.3679							
15											
16			Sum =	22.4047	I =	0.7468						
17											
18											

Ready

Figure 9.7

Figure 9.7 shows an Excel worksheet that incorporates the previously described layout. Column A contains the 11 equally spaced x-values, ranging from 0 to 1. All of these values, beginning with the second, are generated using cell formulas. The value in cell A5, for example, is generated using the cell formula =A4+0.1. This formula is copied to the remaining cells in column A, resulting in the values shown. Figure 9.8 shows the cell formulas used to generate the values that are shown in Fig. 9.7.

Column B contains the corresponding y-values. These values are all generated using cell formulas. For example, the value in cell B4 is generated using the cell formula =EXP(−($A4^2)). This formula is copied to the remaining cells in column B, as shown in Fig. 9.8.

Column D contains modifications of the formulas used to generate the values in column B. These formulas, like those in column B, refer to the x-values in column A.

(Hence the use of absolute addressing with respect to column A.) Now, however, all of the cell formulas, other than the first and the last, are multiplied by either 4 or 2, as required by Simpson's rule. Thus, cell D5 contains the formula =4*EXP(−($A5^2)), cell D6 contains the formula =2*EXP(−($A6^2)), and so on, as shown in Fig. 9.8.

The sum of the entries in column D is computed in cell D16. This value represents the sum of the terms shown in parentheses in Equation (9.15). This value (22.4047) is obtained using the cell formula =SUM(D4:D14).

	A	B	C	D	E	F
F16		=	=D16*(0.1/3)			
1	Numerical Integratio					
2						
3	x	y				
4	0	=EXP(-($A4^2))		=EXP(-($A4^2))		
5	=A4+0.1	=EXP(-($A5^2))		=4*EXP(-($A5^2))		
6	=A5+0.1	=EXP(-($A6^2))		=2*EXP(-($A6^2))		
7	=A6+0.1	=EXP(-($A7^2))		=4*EXP(-($A7^2))		
8	=A7+0.1	=EXP(-($A8^2))		=2*EXP(-($A8^2))		
9	=A8+0.1	=EXP(-($A9^2))		=4*EXP(-($A9^2))		
10	=A9+0.1	=EXP(-($A10^2))		=2*EXP(-($A10^2))		
11	=A10+0.1	=EXP(-($A11^2))		=4*EXP(-($A11^2))		
12	=A11+0.1	=EXP(-($A12^2))		=2*EXP(-($A12^2))		
13	=A12+0.1	=EXP(-($A13^2))		=4*EXP(-($A13^2))		
14	=A13+0.1	=EXP(-($A14^2))		=EXP(-($A14^2))		
15						
16			Sum =	=SUM(D4:D14)	I =	=D16*(0.1/3)
17						
18						

Ready

Figure 9.8

Cell F16 contains the value of the integral, $I = 0.7468$, as determined by Simpson's rule. This value was obtained with the cell formula =D16*(0.1/3), as indicated by Equation (9.15). (Note that the value 0.1 appearing within the parentheses represents the interval width, Δx.) The value obtained agrees with the exact answer to four significant figures. This value is slightly more accurate than the value obtained using Simpson's rule in Example 9.5 because of the greater precision in the spreadsheet solution.

The summation term in Equation (9.15) can be evaluated several other ways in Excel. For example, we might place the end points in one column, two times the interior points in a second column, and two times every other interior point in a third column. The desired summation can be obtained by adding the entries in each column and then adding the three column sums.

Alternatively, we might place the end points in one column, four times every other interior point in a second column, and two times the remaining interior points in a third column. The desired summation can again be obtained by adding the entries in each column and then adding the three column sums.

Example 9.7 Simpson's Rule in Excel — Another Layout

Let us again evaluate the integral

$$I = \int_0^1 e^{-x^2}\, dx$$

using Simpson's rule with 10 equally spaced intervals within an Excel worksheet, as in Example 9.6. Now, however, we will use a different layout to determine the summation term in Equation (9.15).

The Excel worksheet containing the solution is shown in Fig. 9.9. Within this worksheet, columns A and B contain the *x*-values and the corresponding *y*-values, as explained in Example 9.6. Column D contains the end points. Column E contains two times the calculated *y*-values for *all* of the interior points, and column F contains two times the calculated *y*-values for *every other* interior point. Line 16 contains the sums of the values in columns D, E, and F. These three sums are added together in cell D18. Finally, the value of the integral is obtained in cell D20 by multiplying the value in cell D18 by (0.1/3).

	D20		=D18*(0.1/3)								
	A	B	C	D	E	F	G	H	I	J	K	L
1	Numerical Integration Using Simpson's Rule										
2											
3	x	y		y	2y	2y					
4	0.0	1.0000		1.0000							
5	0.1	0.9900			1.9801	1.9801					
6	0.2	0.9608			1.9216						
7	0.3	0.9139			1.8279	1.8279					
8	0.4	0.8521			1.7043						
9	0.5	0.7788			1.5576	1.5576					
10	0.6	0.6977			1.3954						
11	0.7	0.6126			1.2253	1.2253					
12	0.8	0.5273			1.0546						
13	0.9	0.4449			0.8897	0.8897					
14	1.0	0.3679		0.3679							
15											
16		Column Sums =		1.3679	13.5563	7.4805					
17											
18		Overall Sum =		22.4047							
19											
20			I =	0.7468							
21											
22											

Ready

Figure 9.9

Figure 9.10 shows the cell formulas used to generate the values shown in Fig. 9.9.

	D20	▼	=	=D18*(0.1/3)		
	A	B	C	D	E	F
1	Numerical Integratio					
2						
3	x	y		y	2y	2y
4	0	=EXP(-($A4^2))		=EXP(-($A4^2))		
5	=A4+0.1	=EXP(-($A5^2))			=2*EXP(-($A5^2))	=2*EXP(-($A5^2))
6	=A5+0.1	=EXP(-($A6^2))			=2*EXP(-($A6^2))	
7	=A6+0.1	=EXP(-($A7^2))			=2*EXP(-($A7^2))	=2*EXP(-($A7^2))
8	=A7+0.1	=EXP(-($A8^2))			=2*EXP(-($A8^2))	
9	=A8+0.1	=EXP(-($A9^2))			=2*EXP(-($A9^2))	=2*EXP(-($A9^2))
10	=A9+0.1	=EXP(-($A10^2))			=2*EXP(-($A10^2))	
11	=A10+0.1	=EXP(-($A11^2))			=2*EXP(-($A11^2))	=2*EXP(-($A11^2))
12	=A11+0.1	=EXP(-($A12^2))			=2*EXP(-($A12^2))	
13	=A12+0.1	=EXP(-($A13^2))			=2*EXP(-($A13^2))	=2*EXP(-($A13^2))
14	=A13+0.1	=EXP(-($A14^2))		=EXP(-($A14^2))		
15						
16		Column Sums =		=SUM(D4:D14)	=SUM(E4:E14)	=SUM(F4:F14)
17						
18		Overall Sum =		=SUM(D16:F16)		
19						
20			I =	=D18*(0.1/3)		
21						
22						
Ready						

Figure 9.10

Problems

9.10 Repeat Prob. 9.1 using Simpson's rule within an Excel worksheet. Compare the results obtained using Simpson's rule with those obtained earlier using the trapezoidal rule.

9.11 When solving Prob. 9.10, how many equally spaced intervals would be required using Simpson's rule in order to obtain an answer that is within 2 percent of the answer obtained with classical calculus? Use Excel to carry out the integration using Simpson's rule. Compare the results with those obtained in Prob. 9.2 using the trapezoidal rule.

9.12 Repeat Prob. 9.3 using Simpson's rule within an Excel worksheet. Compare the results obtained using Simpson's rule with those obtained earlier using the trapezoidal rule.

9.13 In Prob. 9.5 you were asked to fit a polynomial to the voltage versus time data given in Example 9.1 and then integrate the polynomial over the time interval $0 \leq t \leq 0.5$ sec. using classical calculus. In this problem you are asked to integrate the polynomial using Simpson's rule within an Excel worksheet. Carry out the integration using

(a) Ten equally spaced intervals.

(b) Thirty equally spaced intervals.

(c) One hundred equally spaced intervals.

Compare with the results obtained in Prob. 9.6, using the trapezoidal rule.

9.14 Integrate the data given in Prob. 9.8 using Simpson's rule to determine the average power generation over the given range of wind velocities. Evaluate the integral within an Excel worksheet. Compare with the results obtained in Prob. 9.8 using the trapezoidal rule.

9.15 The charge within a capacitor is given by the expression

$$Q = \int_0^{t_{max}} i\, dt$$

where Q is the charge, in coulombs, i is the current, in amperes, and t is the time, in seconds. Suppose the variation of current with time is given by the equation

$$i = 0.2\, e^{-0.1t}$$

Determine the charge stored within the capacitor during the first 200 seconds using each of the following methods:

(a) Classical calculus.

(b) The trapezoidal rule.

(c) Simpson's rule.

Implement the trapezoidal rule and Simpson's rule within an Excel worksheet.

9.16 The total amount of water discharged from the bottom of a tank is given by the expression

$$V = \int_0^t q\, dt$$

where q, the volumetric flow rate in cubic feet per second, is given by the expression

$$q = 0.1\,(80 - t), \qquad 0 \le t \le 80 \text{ sec.}$$

and t is the time in seconds. Determine

(a) The total amount of water initially in the tank, assuming the entire tank is emptied during the first 80 seconds.

(b) The quantity of water discharged during the first 40 seconds.

(c) The quantity of water discharged between times $t = 15$ and $t = 60$ seconds.

Determine each value using each of the following methods:

(a) Classical calculus.

(b) The trapezoidal rule, within an Excel worksheet.

(c) Simpson's rule, within an Excel worksheet.

9.17 A group of students have designed a small rocket, which they plan to fire from the football field. The rocket contains enough fuel to burn for eight seconds. During this period, the rocket's vertical velocity will be determined by the expression

$$v = 6t^2, \qquad 0 \le t \le 8$$

where v represents the vertical velocity in ft/sec and t represents the time, in seconds.

After the eight-second burn, the rocket will fall to the ground under the force of gravity. Hence,

$$v = v_0 - 32.2\,(t - 8), \qquad t > 8$$

where v_0 is the velocity at the end of the eight-second burn.

Integrate these two expressions to determine the maximum height attained by the rocket and the time required for the rocket to return to the ground.

Calculate each value using each of the following methods:

(a) Classical calculus.

(b) The trapezoidal rule, within an Excel worksheet.

(c) Simpson's rule, within an Excel worksheet.

9.18 The heat absorbed by a solid when its temperature is increased is given by the expression

$$Q = m \int_{T_1}^{T_2} C_p \, dT$$

In this expression, Q represents the total heat absorbed, in BTUs; m is the mass of the solid, in lbs; C_p is the specific heat of the solid, in BTU/(lb)(°F); and T is the temperature, in degrees Fahrenheit.

How much heat is absorbed when a 20-lb mass of copper is heated from 100°F to 500°F?   The specific heat of copper is given by the expression

$$C_p = 0.0909 + 2\times10^{-5}\,T - 1\times10^{-9}\,T^2$$

Determine your answer using each of the following methods:

(*a*)   Classical calculus.
(*b*)   The trapezoidal rule, within an Excel worksheet.
(*c*)   Simpson's rule, within an Excel worksheet.

9.19 When water flows slowly within a circular pipe, its velocity increases with the square of the distance from the pipe wall. Thus, the velocity is zero at the wall and reaches a maximum at the center of the pipe.

Suppose the velocity distribution within a 12-inch diameter pipe is given by the expression

$$v = 0.3\left(1 - \frac{r}{0.5}\right)^2$$

where v is the velocity, in ft/sec, and r is the radial distance from the center of the pipe, in feet. (Hence, $0 \le r \le 0.5$). Then the volumetric flow rate through the pipe, in cu ft/sec, will be given by the expression

$$Q = \int v \, dA = 2\pi \int_0^{0.5} vr \, dr$$

Determine the volumetric flow within the pipe, using each of the following methods:

(*a*)   Classical calculus.
(*b*)   The trapezoidal rule, within an Excel worksheet.
(*c*)   Simpson's rule, within an Excel worksheet.

ADDITIONAL READING

Akai, T. J. *Applied Numerical Methods for Engineers*. New York: Wiley, 1994.

Carnahan, B., H. A. Luther, and J. O. Wilkes. *Applied Numerical Methods*. New York: Wiley, 1969.

Chapra, S. C. and T. P. Canale. *Numerical Methods for Engineers, with Personal Computer Applications*. New York: McGraw-Hill, 1985.

Mayo, W. E. and M. Cwiakala. *Introduction to Computing for Engineers*. New York: McGraw-Hill, 1991.

Nakamura, S. *Applied Numerical Methods in C*. Englewood Cliffs, NJ: Prentice-Hall, 1993.

CHAPTER 10

COMPARING ECONOMIC ALTERNATIVES

A new engineering design frequently results in a potential investment opportunity for the sponsoring company. Thus, the desirability of the proposed design is often measured in economic rather than technical terms. Furthermore, if several different designs have been proposed, the desirability of one investment opportunity over another is generally based upon certain economic criteria. Engineers must therefore have some knowledge of engineering economic analysis as well as a mastery of technical fundamentals.

Spreadsheets provide an excellent means for carrying out engineering economic analysis since they can accommodate complex investment patterns projected over future time periods. They also include library functions for carrying out the required calculations. Excel is particularly effective in this area.

In this chapter, we will present some fundamental concepts in engineering economic analysis. In particular, this chapter discusses compound interest and the time value of money, illustrates how these concepts are used to analyze complicated cash flows, and discusses how Excel can conveniently be used to carry out these studies.

10.1 COMPOUND INTEREST

Money is a commodity that is used to purchase goods and services. It therefore has *value*. If a person or an organization loans, deposits, or invests a sum of money, the lender is entitled to a payment for the use of that money. Similarly, if a person or an organization borrows a sum of money, the borrower must pay for the use of the money. A sum of money that is loaned or borrowed is called *principal*; a payment that is made for the use of someone else's money in known as *interest*.

Interest calculations are always based upon an *interest rate*. The interest rate is ordinarily expressed as a percentage, but it must be converted to decimal form in order to carry out an economic analysis calculation. To do so, we first express any fractional percentage points in decimal form and then divide the entire interest rate by 100. For example, an annual interest rate of 5-1/2 percent would be expressed in fractional form as 0.055. (Notice that we have not stated how often the interest is paid; for example, annually, quarterly, monthly, etc.)

The determination of an appropriate interest rate is normally based upon two factors: the degree of risk associated with the intended investment and general economic conditions. Risky investments are associated with higher interest rates than conservative, safe investments.

Economic calculations are generally based upon the use of *compound interest*. When carrying out compound interest calculations, it is assumed that the overall time span is broken up into several consecutive interest periods (e.g., several consecutive years) and that interest accumulates from one interest period to the next.

Suppose we invest a given amount of principal (P), which earns interest at a constant rate (i) during each of several consecutive interest periods. Within a given interest period, the current interest (I) is determined as a fraction of the total accumulation (i.e., the principal plus the previously accumulated interest). Thus, for the first interest period, the interest earned is determined as

$$I_1 = iP \tag{10.1}$$

and the total amount of money that has accumulated at the end of the first interest period is

$$F_1 = P + I_1 = P + iP = P(1 + i) \tag{10.2}$$

For the second interest period, the interest is determined as

$$I_2 = iF_1 = iP(1 + i) \tag{10.3}$$

and the total amount accumulated at the end of the second interest period is

$$F_2 = P + I_1 + I_2 = P + iP + iP(1 + i)$$

$$= P[(1 + i) + i(1 + i)] = P(1 + i)^2 \tag{10.4}$$

For the third interest period, we obtain

$$I_3 = iP(1 + i)^2 \tag{10.5}$$

$$F_3 = P(1 + i)^3 \tag{10.6}$$

and so on.

In general, if there are n interest periods, the total amount of money accumulated at the end of the last interest period is given by

$$F = F_n = P(1 + i)^n \tag{10.7}$$

This is the so-called "law" of compound interest. Equation (10.7) can be used to determine either the amount of money accumulated, in the case of an investment, or the amount of money owed, in the case of a loan. It is easily implemented within a spreadsheet.

Example 10.1 Accumulating Compound Interest

A student invests $2000 in a bank account earning 5 percent interest, compounded annually. If the student does not make any subsequent deposits or withdrawals, how much money will accumulate after 20 years? Calculate and plot the amount of money accumulated on a year-by-year basis.

In this problem, we know that $P = \$2000$, $i = 0.05$, and $n = 20$. Therefore, we can determine the amount of money accumulated after 20 years using Equation (9.7):

$$F = \$2000\,(1 + 0.05)^{20} = \$2000\,(1.05)^{20} = \$5306.60$$

Note that the interest accumulated over the 20-year period is

$$F - P = \$5306.60 - \$2000.00 = \$3306.60$$

which is substantially more than the original deposit.

The easiest way to determine the year-by-year accumulation is to construct an Excel worksheet based upon the repeated use of Equation (10.7), as shown in Fig. 10.1. The amount of money accumulated at the end of each year is shown in column E, beginning with the end of year 0 (the time of the initial deposit) and ending with the end of year 20. Each value in column E is obtained by applying Equation (10.7) to the principal entered in cell B3, the interest rate in cell B5, and the number of years in the adjoining cell in column D. (Notice, for example, the formula used to obtain the final value, 5306.60, which is shown in the highlighted cell E25.) Thus, changing a value in cell B3 or cell B5 will result in different values for all of the accumulations shown in column E.

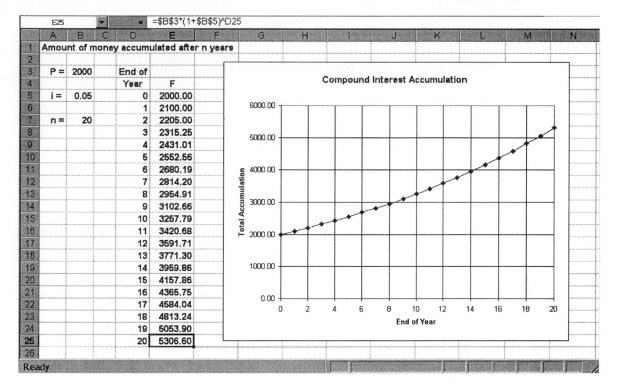

E25   =$B$3*(1+$B$5)^D25

Figure 10.1

Equation (10.7) can be used to determine the cost of a single-payment loan as well as the value of an accumulation. When dealing with a loan of this type, the principal (*P*) represents the amount of money originally borrowed, and the future accumulation (*F*) represents the amount of money that must be repaid (principal plus accumulated interest) after *n* interest periods.

Example 10.2 A Single-Payment Loan

A small manufacturing firm plans to borrow $500,000 in order to expand its loading dock. Interest will be charged at the rate of 7 percent per year, compounded annually. The company plans to allow the interest to accumulate for four years, at the end of which the company will repay the principal plus all of the accumulated interest. How much must the company repay at the end of the loan period? How much money will this loan cost the company, in terms of accumulated interest?

This problem can be solved using Equation (10.7), with *P* = $500,000, *i* = 0.07, and *n* = 4. Substituting these values into Equation (10.7), we obtain

$$F = \$500{,}000\,(1 + 0.07)^4 = \$500{,}000\,(1.07)^4 = \$655{,}398.$$

Hence, the company must repay \$655,398 at the end of four years. The cost of the loan will be \$655,398 − \$500,000 = \$155,398 in accumulated interest.

Although interest rates are typically specified on an annual basis, the interest is often compounded more frequently. Thus, if i is the interest rate, expressed on an annual basis; m is the number of compounding periods per year (e.g., $m = 12$ for monthly compounding); and n is the total number of compounding periods (i.e., $n = m \times$ number of years), Equation (10.7) can be modified to read

$$F = P\,(1 + i/m)^n \tag{10.8}$$

Thus, if the interest is compounded quarterly, Equation (10.8) would be written as

$$F = P\,(1 + i/4)^n \tag{10.9}$$

where n is the total number of quarters. Similarly, if the interest is compounded monthly, Equation (10.8) becomes

$$F = P\,(1 + i/12)^n \tag{10.10}$$

where n now represents the total number of months.

The interest can even be compounded daily. In such situations, Equation (10.8) becomes

$$F = P\,(1 + i/365)^n \tag{10.11}$$

where n represents the total number of days. (The banking industry sometimes uses a factor of 360 rather than 365 for daily compounding since 360 is a multiple of both 4 and 12. This results in a simple relationship between the annual interest rate, the quarterly interest rate, the monthly interest rate, and the daily interest rate. Obviously, this practice precedes the use of computers in the banking industry.)

Example 10.3 Compound Interest: Frequency of Compounding

In Example 10.1, we considered a \$2000 deposit earning interest at 5 percent per year, compounded annually. After 20 years, we found that \$5306.60 had accumulated. Let us now recalculate the amount of money accumulated, based upon an annual interest rate of 5 percent and

(*a*) Quarterly compounding

(*b*) Monthly compounding

(*c*) Daily compounding

The accumulation resulting from quarterly compounding is determined using Equation (10.9); thus,

$$F = \$2000 \, (1 + 0.05 / 4)^{4 \times 20} = \$5402.97$$

If the compounding is monthly, we use Equation (10.10) to obtain

$$F = \$2000 \, (1 + 0.05 / 12)^{12 \times 20} = \$5425.28$$

Similarly, if the compounding is daily, we use Equation (10.11) to obtain

$$F = \$2000 \, (1 + 0.05 / 365)^{365 \times 20} = \$5436.19$$

The results are summarized below:

Frequency of Compounding	*Accumulation*
Annual	$5306.60
Quarterly	$5402.97
Monthly	$5425.28
Daily	$5436.19

Comparing annual compounding with daily compounding, we see that an additional $130 (approximately) is obtained as a result of the more frequent compounding.

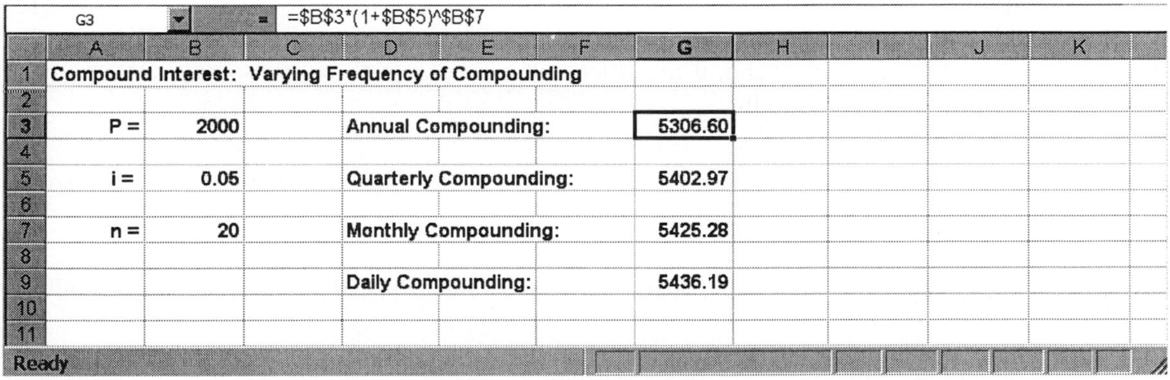

G3		=	=$B$3*(1+$B$5)^$B$7						
	A	B	C	D	E	F	G	H	I	J	K
1	Compound Interest: Varying Frequency of Compounding								
2									
3	P =	2000		Annual Compounding:			5306.60		
4									
5	i =	0.05		Quarterly Compounding:			5402.97		
6									
7	n =	20		Monthly Compounding:			5425.28		
8									
9				Daily Compounding:			5436.19		
10									
11									
Ready									

Figure 10.2

The calculations can be carried out within an Excel worksheet, as shown in Fig. 10.2. The accumulated results, which appear in column G, were obtained using the appropriate compound interest equation for the given compounding period. Notice, for example, the formula used to obtain the value within the highlighted cell G3. This formula corresponds to Equation (10.7), which is the appropriate equation for annual compounding.

Figure 10.3 contains the cell formulas used to obtain the values shown in Fig. 10.2. (The cell widths have been modified to display their entire contents.) Notice the cell formulas used to generate the values appearing in column G. These formulas correspond to Equations (10.7), (10.9), (10.10), and (10.11).

	G3			▼		=	=$B$3*(1+$B$5)^$B$7		
	A	B	C	D	E	F	G		H
1	Compound Interest:								
2									
3		P =	2000		Annual Compounding:			=$B$3*(1+$B$5)^$B$7	
4									
5		i =	0.05		Quarterly Compounding:			=$B$3*(1+$B$5/4)^(4*$B$7)	
6									
7		n =	20		Monthly Compounding:			=$B$3*(1+$B$5/12)^(12*$B$7)	
8									
9					Daily Compounding:			=$B$3*(1+$B$5/365)^(365*$B$7)	
10									
11									
Ready									

Figure 10.3

Sometimes F and P are both specified, and it is necessary to solve for either an interest rate (i) or the number of interest periods (n). If n is given and i is unknown, we can rearrange Equation (10.7) to read

$$i = (F/P)^{1/n} - 1 \qquad (10.12)$$

On the other hand, if i is given and n is unknown, we can solve Equation (10.7) for n, resulting in

$$n = \frac{\log(F/P)}{\log(1+i)} \qquad (10.13)$$

Equations (10.9), (10.10), and (10.11) can also be solved for i and n, with similar results.

Example 10.4 A Single-Payment Loan with Monthly Compounding

In Example 10.2, we determined the cost of a four-year $500,000 loan at an interest rate of 7 percent per year, compounded annually. At the end of four years, we found that the total amount of money that must be repaid is $655,398.

(a) What annual interest rate would the company be charged if the total amount due is $700,000 at the end of four years, assuming monthly compounding?

(b) Suppose the company will be charged 7 percent per year, compounded monthly, but the company cannot afford to repay more than $625,000. For how many months should the loan be taken out?

The first part of this question can be answered by solving Equation (10.10) for the annual interest rate, resulting in

$$i = 12\,[(F/P)^{1/n} - 1]$$

Substituting numerical values into this expression,

$$i = 12\,[(700,000/500,000)^{1/(12 \times 4)} - 1] = 12\,[(1.4)^{1/48} - 1] = 0.0844$$

Hence, the interest rate is 8.44 percent per year, based upon monthly compounding.

To answer the second part of the question, we solve Equation (10.10) for n, resulting in

$$n = \frac{\log(F/P)}{\log(1+i/12)}$$

Substituting numerical values into this expression and taking logarithms to the base 10, we obtain

$$n = \frac{\log(620,000/500,000)}{\log(1+0.07/12)} = \frac{\log(1.250)}{\log(1.0058)} = \frac{0.0969100}{0.00252602} = 38.36$$

Therefore, the loan should be taken out for 38 months. (Note that we could have obtained this same result using natural logarithms.)

As an alternative, we can use Excel's Goal Seek feature (see Chap. 7) to solve for either i or n when the other three parameters are known. The details are provided in the following example.

Example 10.5 A Single-Payment Loan with Monthly Compounding in Excel

Repeat the calculations in Example 10.4 within an Excel worksheet. Use Excel's Goal Seek feature to solve for the interest rate and the number of months, as required.

Figure 10.4 shows an Excel worksheet containing the necessary information. Note that the values $P = 500,000$, $i = 0.07$, and $n = 48$ are entered directly into cells B3, B5, and

B7, respectively. The corresponding value $F = 661{,}026.94$ shown in cell B9 is obtained by a cell formula, which is an expression of Equation (10.9).

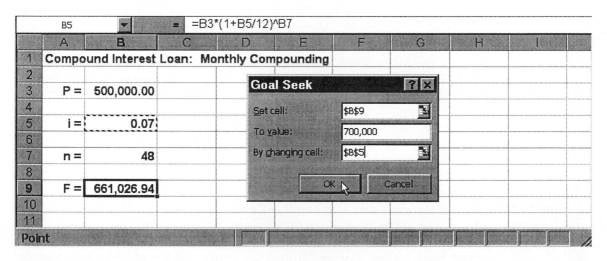

Figure 10.4

Figure 10.5 shows the Goal Seek dialog box, obtained by selecting Goal Seek from the Tools menu. The dialog box specifies that the formula in cell B9 must result in a value of 700,000. This value will be obtained by altering the value of the interest rate in cell B5.

Figure 10.5

Figure 10.6 shows the resulting solution. We see that the calculated accumulation is $700,000, as required. To obtain this result, Goal Seek selected a value of 0.0844 (i.e., an annual interest rate of 8.44 percent) in cell B5. This corresponds to the value obtained in Example 10.4.

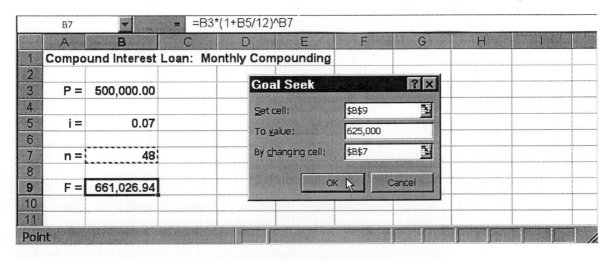

Figure 10.6

To solve the second part of the problem, we reset the annual interest rate in cell B5 to 0.07 and again select Goal Seek from the Tools menu. Now, however, we require that the formula in cell B9 result in a value of 625,000. To obtain this value, we require that the value in cell B7 (the number of months) be altered, as required. Figure 10.7 shows the Goal Seek dialog box for this problem.

Figure 10.7

The solution is shown in Fig. 10.8. Notice that the calculated accumulation is $625,000, as required. This value was obtained by changing the length of the loan from 48 months (the value originally entered into cell B7) to 38.36 months. This result is consistent with the value obtained in Example 10.4. Rounding to the nearest lower integer, we conclude that the length of the loan should not exceed 38 months.

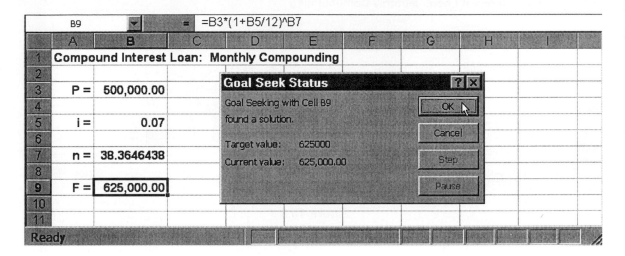

Figure 10.8

Problems

10.1 In Example 10.1, suppose the student seeks a bank that will pay interest more often than once a year. Using Excel, determine the additional amount of money that will accumulate if the interest is compounded

(a) Quarterly

(b) Monthly

(c) Daily

Assume an annual interest rate of 5 percent in each case. Compare the total accumulation with that obtained in Example 9.1 for annual compounding.

10.2 A student received some cash gifts when she graduated from college. She hopes to accumulate $50,000 by the time she retires, 40 years later. If the local bank pays interest at 4-1/2 percent per year, compounded monthly, how much money must the student deposit when she graduates in order to meet her goal? Solve using Excel.

10.3   Suppose a student deposits $5000 in a bank account when she graduates from college. Using Excel, determine the required annual interest rate for the student to accumulate $50,000 after 40 years. Assume the interest is compounded

(*a*)   Annually

(*b*)   Quarterly

(*c*)   Monthly

(*d*)   Daily

10.4   Suppose a student deposits $5000 in a bank account when she graduates from college. The bank pays interest at 5 percent per year. Using Excel, determine how long will it take for for her money to double if the interest is compounded

(*a*)   Aannually

(*b*)   Quarterly

(*c*)   Monthly

(*d*)   Daily

10.5   Suppose a student deposits $5000 in a bank account when she graduates from college. She plans to keep the money in the bank, accumulating interest, for 40 years. Using Excel, determine how much money will accumulate after 40 years if the annual interest rate is

(*a*)   4 percent

(*b*)   5 percent

(*c*)   8 percent

Assume quarterly compounding in each case.

10.6   Suppose you borrow $3000 at the beginning of your senior year to meet college expenses. If you make no payments for 10 years and then repay the entire amount of the loan, including accumulated interest, how much money will you owe? Use Excel to determine your solution. Assume interest is 6 percent per year, compounded

(*a*)   Annually

(*b*)   Quarterly

(*c*)   Monthly

(*d*)   Daily

How significant is the frequency of compounding?

10.7   A company plans to borrow $150,000 to equip a product-testing lab. If the company must repay $275,000 after seven years, what interest rate is being charged? Assume annual compounding. Use Excel to determine a solution.

10.8 A company plans to borrow $250,000 to promote a new product. The current interest rate is 8 percent per year, compounded monthly. Suppose the company does not plan to repay any of the loan until the end of the loan period, at which time the entire loan will be repaid. If the company cannot repay more than $325,000, what is the maximum permissible loan period?

10.9 A company has recently developed an inexpensive, battery-powered lawn mower. At present, the company's annual production rate is only 40 percent of total production capacity. However, the company expects its annual sales (and hence its annual production rate) to increase by 9 percent each year, as the demand for battery-powered lawn mowers increases. If so, how long will it take for the company to be producing lawn mowers at its full capacity?

10.10 A local beach is monitored each year for biological pollutants. Currently, the concentration of a certain hazardous bacterium is 6 ppm (parts per million). The maximum safe level of this bacterium is 20 ppm. If the concentration increases at 5 percent per year, how long will it take before the beach must be closed?

10.2 THE TIME VALUE OF MONEY

Because money has the ability to earn interest, its value increases with time. Thus, $100 today is equivalent to $105 one year from now if the interest rate is 5 percent per year, compounded annually. We therefore say that the *future value* of $100 is $105 if i = 5 percent, compounded annually, and n = 1.

If money *increases* in value as we move from the present to the future, it must *decrease* in value as we move from the future to the present. Thus, $100 one year from now is equivalent to $95.24 today if the interest rate is 5 percent per year, compounded annually. (Note that $95.24 × 1.05 = $100.) Thus, we say that the *present value* of $100 is $95.24 if i = 5 percent, compounded annually, and n = 1. The present value is also referred to as the *net present value*.

Example 10.6 Present Value of a Future Sum of Money

A student will inherit $10,000 in three years. The student has a savings account that pays 5-1/2 percent per year, compounded annually. What is the present value of the student's inheritance?

This problem can be solved by rearranging Equation (10.7) into the following form:

$$P = F / (1 + i)^n$$

Substituting numerical values into this expression,

$$P = \$10,000 / (1 + 0.055)^3 = \$8516.14$$

Thus, the present value of the student's inheritance is $8516.14. (If the student were to deposit $8516.14 in a savings account that pays 5-1/2 percent per year, compounded annually, he would accumulate $10,000 at the end of three years.)

Engineers frequently compare the present value of one investment strategy with the present value of another. Thus, the present value provides a basis of comparison for various economic alternatives. This concept is particularly useful when the time frame of one investment strategy differs from that of another.

Example 10.7 Comparing Two Economic Alternatives

The services of an engineering consulting firm are being sought by two different clients. The first client wants the firm to carry out a study on the design of a new water purification system. The study will result in a payment of $700,000 in two years. The second client wants the firm to conduct a series of tests on an existing water purification system and then make recommendations for redesigning the system, based upon the results of the tests. This study will result in a payment of $800,000 in five years. If the consulting firm can accept only one additional client, which one should it be? Solve using Excel, assuming the firm normally earns 8 percent per year on its assets.

We wish to determine the present value for each of the proposed payments, using the 8 percent interest rate. This allows us to compare one possible payment received two years in the future with another possible payment received five years in the future.

Figure 10.9 contains an Excel worksheet showing the present value of each of the proposed payments. The numerical values associated with the two proposals are given in columns B and D, respectively. Thus, the present value associated with the first client is shown in cell B9 as $P_1 = \$600,127.17$, and the present value associated with the second client is shown in cell D9 as $P_2 = \$544,466.56$. Since P_1 exceeds P_2, the consulting firm should choose the first client.

The values in cells B9 and D9 were obtained by solving Equation (10.7) for P, using the values for F, i, and n given in each column. The cell formulas are shown in Fig. 10.10.

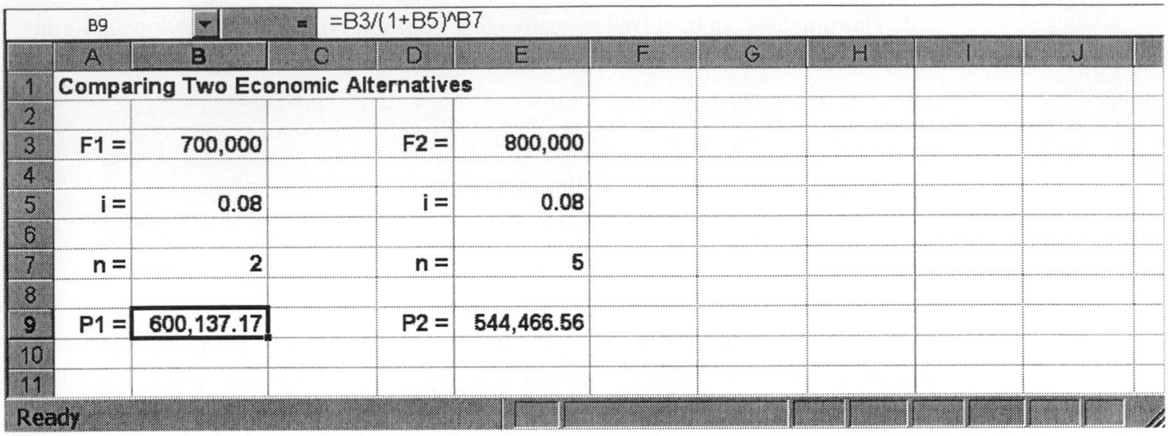

B9		▼		■	=B3/(1+B5)^B7					

	A	B	C	D	E	F	G	H	I	J
1	Comparing Two Economic Alternatives									
2										
3	F1 =	700,000		F2 =	800,000					
4										
5	i =	0.08		i =	0.08					
6										
7	n =	2		n =	5					
8										
9	P1 =	600,137.17		P2 =	544,466.56					
10										
11										

Ready

Figure 10.9

B9		▼		■	=B3/(1+B5)^B7	

	A	B	C	D	E
1	Comparing Tw				
2					
3	F1 =	700000		F2 =	800000
4					
5	i =	0.08		i =	0.08
6					
7	n =	2		n =	5
8					
9	P1 =	=B3/(1+B5)^B7		P2 =	=E3/(1+E5)^E7
10					
11					

Ready

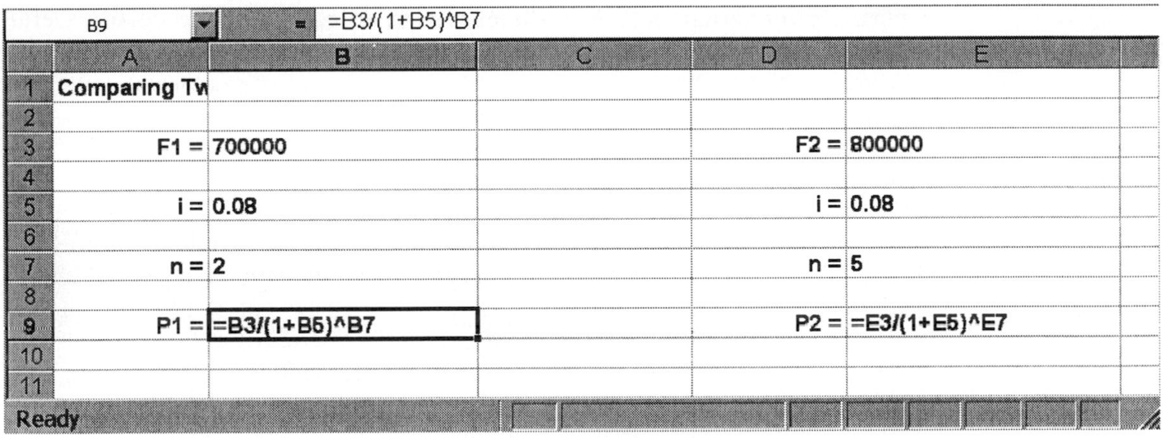

Figure 10.10

Problems

10.11 Which would have a greater present value?

(a) $10,000 five years from now.

(b) $12,500 eight years from now.

Assume an interest rate of 7 percent, compounded annually.

10.12 Repeat Problem 10.11 using an interest rate of 8 percent, compounded annually. Compare your answer with that obtained for Prob. 10.11.

10.13 Repeat Problem 10.11 using an annual interest rate of 7 percent, compounded daily. Compare your answer with that obtained for Prob. 10.11.

10.14 Which would have a greater present value?

(*a*) $10,000 five years from now.

(*b*) $12,500 nine years from now.

Assume an interest rate of 7 percent, compounded annually. Compare your answer with that obtained for Prob. 10.11.

10.3 UNIFORM, MULTIPAYMENT CASH FLOWS

Most realistic economic alternatives involve a number of cash inflows (*receipts*) and/or cash outflows (*disbursements*) over a period of several years. These items may all be identical, they may all be different, or they may involve some repeated items within an overall irregular pattern. A particularly common investment pattern involves an initial cash outflow (i.e., an initial investment) followed by a series of cash inflows at later times.

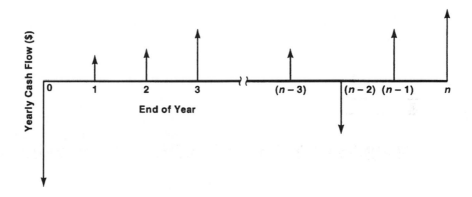

Figure 10.11

Collectively, the cash inflows and cash outflows associated with a proposed investment are called *cash flows*. It is often convenient to represent a series of cash flows graphically, by means of a *cash flow diagram*, as illustrated in Fig. 10.11. In a cash flow diagram, we plot the individual cash flow items as vertical arrows along the time axis. Cash *inflows* are represented as *upward*-pointing arrows, and cash *outflows* are represented as *downward*-pointing arrows.

Of particular interest is a cash flow pattern consisting of a single initial investment followed by *n* uniform payments, where each payment is made at the end of a compounding period. This cash flow pattern might represent a loan, where the initial investment is the amount of the loan (an investment, from the lender's perspective), and the uniform periodic payment is the amount repaid at the end of each compounding period (e.g., at the end of each month). Fig. 10.12 shows a diagram of this cash flow from the lender's perspective.

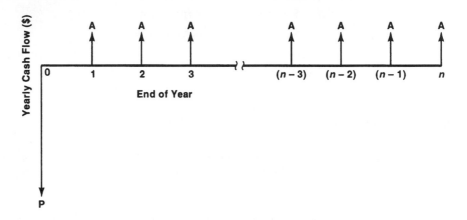

Figure 10.12

This cash flow pattern can be represented mathematically by the expression

$$A = P\left[\frac{(i/m)(1+i/m)^n}{(1+i/m)^n - 1}\right] \tag{10.14}$$

where P represents the initial investment, A is the amount of each payment, i is the annual interest rate, m is the number of compounding periods per year and n is the total number of payments (i.e., the total number of compounding periods).

Example 10.8 Determining the Cost of a Loan

Suppose you want to borrow $10,000 to buy a car, which you plan to repay on a monthly basis over three years (i.e., 36 equal monthly payments). If the current interest rate is 8 percent per year, compounded monthly, how much will you have to repay each month?

This question can be answered through the use of Equation (10.14). Thus,

$$A = \$10,000\left[\frac{(0.08/12)(1+0.08/12)^{12\times3}}{(1+0.08/12)^{12\times3} - 1}\right] = \$313.36$$

We therefore conclude that the amount of each monthly payment is $313.36. (Note that the total amount of money repaid, including interest, is $313.36 × 36 = $11,280.96. Hence, the total interest paid is $11,280.96 – $10,000 = $1280.96.)

Equation (10.14) can easily be evaluated in Excel. However, Excel includes two library functions that simplify uniform payment loan calculations even further. In particular, the PMT function, written as PMT(i, n, P), returns the periodic payment (A) for an n-payment loan of P dollars at interest rate i. (Note that i represents the interest rate for each *payment period*, not the *annual* interest rate.) Similarly, the PV function, written as PV(i, n, A), returns the present value (P) of a series of n payments of A dollars each, at interest rate i. The direct use of Equation (10.14) and the use of the PV function are illustrated in the next example.

Example 10.9 Present Value of a Proposed Investment in Excel

A company is considering investing $1 million in a new product. This investment will result in a return of $140,000 at the end of each year for the next 12 years. The company normally expects an 8 percent return per year, compounded annually, on its investments. Should the company invest in this product? Solve using Excel.

We can solve this problem by calculating the present value of the 12 yearly payments of $140,000 each and then comparing this value with the required initial cost of $1 million. If the present value of the 12 payments is higher than the actual initial cost, the investment would be desirable.

	B11		=	=PV(B5,B7,B3)						
	A	B	C	D	E	F	G	H	I
1	Present Value of an Investment								
2									
3	A =	140,000							
4									
5	i =	0.08							
6									
7	n =	12							
8									
9	P =	1,055,050.92							
10									
11	P =	-1,055,050.92							
12									
Ready									

Figure 10.13

Figure 10.13 contains an Excel worksheet showing the solution, $P = \$1,055,051$ (rounded to the nearest dollar). For illustrative purposes, the solution has been obtained two different ways: using Equation (10.14) directly (shown in cell B9) and using the PV function (cell B11). The value obtained using the PV function is negative, to indicate a cash outflow. (When using Excel financial functions, the signs of the cash flow components are consistent with the sign convention in cash flow diagrams, as in Fig. 10.11.) Since the present value of the proposed investment (\$1,055,051) exceeds its cost (\$1 million), we conclude that the company *should* invest in this product.

Figure 10.14 shows the cell formulas used to obtain the results in Fig. 10.13.

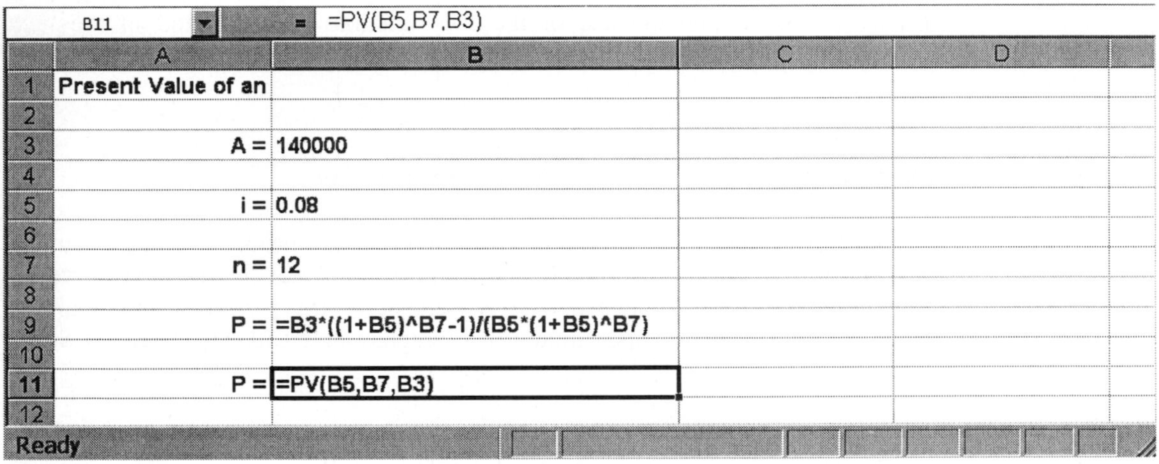

B11		=	=PV(B5,B7,B3)		
	A	B		C	D
1	Present Value of an				
2					
3	A =	140000			
4					
5	i =	0.08			
6					
7	n =	12			
8					
9	P =	=B3*((1+B5)^B7-1)/(B5*(1+B5)^B7)			
10					
11	P =	=PV(B5,B7,B3)			
12					
Ready					

Figure 10.14

Another cash flow pattern that has been studied extensively is the future value of n uniform payments. The payments can be viewed as cash outflows, and the accumulation as a cash inflow. This cash flow pattern represents a bank savings plan, where a fixed amount of money, A, is deposited at the end of each month, accumulating interest at an annual rate, i, compounded m times per year. The accumulation, F, is then withdrawn at the end of the year. Figure 10.15 shows the cash flow diagram.

Equation (10.15) provides the relationship between F and A. This equation can be implemented directly in Excel, though Excel includes an equivalent library function, $FV(i, n, A)$. The use of Equation (10.15) and the FV function are illustrated in the next example.

$$F = A\left[\frac{(1+i/m)^n - 1}{(i/m)}\right]$$

(10.15)

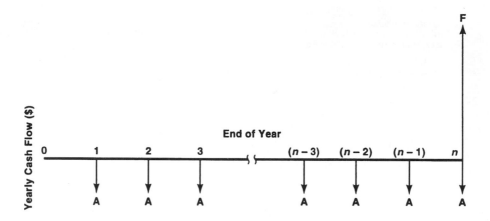

Figure 10.15

Example 10.10 Future Value of a Series of Uniform Payments

A student has opened a savings plan at a local bank. The bank requires the student to deposit $100 into the savings plan at the end of each month (January through November) for 11 months. The student then receives the total amount of the accumulation at the end of the calendar year (end of December). How much will the student accumulate, if the annual interest rate is 5 percent, compounded monthly?

Equation (10.15) can be used to answer this question. Note, however, that Equation (10.15) applied to this problem would provide the amount of money accumulated from 12 uniform monthly payments. The last payment would occur at the end of December, at the same time as the withdrawal (a mathematical artifice). In reality, only 11 payments will be made. Hence, we must subtract the amount of one monthly payment ($100) from the calculated value of F obtained from Equation (10.15).

Substituting into Equation (10.15), we obtain

$$F = \$100 \left[\frac{(1+0.05/12)^{12}}{(0.05/12)} \right] = \$1227.89$$

Subtracting $100 for the nonexistent December payment, we conclude that the accumulation will be $1127.89.

Figure 10.16 shows an Excel worksheet containing the solution. The value shown in cell B11 was obtained directly from Equation (10.15), whereas the value in cell B13 was obtained using the FV function. Both methods result in a value of $F = \$1127.89$, which is consistent with our earlier result.

	B13	▼		=	=FV(B5/B7,B9,-B3)-B3						
	A	B	C	D	E	F	G	H	I	J	
1	Future Value of a Monthly Savings Plan										
2											
3	A =	100									
4											
5	i =	0.05									
6											
7	m =	12									
8											
9	n =	12									
10											
11	F =	1,127.89									
12											
13	F =	1,127.89									
14											
Ready											

Figure 10.16

	B13	▼	=	=FV(B5/B7,B9,-B3)-B3		
	A		B	C	D	E
1	Future Value of a M					
2						
3	A =		100			
4						
5	i =		0.05			
6						
7	m =		12			
8						
9	n =		12			
10						
11	F =		=B3*((1+B5/B7)^B9.			
12						
13	F =		=FV(B5/B7,B9,-B3)-			
14						
Ready						

Figure 10.17

Figure 10.17 shows the corresponding cell formulas. The formula shown in cell B11 corresponds to Equation (10.15). Note that one monthly payment has been subtracted to account for the nonexistent December payment.

Cell B13 shows the formula using the FV function. The function arguments represent the monthly interest rate, the number of compounding periods, and the monthly payment, respectively. Notice that the monthly payment is entered as a negative quantity since it represents a cash outflow. Also, note that one monthly payment has been subtracted from the FV function to correct for the nonexistent December payment.

Problems

10.15 Redraw the cash flow shown in Fig. 10.12 from a borrower's perspective.

10.16 Redraw the cash flow shown in Fig. 10.15 from the bank's perspective.

10.17 For the investment described in Example 10.9, how much would the company have to recover each year in order to break even after 12 years? Solve using Excel.

10.18 An engineer has accumulated $300,000 in a retirement plan and is now considering an early retirement. To do so, the engineer plans to withdraw a fixed amount of money at the end of each year for 20 years. If the accumulated money earns interest at 6 percent per year, compounded annually, what is the maximum amount that can be withdrawn each year? Solve using Excel.

10.19 An electronics firm is planning to invest $1,500,000 to develop a new product. It is assumed that the company will receive a return of $250,000 a year for 10 years from the sale of this product, beginning at the end of the first year. To what interest rate does this correspond, assuming annual compounding? Is this a good investment? Solve using Excel.

10.20 A company plans to borrow $150,000 at 9 percent per year, compounded annually, in order to finance the installation of pollution reduction equipment. The loan is to be repaid in 8 equal annual payments. Using Excel, determine the amount of each payment.

10.21 Suppose the company described in Prob. 10.20 cannot repay more than $15,000 each year. For how long should the loan be taken, assuming an interest rate of 9 percent per year, compounded annually?

10.22 An engineer plans to borrow $80,000 in order to buy a house. Suppose the interest rate is 9 percent per year, compounded monthly.
- (*a*) What will be the engineer's monthly payment if the payback period is 30 years?
- (*b*) How much interest will be paid on the loan over the entire 30-year period?

Solve using Excel.

10.23 A recent engineering graduate is planning to buy a sports car. To do so, he must borrow $12,000, which he will repay monthly over a four-year period. Using Excel, determine (*a*)   The monthly payment if the bank charges interest at the rate of 10 percent per year, compounded monthly.

(*b*)   The total amount of interest paid over the entire life of the loan.

10.24 An engineer deposits $3000 into a retirement account at the end of each year.   The account pays interest at the rate of 6 percent per year, compounded annually. Using Excel, determine

(*a*)   How much money will have accumulated over 30 years.

(*b*)   How long it will take for the engineer to accumulate $300,000.

10.25 An engineer deposits a predetermined amount of money into a savings account at the end of each year. The bank pays interest at the rate of 6 percent per year, compounded annually. How much money must the engineer save each year in order to accumulate $300,000 after 30 years? Solve using Excel.

10.26 Resolve Prob. 10.25 assuming that the money is deposited at the *beginning* of each year. Solve using Excel. Compare your answer with the result obtained in Prob. 10.25.

10.27 A country has sufficient coal reserves to last for 400 years at the current rate of consumption. It is expected, however, that the rate of consumption will increase by 5 percent each year. If this is so, how long will it take for the country's coal reserves to be depleted? Solve using Excel.

10.28 A chemical company has developed a new lightweight, high-strength plastic.   The company is planning to spend $20 million for a manufacturing plant in order to begin full-scale production. As a result, the company expects to receive $2.4 million a year for 10 years, beginning at the end of the first year. Use Excel to determine the corresponding interest rate, assuming annual compounding.

10.29 A student has just won $1 million in the state lottery. She is given the choice of receiving $50,000 a year for 20 years or accepting a lump-sum payment of $650,000. Suppose the student is able to deposit the money into a money market fund that pays 5-1/2 percent per year, compounded monthly. Using Excel, determine the best investment policy.

10.30 A company is considering two investment alternatives. Each involves an initial investment of $10,000,000. It is estimated that the first alternative will return $1,800,000 at the end of each year for 10 years. The second alternative should return $1,500,000 at the end of each year for 12 years. If the company normally receives 10 percent per year on its investments, which alternative should the company select? Solve using Excel.

10.31 Suppose the company described in Prob. 10.30 anticipates that it can receive 13 percent per year on its investments during the next 12 years. If so, should the company invest in either of the proposed alternatives? If so, which one? If not, why not? Solve using Excel.

10.4 IRREGULAR CASH FLOWS

The cash flows associated with engineering investments usually do not follow the simple cash flow patterns described in the last section. Rather, they tend to be irregular, with a different payment corresponding to each payment period. Thus, it is much more difficult to evaluate the present value of an irregular cash flow since the simple mathematical relationships presented in the last section do not apply. The use of a spreadsheet is particularly helpful in situations of this type.

One way to evaluate the present value of an irregular cash flow within an Excel worksheet is to enter the cash flow components in one column and the present value of each component in an adjoining column. Equation (10.7), or a related equation [i.e., Equations (10.9) through (10.11)] can be used to determine the present value of each component. The present value of the entire cash flow can then be obtained by summing the present value of the individual components. The following example illustrates the procedure.

Example 10.11 Present Value of an Irregular Cash Flow in Excel

A manufacturing company is considering bringing out a new product. To do so, the company must invest $1 million at the beginning of year 1 (i.e., at the end of year 0) and another $8 million one year later. Marketing studies suggest that this investment will generate an irregular series of revenues from years 3 through 12. The anticipated cash flow components are summarized below:

End of Year	Revenue	End of Year	Revenue
0	−$10,000,000	7	$5,000,000
1	−8,000,000	8	6,000,000
2	0	9	5,000,000
3	1,000,000	10	4,000,000
4	2,000,000	11	3,000,000
5	3,000,000	12	2,000,000
6	4,000,000	13	1,000,000

Enter this cash flow into an Excel worksheet and determine the present value of this proposed investment, assuming the company normally receives a return of 8 percent per year, compounded annually. (Note that the initial investments are shown as negative cash flow components at the end of years 0 and 1.) Based upon the present value of the cash flow, determine whether or not this new product would represent an attractive investment strategy.

Figure 10.18 shows an Excel worksheet containing the cash flow components in column B and the present value of the cash flow in column D. A bar graph of the cash flow components (column D) is also shown.

Cell D21 contains the sum of the present values of the cash flow components. This value ($2,380,571) is positive, indicating that the present value of the future cash inflows exceeds the present value of the initial investments. Hence, this new product represents an attractive investment opportunity.

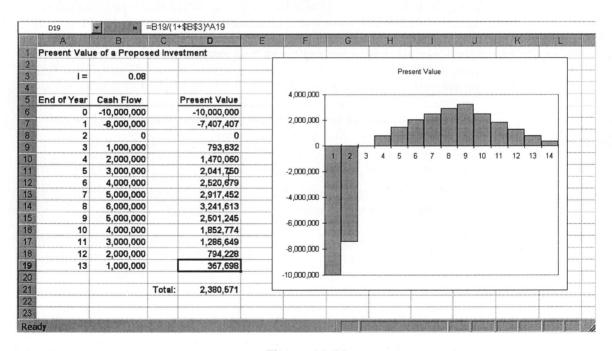

Figure 10.18

Figure 10.19 shows the cell formulas that were used to generate the present values in Fig. 10.18. The formulas shown in cells D7 through D19 each represent Equation (10.7), solving for P as a function of F (in column B), i (cell B3), and n (column A). The formula in cell D21 makes use of the SUM function to determine the sum of the individual present values.

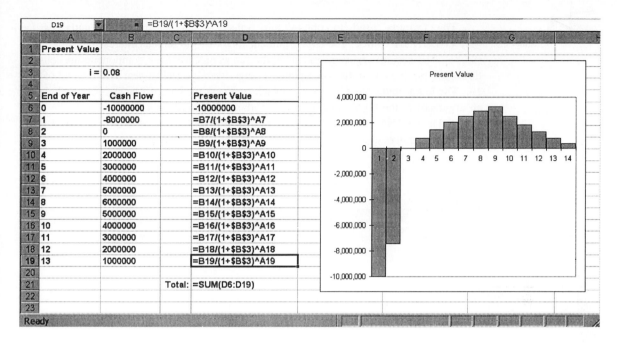

Figure 10.19

The NPV function can also be used to determine the present value of an irregular cash flow in Excel. This function is written as NPV(i, $A1$, $A2$, . . .), where i represents the interest rate, $A1$ represents the cash flow component at the end of the first period, $A2$ represents the cash flow component at the end of the second period, and so on. The total number of cash flow components (n) need not be specified.

Note that the arguments for the NPV function do not include a cash flow component at the beginning of the first year (i.e., at the end of year 0). Hence, if a cash flow includes an initial investment at the beginning of the first year (as many do), this value must be subtracted from the value that is returned by the NPV function. The next example illustrates this procedure.

Example 10.12 Use of the NPV Function in Excel

Use Excel's NPV function to determine the present value of the cash flow given in Example 10.11.

Figure 10.20 shows an Excel worksheet containing the desired solution. The present value of the cash flow appears in cell B21 as $2,380,571. This value agrees with that obtained in Example 10.11, as expected.

The formula used to obtain this value is shown in the formula bar as =NPV(B3,B7:B19)+B6. The first argument refers to the annual interest rate in cell B3 (0.08). The second argument indicates the range of cash flow components in cells B7 through B19. Note that these components range from the end of year 1 to the end of year 13. The initial investment in cell B6, which occurs at the end of year 0, is not included as an argument to the NPV function. Hence, this value must be subtracted from the value returned by the NPV function. (The cell address B6 is *added* to the cell formula because it contains a negative quantity, resulting in a decrease in the present value of the cash flow.)

	A	B	C	D	E	F	G	H	I	J	K
A1		■ Present Value of a Proposed Investment									
1	Present Value of a Proposed Investment										
2											
3	i =	0.08									
4											
5	End of Year	Cash Flow									
6	0	-10,000,000									
7	1	-8,000,000									
8	2	0									
9	3	1,000,000									
10	4	2,000,000									
11	5	3,000,000									
12	6	4,000,000									
13	7	5,000,000									
14	8	6,000,000									
15	9	5,000,000									
16	10	4,000,000									
17	11	3,000,000									
18	12	2,000,000									
19	13	1,000,000									
20											
21	NPV =	2,380,571									
22											
23											
Ready											

Figure 10.20

Use of the NPV function is particularly convenient when comparing one investment opportunity with another. The cash flow associated with each investment can be entered into a separate column in an Excel worksheet, with the present value of the cash flow at the bottom of the column. The investment with the largest positive present value is selected as the most desirable. If none of the proposed investments result in a positive present value, none of the proposals are considered desirable. (The company should then invest in whatever it normally does to earn the interest rate used in the calculations.)

Example 10.13 Comparing Two Investment Opportunities

A company has developed two promising new products but can afford to manufacture and market only one of them. Each product requires an initial investment of $3,500,000, and each is expected to provide a return of $7,200,000 over a six-year period. The expected revenues are distributed differently with respect to time, however, as indicated by the following cash flows. If the company can normally earn the equivalent of 10 percent per year, compounded annually, in which product should the company invest? Solve using Excel.

End of Year	Revenue, Proposal A	Revenue, Proposal B
0	-$3,500,000	-$3,500,000
1	1,200,000	600,000
2	1,200,000	900,000
3	1,200,000	1,100,000
4	1,200,000	1,300,000
5	1,200,000	1,500,000
6	1,200,000	1,800,000

C15   =NPV($B$3,C8:C13)+C7

	A	B	C	D	E	F	G	H	I	J	K
1	Comparing Two Investment Opportunities										
2											
3	i =	0.1									
4											
5	End of		Cash Flow,		Cash Flow,						
6	Year		Proposal A		Proposal B						
7	0		-3,500,000		-3,500,000						
8	1		1,200,000		600,000						
9	2		1,200,000		900,000						
10	3		1,200,000		1,100,000						
11	4		1,200,000		1,300,000						
12	5		1,200,000		1,500,000						
13	6		1,200,000		1,800,000						
14											
15	NPV =		1,726,313		1,451,055						
16											
17											

Ready

Figure 10.21

Figure 10.21 contains the solution. The cash flow associated with Proposal A is entered in Column C, and the cash flow associated with Proposal B is entered in Column E. Beneath each cash flow is the present worth, as determined by the NPV function.

Thus, cell C15 shows that the present value for Proposal A is $1,726,313. This value was obtained using the formula =NPV($B$3,C8:C13)+C7. Similarly, cell E15 shows a value of $1,451,055 for Proposal B, obtained from the formula =NPV($B$3,E8:E13)+E7. In each formula, the first argument refers to the annual interest rate in cell B3, and the second argument indicates the range of cells containing the cash flow components. Note that the initial investment (at end of year 0) is not included in the range of cash flow components and must therefore be added separately at the end of each formula.

The present value of each proposal is seen to be positive; hence, either proposal represents a viable investment opportunity. Since the present value of Proposal A exceeds the present value of Proposal B, the correct decision is to select proposal A.

Problems

10.32 Determine the present value of the following cash flow, using an interest rate of 6 percent per year, compounded annually. Solve with Excel, using

(*a*)   Equation (10.7).

(*b*)   The NPV function.

End of Year	Revenue
0	−$1,00,000
1	0
2	200,000
3	250,000
4	300,000
5	350,000
6	400,000

10.33 Resolve Prob. 10.32 using an interest rate of 12 percent per year, compounded annually. Compare your answer with that obtained in Prob. 10.32. Explain why the two solutions are fundamentally different.

10.34 Using Excel, determine which of the following two investment opportunities is more desirable, based upon their present values. Assume an interest rate of 12 percent per year, compounded annually.

End of Year	Revenue, Proposal A	Revenue, Proposal B
0	−$500,000	0
1	200,000	−$800,000
2	200,000	0
3	200,000	400,000
4	200,000	400,000
5	0	400,000

Compare with the conclusion that would be reached if the time value of money were not taken into account.

10.35 Use Excel to determine which of the following two investment opportunities is more desirable, based upon their present values. Assume an interest rate of 10 percent per year, compounded annually. (Note that both cash flows sum to the same value, though the order of the components is different for each proposal.)

End of Year	Revenue, Proposal A	Revenue, Proposal B
0	-$25,000	-$25,000
1	20,000	10,000
2	12,000	12,000
3	10,000	15,000
4	15,000	20,000
5	20,000	20,000

10.36 Use Excel to determine which of the following two investment opportunities is the most desirable.

End of Year	Revenue, Proposal A	Revenue, Proposal B
0	-$500,000	-$500,000
1	150,000	50,000
2	150,000	70,000
3	150,000	100,000
4	150,000	200,000
5	150,000	400,000

Assume annual compounding, using an annual interest rate of

(*a*)   0 percent

(*b*)   6 percent

(*c*)   12 percent

Is the choice influenced by the interest rate?

10.37 Use Excel to determine which of the following two investment opportunities is the most desirable, based upon an interest rate of 8 percent per year, compounded annually.

End of Year	Revenue, Proposal A	Revenue, Proposal B
0	−$750,000	−$450,000
1	0	−400,000
2	150,000	100,000
3	150,000	150,000
4	200,000	200,000
5	200,000	250,000
6	200,000	250,000
7	200,000	250,000

How would the two investments compare if the time value of money were not considered?

10.38 A car manufacturer plans to invest $27 million in order to expand its production facilities. Two different proposals are being considered. The first proposal is to expand a current midwestern assembly plant. This will allow the manufacturer to produce an additional 50,000 cars a year, at an average net profit of $200 per car. The second proposal is to build a new assembly plant on the West Coast. This facility will have an annual output of only 40,000 cars per year, but the profit is expected to be $300 per car.

The midwestern plant can begin production during the first year (i.e., the same year that the initial investment is made). However, the West Coast plant requires a one-year startup, so that production will not begin until the second year. For planning purposes, each assembly plant is assumed to have a lifetime of five years of actual production.

If the company uses an annual interest rate of 12 percent, compounded annually, which is the better proposal? Solve using Excel.

10.39 In the preceding problem, suppose the expected profit is reduced to $265 per car at the west coast plant because of higher labor and utility costs. Which proposal will now be more desirable?

10.40 A large regional office of a computer company requires an additional 400,000 square feet of storage space. One proposal is to erect a 400,000 square foot building now, at a cost of $25 per square foot. Another proposal is to erect a 250,000 square foot building now at a cost of $26 per square foot, and to add another 150,000 square feet three years from now at a cost of $30 per square foot. Which is the better proposal, based upon an annual interest rate of 8 percent? Solve using Excel.

10.41 For the situation described in the previous problem, suppose the annual interest rate is 12 percent rather than 8 percent. Which would be the better proposal?

10.5 INTERNAL RATE OF RETURN

The *internal rate of return* (IRR) is another criterion that is often used to compare one investment opportunity with another. Like net present value (NPV), it is derived from the cash flow associated with an investment. Unlike net present value, however, it is not necessary to specify an interest rate when determining the internal rate of return. The internal rate of return is a popular criterion for this reason.

To understand the internal rate of return, let us begin with a cash flow consisting of an initial cash outflow (i.e., an initial investment), followed by a series of cash inflows, not necessarily uniform, whose sum exceeds the initial cash outflow. Hence, if we neglect the time value of money (i.e., if we select an interest rate of zero), the cash flow will have a positive net value.

Now suppose we plot the net present value of the cash flow as a function of interest rate, as shown in Fig. 10.22. We see that the net present value decreases with increasing interest rate, eventually crossing the abscissa and becoming negative. The internal rate of return is the *crossover point*; that is, the value of the interest rate at which the net present value is zero. Fig. 10.22, for example, shows an internal rate of return slightly in excess of 15 percent.

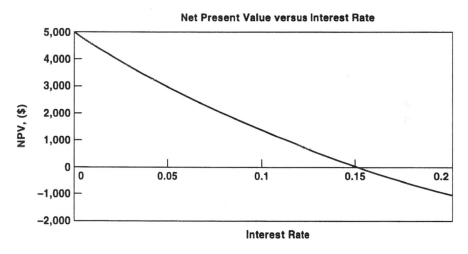

Figure 10.22

When the interest rate is zero, the net present value is equal to the algebraic sum of the cash flow components. This value will be positive when the sum of the cash inflows exceeds the initial cash outflow. When the interest rate becomes very large, however, the net present value will asymptotically approach the (negative) value of the initial cash outflow.

Example 10.14 Calculating the Internal Rate of Return

Determine the internal rate of return for the following cash flow within an Excel worksheet. To do so, proceed as follows:

(a) Determine the present value of the cash flow for each of several different interest rates.

(b) Plot the net present value of the cash flow as a function of interest rate.

(c) From the resulting graph, determine the internal rate of return.

End of Year	Revenue
0	−$100,000
1	15,000
2	20,000
3	25,000
4	30,000
5	35,000
6	40,000

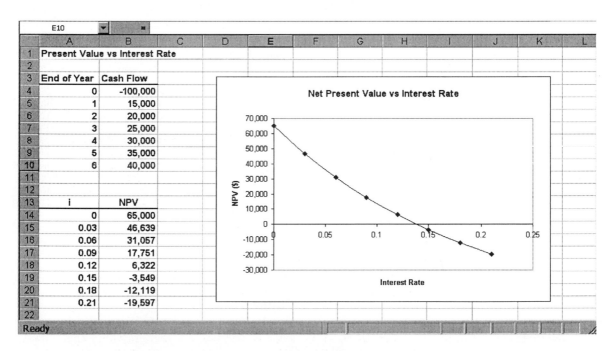

Figure 10.23

Figure 10.23 contains the desired information. The cash flow components are entered into cells B4 through B10. Cells A14 through A21 contain various interest rates. The corresponding NPV values, obtained using the NPV function, are shown in Cells B14 through B21. Notice that the value in cell B14, which corresponds to an interest rate of zero, is simply the sum of the individual cash flow components. As the interest rate increases, the NPV value decreases, becoming negative for interest rate values above 0.12.

The figure also contains a plot of NPV versus interest rate. The NPV value decreases with increasing interest rate, as expected, crossing the abscissa at about $i = 0.14$. Hence, we conclude that the internal rate of return is about 0.14 (14 percent).

Figure 10.24 shows the cell formulas that were used to obtain the NPV values shown in Fig. 10.23.

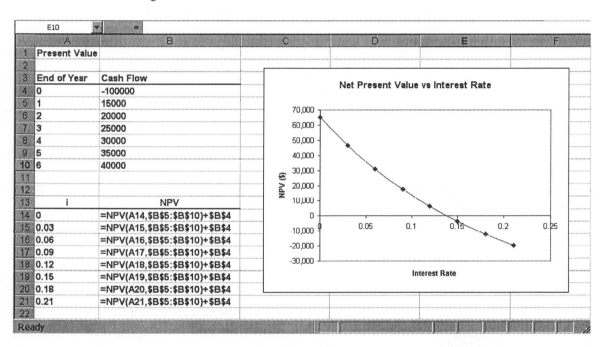

Figure 10.24

The internal rate of return, like the present value, is particularly useful when selecting among multiple investment opportunities. The cash flow having the largest internal rate of return is considered the more desirable alternative. In most situations, this is equivalent to selecting the cash flow that has the greatest net present value.

Excel includes a library function (IRR) that determines the internal rate of return directly, thus avoiding the need to calculate the present value of the cash flow as a function of interest rate. This function is written as IRR$(P, A1, A2, \ldots)$, where P represents the initial cash outflow and $A1, A2, \ldots$ represent the various cash inflows. Note that this function, in contrast to the NPV function, *does* include the initial cash outflow in its list of arguments. All cash outflows, including the initial cash outflow, must be expressed as negative quantities.

Example 10.15 Comparing Two Investment Opportunities Using IRR

Example 10.13 presented the cash flows associated with two different investment opportunities and compared them on the basis of their present value. Let us now compare these alternatives on the basis of their internal rate of return. To do so, we will modify the Excel worksheet shown in Fig. 10.21 to include the use of the IRR function as well as the NPV function. The cash flows are reproduced below for your convenience.

End of Year	Revenue, Proposal A	Revenue, Proposal B
0	−$3,500,000	−$3,500,000
1	1,200,000	600,000
2	1,200,000	900,000
3	1,200,000	1,100,000
4	1,200,000	1,300,000
5	1,200,000	1,500,000
6	1,200,000	1,800,000

	G21			▼	■						
	A	B	C	D	E	F	G	H	I	J	K
1	Comparing Two Investment Opportunities										
2											
3	i =	0.1									
4											
5	End of		Cash Flow,		Cash Flow,						
6	Year		Proposal A		Proposal B						
7	0		-3,500,000		-3,500,000						
8	1		1,200,000		600,000						
9	2		1,200,000		900,000						
10	3		1,200,000		1,100,000						
11	4		1,200,000		1,300,000						
12	5		1,200,000		1,500,000						
13	6		1,200,000		1,800,000						
14											
15	NPV =		1,726,313		1,451,055						
16											
17	IRR =		26%		21%						
18											
Ready											

Figure 10.25

Figure 10.25 shows an Excel worksheet containing the two cash flows. Both the present value and the internal rate of return are shown for each of the cash flows. The present values were determined using the NPV function, and the internal rates of return were determined with the IRR function. The conclusions based upon each criterion are consistent, with Proposal A appearing more attractive in each case. That is, the present valuc of Proposal A ($1,726,313) is higher than that of of Proposal B ($1,451,055), and the internal rate of return of proposal A (26 percent) exceeds that of Proposal B (21 percent).

Figure 10.26 reveals the cell formulas used to generate the NPV and IRR values shown in Fig. 10.25. Notice that the cash flow components are entered into the two formulas differently. In particular, the initial cash outflow (end of year 0) is not included in the list of arguments in the NPV function. Instead, this value is added to the value of the future cash inflows that is returned by the NPV function. (Since the initial cash outflow is negative, we are really subtracting this dollar amount rather than adding it.) The IRR function, on the other hand, accepts *all* of the cash flow components, including the initial cash outflow, as arguments. Thus, there is no need to treat this cash flow component separately.

	A	B	C	D	E	
1	Comparing Tw					
2						
3	i =	0.1				
4						
5	End of		Cash Flow,		Cash Flow,	
6	Year		Proposal A		Proposal B	
7	0		-3500000		-3500000	
8	1		1200000		600000	
9	2		1200000		900000	
10	3		1200000		1100000	
11	4		1200000		1300000	
12	5		1200000		1500000	
13	6		1200000		1800000	
14						
15	NPV =		=NPV($B$3,C8:C13)+C7		=NPV($B$3,E8:E13)+E7	
16						
17	IRR =		=IRR(C7:C13)		=IRR(E7:E13)	
18						

Figure 10.26

The internal rate of return is actually a root of a polynomial equation in terms of the interest rate (*i*). Thus, in order to determine the internal rate of return, we must solve a polynomial equation for a positive, real root. This is carried out automatically when using the IRR function. Remember, however, that some polynomials have no real roots, while others have multiple real roots. Therefore, there are some cash flows for which the internal rate of return cannot be determined. Moreover, other cash flows have multiple values for the internal rate of return. (In this situation, the usual procedure is to select the smallest positive value.) This places a limitation on the use of the internal rate of return as a basis for comparing cash flows. The present value, on the other hand, can always be calculated for any cash flow.

Convergence can also be a problem when solving for the internal rate of return. Since the governing polynomial equation is generally solved iteratively (see Sec. 7.6), there is no assurance that the solution will converge to the smallest positive real root or to any real root at all. This limitation applies to the use of the IRR function as well as a solution that might be attempted using Excel's Goal Seek feature. On the other hand, a graphical solution, though tedious, can always be obtained if a real root exists.

Example 10.16 Internal Rate of Return as the Root of a Polynomial

In Example 10.14, we calculated the internal rate of return for the following cash flow:

End of Year	Revenue
0	−$100,000
1	15,000
2	20,000
3	25,000
4	30,000
5	35,000
6	40,000

Show that the equation for the present value of this cash flow is a polynomial in i and that the internal rate of return is a positive real root of this polynomial.

Let us make use of Equation (10.7) to determine the present value of this cash flow. Thus, we can write the present value as

$$NPV = -100,000 + 15,000\frac{1}{(1+i)} + 20,000\frac{1}{(1+i)^2} + 25,000\frac{1}{(1+i)^3} +$$

$$30,000\frac{1}{(1+i)^4} + 35,000\frac{1}{(1+i)^5} + 40,000\frac{1}{(1+i)^6}$$

The internal rate of return is the value of i that causes NPV to equal zero. Hence,

$$-100,000 + \frac{15,000}{(1+i)} + \frac{20,000}{(1+i)^2} + \frac{25,000}{(1+i)^3} + \frac{30,000}{(1+i)^4} + \frac{35,000}{(1+i)^5} + \frac{40,000}{(1+i)^6} = 0$$

If we multiply this equation by $(1 + i)^6$, we obtain

$$-100{,}000(1+i)^6 + 15{,}000(1+i)^5 + 20{,}000(1+i)^4 + 25{,}000(1+i)^3 +$$

$$30{,}000(1+i)^2 + 35{,}000(1+i) + 40{,}000 = 0$$

This equation is a sixth-degree polynomial in terms of the interest rate, i. It can be solved several different ways, including the techniques described in Chap. 7 (see Prob. 10.44 below).

Problems

10.42 Enter the cash flow given in Example 10.14 into an Excel worksheet. Determine the internal rate of return using the IRR function, as described in Example 10.15. Compare your answer with that obtained in Example 10.14.

10.43 The following cash flow was originally given in Example 10.11. It is reproduced below for your convenience.

End of Year	Revenue	End of Year	Revenue
0	-$10,000,000	7	$5,000,000
1	-8,000,000	8	6,000,000
2	0	9	5,000,000
3	1,000,000	10	4,000,000
4	2,000,000	11	3,000,000
5	3,000,000	12	2,000,000
6	4,000,000	13	1,000,000

(a) Enter this cash flow into an Excel worksheet. Using the NPV function, determine the present value of the cash flow for several different interest rates, as in Example 10.14 (see Fig. 10.23). Choose a large enough range of interest rates (beginning with $i = 0$) so that the present value becomes negative for large interest rates.

(b) Plot the present value against interest rate, as in Fig. 10.23.

(c) Determine the internal rate of return by observing where the curve crosses the abscissa (i.e., at what interest rate the present value of the cash flow is zero).

(d) Determine the internal rate of return using the IRR function. Compare your answer with that obtained in part (c).

10.44 Using Excel's Goal Seek feature, determine the smallest positive real root of the polynomial given in Example 10.16. Compare this value with the internal rate of return obtained in Example 10.14 and in Prob. 10.42.

Hint: Rewrite the polynomial in terms of the variable u, where $u = (1 + i)$. Solve for u and then determine the internal rate of return as $i = (u - 1)$. Is the root of the polynomial equivalent to the internal rate of return?

10.45 Each of the following previous problems contains cash flows for two different investment opportunities. For each problem, construct an Excel worksheet to determine the more desirable investment opportunity, based upon the internal rate of return. Compare with the results obtained previously, based upon present value.

(*a*) Prob. 10.34

(*b*) Prob. 10.35

(*c*) Prob. 10.37

10.46 The following cash flows describe two competing investment opportunities. (These cash flows were presented previously, in Prob. 10.36.)

End of Year	Revenue, Proposal A	Revenue, Proposal B
0	−$500,000	−$500,000
1	150,000	50,000
2	150,000	70,000
3	150,000	100,000
4	150,000	200,000
5	150,000	400,000

(*a*) Enter both cash flows into an Excel worksheet.

(*b*) Determine which investment is more attractive, based upon present value using an interest rate of 5 percent, compounded annually.

(*c*) Determine which investment is more attractive, based upon the internal rate of return.

(*d*) Explain any apparent anomalies by plotting present worth against interest rate for each of the cash flows.

10.47 Using Excel, solve Prob. 10.38 on the basis of internal rate of return. Compare with the results obtained earlier, using the present value and an annual interest rate of 12 percent.

ADDITIONAL READING

Blank, L. T. and A. J. Tarquin. *Engineering Economy.* 3d ed. New York: McGraw-Hill, 1989.

Degarmo, E. P., W. G. Sullivan, and J. A. Bontadelli. *Engineering Economy.* 9th ed. New York: Macmillan, 1993.

Newnan, D. G. *Engineering Economic Analysis.* 4th ed. San Jose, CA: Engineering Press, Inc., 1991.

Park, C. S. *Contemporary Engineering Economics.* Reading, MA: Addison-Wesley, 1993.

Riggs, J. L. and T. M. West. *Engineering Economics.* 3d ed. New York: McGraw-Hill, 1986.

Thuesen, G. J. and W. J. Fabrycky. *Engineering Economy.* 8th ed. Englewood Cliffs, NJ: Prentice-Hall, 1993.

CHAPTER 11

FINDING AN OPTIMUM SOLUTION

Many problems in engineering have multiple solutions, and selecting among them can be a major task. Normally, there is some criterion, such as cost, profit, weight or yield, that distinguishes one solution from another. When expressed mathematically, this is known as an *objective function*. (It may also be called the *cost function* or the *performance criterion*.) In addition, there are certain conditions, such as conservation laws, capacity restrictions or other technical relationships, that must always be satisfied. In mathematical form, these conditions are known as *constraints*. (They are also referred to as *auxiliary conditions* or *design requirements*.)

When selecting an optimum solution, the goal is to determine the particular solution that causes the objective function to be maximized or minimized while satisfying all of the constraints. Suppose, for example, we want to determine the lightest-weight bridge that will satisfy a given set of design requirements. The objective function will be the weight of the bridge, expressed in terms of the dimensions of the major structural members. The constraints will be the various force and moment relationships that must be satisfied. The problem is to determine the dimensions of the major structural members that will minimize the weight, subject to the applicable constraints. Problems of this type are known as *optimization problems*.

Excel includes a provision for solving a wide variety of optimization problems. Both maximization and minimization problems, linear or nonlinear, with or without constraints, can be solved very easily. However, the *formulation* of an optimization problem generally requires considerable care. In this chapter, we will see how optimization problems are formulated and then solved using Excel's **Solver** feature.

11.1 OPTIMIZATION PROBLEM CHARACTERISTICS

In general terms, an optimization problem can be written in the following manner: Determine the values of the independent variables x_1, x_2, $\cdots$, x_n that will maximize or minimize the *objective function*

$$y = f(x_1, x_2, \cdots, x_n) \tag{11.1}$$

The independent variables are often referred to as *decision variables* or *policy variables*. Also, the objective function is sometimes called the *performance criterion*, *performance index*, *profit function* (for a maximization problem), or *cost function* (for a minimization problem).

In many problems, the optimum solution must satisfy one or more auxiliary conditions, called *constraints*. Each constraint may be expressed as an actual equation or as an inequality; that is,

$$g_j(x_1, x_2, \cdots, x_n) = 0 \tag{11.2}$$

or as

$$g_j(x_1, x_2, \cdots, x_n) \leq 0 \tag{11.3}$$

or

$$g_j(x_1, x_2, \cdots, x_n) \geq 0 \tag{11.4}$$

for $j = 1, 2, \cdots, m$, where m is the total number of constraints.

In addition, the permissible values of the independent variables are usually restricted to being nonnegative. Thus,

$$x_i \geq 0, \qquad i = 1, 2, \cdots, n. \tag{11.5}$$

For now, we will confine ourselves to problems that involve only two independent variables; that is, for which $n = 2$. Problems of higher dimensionality will be considered in the next section, where we discuss solution procedures based upon the use of Excel.

These concepts will become clearer in the following two examples, which present two different types of optimization problems and their respective solution characteristics.

Example 11.1 Scheduling Production to Maximize Profit

Suppose a company manufactures two products, A and B, which can be sold for $120 per unit and $80 per unit, respectively. Management requires that at least 1000 units be manufactured each month. Product A requires five hours of labor per unit, and product B requires three hours. The cost of labor is $12 per hour, and a total of 8000 hours are available per month. Determine a monthly production schedule that will maximize the company's profit.

To express this problem mathematically, we begin by defining the following variables:

x_1 = units of product A manufactured per month

x_2 = units of product B manufactured per month

These variables are referred to as decision variables or policy variables.

Now consider the objective function, which will represent the profit. Let us write the objective function as

$$y = [120 - (5 \times 12)] \, x_1 + [80 - (3 \times 12)] \, x_2$$

or

$$y = 60 \, x_1 + 44 \, x_2 \tag{11.6}$$

Next, we develop the constraints. The minimum production requirement can be expressed as

$$x_1 + x_2 \geq 1000 \tag{11.7}$$

Notice that this condition is expressed as an *inequality* rather than an actual equation (i.e., a strict equality) since the problem states that *at least* 1000 units must be manufactured per month. If the problem required a production schedule of *exactly* 1000 units per month, Equation (11.7) would be written as a strict equality rather than an inequality.

Similarly, the restriction on labor availability can be expressed as

$$5x_1 + 3x_2 \leq 8000 \tag{11.8}$$

Finally, we must restrict the decision variables to being nonnegative; that is,

$$x_1, x_2 \geq 0 \tag{11.9}$$

We can now state the optimization problem in the following concise manner: Determine the values of x_1 and x_2 that will maximize Equation (11.6), subject to the auxiliary conditions (constraints) expressed by Equations (11.7) through (11.9).

It is instructive to view this problem graphically. Thus, Fig. 11.1 shows a plot of x_2 versus x_1. Only the upper right quadrant is shown since this is the region that satisfies the nonnegativity conditions expressed by Equation (11.9). The lower diagonal line represents Equation (11.7). Any point on or above this line will satisfy the minimum production requirement. Similarly, the upper diagonal line represents Equation (11.8). Any point on or below this line will represent the labor availability restriction.

Notice that the three constraints form a polyhedron, which is shown as a shaded area in Fig. 11.1 This is called the *feasible region*. Any point within the feasible region will satisfy all of the constraints.

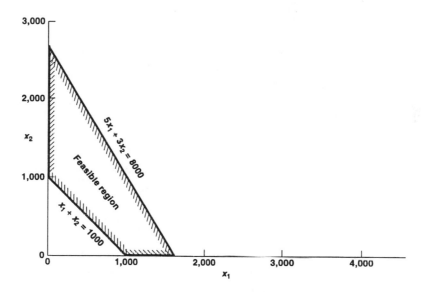

Figure 11.1

Now let us superimpose several lines representing different values of the objective function over the feasible region, as shown in Fig. 11.2. The different values of the objective function are shown as a family of parallel lines. Note that the value of the objective function increases as we move from lower left to upper right. We seek the line representing the largest possible value of the objective function that is within or on the feasible region.

The solution is the line representing the value $y = 11.7 \times 10^4$, which touches the upper left corner of the feasible region at $x_1 = 0$ and $x_2 = 2667$. Any point beneath this line would represent a lower value of y. Similarly, any point above this line would not lie within the feasible region, and the constraints would not be satisfied.

Notice that the solution falls on the line represented by Equation (11.8). Hence, the constraint represented by Equation (11.8) is a strict equality at the optimum. Therefore, we say that this constraint is binding (or is active). On the other hand, the constraint represented by Equation (11.7) is not binding (i.e., is inactive) since the optimum does not fall on this line.

We conclude, then, that the most profitable strategy is to manufacture 0 units of product A and 2667 units of product B per month. The resulting profit will be $117,000.

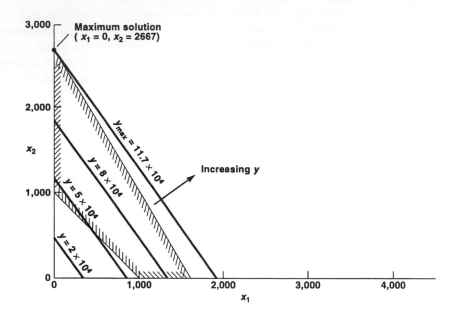

Figure 11.2

Notice that the objective function and the constraints were represented by linear (i.e., straight-line) relationships in the preceding example. Optimization problems of this type are called *linear programming* problems. Linear programming has been studied extensively, and a great deal is known about the nature of the feasible region and the optimal solution. In particular:

1. The feasible region will always be a polyhedron, if it exists (i.e., if the constraints are not inconsistent).

2. The feasible region may be closed, or it may be open (unbounded). If the feasible region is closed, the optimal solution will always occur at one of the vertices.

3. If the feasible region is open, the optimal solution may be unbounded (infinite). Note, however, that an open feasible region does *not* necessarily *require* that the objective function be unbounded.

4. The optimal solution, if it exists, may not be unique. That is, the line (or plane) representing the objective function may be parallel to one of the lines (planes) that bound the feasible region. Hence, the same solution will be obtained at two or more vertices. In such situations, the *optimal value of the objective function* will be the same at each vertex, but the optimal *policy*, that is, the values of the independent variables, will be different.

Linear programming problems can be solved by a precise, though complex, solution procedure known as the *simplex algorithm* (there are several variations). This procedure is included in Excel's Solver feature. We will see how the Solver feature is used in the next section.

In the next example, we will examine another optimization problem, which involves nonlinearities. Nonlinear problems have solution characteristics that may differ markedly from linear programming problems. Hence the procedures used to obtain solutions are more complicated than the procedures used to solve linear programming problems.

Example 11.2 A Minimum-Weight Structure

Consider a structure that consists primarily of two cylindrical load-bearing columns whose radii, in feet, are x_1 and x_2, respectively. The weight of the structure, in thousands of pounds, is given by the expression

$$y = 10 + (x_1 - 0.5)^2 + (x_2 - 0.5)^2 \tag{11.10}$$

Determine the values of x_1 and x_2 that will minimize the weight of the structure, subject to the following requirements:

1. The combined cross-sectional area of the two columns must be at least 10 square feet; that is,

$$\pi (x_1^2 + x_2^2) \geq 10 \tag{11.11}$$

2. Structural considerations require that the radius of the first column not exceed 1.25 times the radius of the second column; that is,

$$x_1 \leq 1.25\, x_2 \tag{11.12}$$

In addition, the radii must be nonnegative; that is,

$$x_1, x_2 \geq 0 \tag{11.13}$$

Figure 11.3 shows the feasible region defined by the constraints. Any point within the shaded region will satisfy all of the constraints, as expressed by Equations (11.11) through (11.13). Notice that the feasible region now has a curved boundary because of the nonlinear constraint expressed by Equation (11.11). In addition, note that the feasible region is not bounded (some linear problems also exhibit this characteristic).

The objective function represents a family of circles whose center is ($x_1 = 0.5$, $x_2 = 0.5$). Let us superimpose several circular arcs representing different values of the objective function over the feasible region, as shown in Fig. 11.4. Notice that the value of the objective function increases as we move radially outward from the center. The smallest value of the objective function that satisfies all of the constraints is the circle that is tangential to the feasible region at $x_1 = x_2 = 1.26$. This circle has a value of 11.16. Hence, the minimum weight structure that satisfies all of the imposed conditions is 11,160 lb_f. Each of the supporting columns will have a radius of 1.26 ft.

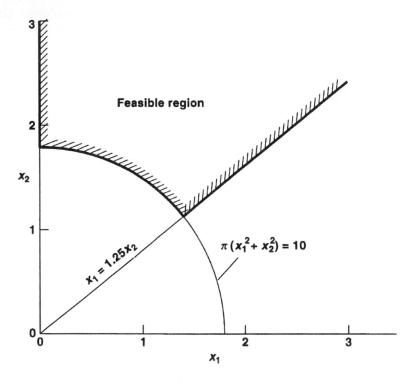

Figure 11.3

Note that the constraint represented by Equation (11.11) is active at the optimum, but the other constraint, represented by Equation (11.12), is inactive. If the family of objective functions were oriented differently, however (i.e., if the center of the circles were located elsewhere), the optimum solution might be at one of the corners, where Equation (11.11) intersects the x_2 axis or where Equation (11.11) and Equation (11.12) intersect. Moreover, the optimum solution might occur *within* the feasible region, where none of the constraints would be active.

This last example illustrates some fundamental differences between linear and nonlinear optimization problems. Namely, in nonlinear problems, the feasible region is not necessarily a polyhedron, and the optimum solution does not necessarily occur at a corner. In fact, the optimum solution may occur along one of the bounding surfaces or within the feasible region.

Moreover, some nonlinear problems exhibit multiple optimum points (called *local optima*). Thus, a maximum point obtained for this type of problem may represent the height of the nearest hill rather than the height of the highest hill around. These characteristics place greater demands on the mathematical procedures used to find the optimum of a nonlinear problem.

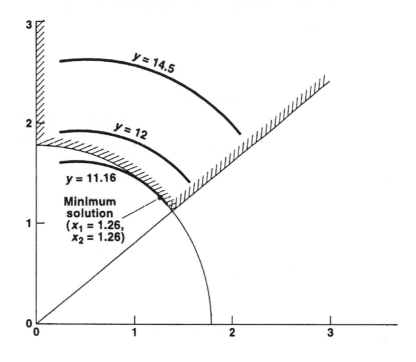

Figure 11.4

Before leaving this section, we note that engineering optimization problems generally have two distinct characteristics. First, most realistic optimization problems involve many independent variables. Such multidimensional problems cannot be represented graphically; hence, we must depend upon complex mathematical procedures to obtain an optimum solution. Our use of simple two-dimensional problems was intended only to provide insight and understanding into the nature of optimization problems.

Secondly, we should understand that engineering optimization problems tend to fall into one of the following two categories:

1. Large linear optimization problems, which may have hundreds, perhaps thousands, of independent variables. Problems involving production scheduling, inventory control, product blending, etc. often fall into this category.

2. Problems involving one or more nonlinearities. As a rule, these problems have fewer independent variables than purely linear problems, though they may be much more difficult to solve because of the nonlinearities. Engineering design problems usually fall into this category.

Excel includes mathematical solution procedures for both problem categories. We will see how they are used in the next section.

Problems

11.1 Each of the following problems is a variation of the linear optimization problem given in Example 11.1. For each problem, draw the feasible region. Then determine the optimum solution by superimposing several lines or curves that represent different values of the objective function over the feasible region. Determine both the optimum value of the objective function and the corresponding optimum policy.

(a) Minimize

$$y = 60\,x_1 + 44\,x_2$$

subject to the following constraints:

$$x_1 + x_2 \geq 1000$$
$$5\,x_1 + 3\,x_2 \leq 8000$$
$$x_1, x_2 \geq 0$$

(b) Maximize

$$y = 60\,x_1 + 30\,x_2$$

subject to the following constraints:

$$x_1 + x_2 \geq 1000$$
$$5\,x_1 + 3\,x_2 \leq 8000$$
$$x_1, x_2 \geq 0$$

(c) Maximize

$$y = 50\,x_1 + 30\,x_2$$

subject to the following constraints:

$$x_1 + x_2 \geq 1000$$
$$5\,x_1 + 3\,x_2 \leq 8000$$
$$x_1, x_2 \geq 0$$

(*d*)   Maximize

$$y = 60\,x_1 + 44\,x_2$$

subject to the following constraints:

$$x_1 + x_2 \geq 1000$$
$$5\,x_1 + 3\,x_2 \geq 8000$$
$$x_1, x_2 \geq 0$$

(*e*)   Minimize

$$y = 60\,x_1 + 44\,x_2$$

subject to the following constraints:

$$x_1 + x_2 \geq 1000$$
$$5\,x_1 + 3\,x_2 \geq 8000$$
$$x_1, x_2 \geq 0$$

(*f*)   Maximize

$$y = 60\,x_1 + 44\,x_2$$

subject to the following constraints:

$$x_1 + x_2 \leq 1000$$
$$5\,x_1 + 3\,x_2 \geq 8000$$
$$x_1, x_2 \geq 0$$

11.2   The following linear optimization problems are variations of one another. For each problem, draw the feasible region. Then determine the optimum solution by superimposing several lines or curves that represent different values of the objective function over the feasible region. Determine both the optimum value of the objective function and the corresponding optimum policy.

(*a*)   Maximize

$$y = x_1 + x_2$$

subject to the following constraints:

$$x_1 + 2\,x_2 \leq 6$$
$$2\,x_1 + x_2 \leq 8$$
$$x_1, x_2 \geq 0$$

(*b*) Minimize

$$y = x_1 + x_2$$

subject to the following constraints:

$$x_1 + 2x_2 \le 6$$
$$2x_1 + x_2 \le 8$$
$$x_1, x_2 \ge 0$$

(*c*) Minimize

$$y = x_1 + x_2$$

subject to the following constraints:

$$x_1 + 2x_2 \ge 6$$
$$2x_1 + x_2 \ge 8$$
$$x_1, x_2 \ge 0$$

(*d*) Maximize

$$y = x_1 + x_2$$

subject to the following constraints:

$$x_1 + 2x_2 \ge 6$$
$$2x_1 + x_2 \ge 8$$
$$x_1, x_2 \ge 0$$

(*e*) Minimize

$$y = 2x_1 + 4x_2$$

subject to the following constraints:

$$x_1 + 2x_2 \ge 6$$
$$2x_1 + x_2 \ge 8$$
$$x_1, x_2 \ge 0$$

11.3 Each of the following problems is a variation of the nonlinear optimization problem given in Example 11.2. For each problem, draw the feasible region. Then determine the optimum solution by superimposing several lines or curves that represent different values of the objective function over the feasible region. Determine both the optimum value of the objective function and the corresponding optimum policy.

(a) Minimize
$$y = 10 + (x_1 - 0.5)^2 + (x_2 + 2)^2$$
subject to the following constraints:
$$\pi (x_1^2 + x_2^2) \geq 10$$
$$x_1 - 1.25\, x_2 \leq 0$$
$$x_1, x_2 \geq 0$$

(b) Minimize
$$y = 10 + (x_1 + 2)^2 + (x_2 - 0.5)^2$$
subject to the following constraints:
$$\pi (x_1^2 + x_2^2) \geq 10$$
$$x_1 - 1.25\, x_2 \leq 0$$
$$x_1, x_2 \geq 0$$

(c) Minimize
$$y = 10 + (x_1 - 0.5)^2 + (x_2 - 0.5)^2$$
subject to the following constraints:
$$\pi (x_1^2 + x_2^2) \leq 10$$
$$x_1 - 1.25\, x_2 \leq 0$$
$$x_1, x_2 \geq 0$$

(d) Maximize
$$y = 10 + (x_1 - 0.5)^2 + (x_2 - 0.5)^2$$
subject to the following constraints:
$$\pi (x_1^2 + x_2^2) \geq 10$$
$$x_1 - 1.25\, x_2 \leq 0$$
$$x_1, x_2 \geq 0$$

(e) Minimize
$$y = 10 + (x_1 - 0.5)^2 + (x_2 - 0.5)^2$$
subject to the following constraints:
$$\pi (x_1^2 + x_2^2) \leq 10$$
$$x_1 - 1.25\, x_2 \leq 0$$
$$x_1, x_2 \geq 0$$

(f) Maximize

$$y = 10 + (x_1 - 0.5)^2 + (x_2 - 0.5)^2$$

subject to the following constraints:

$$\pi(x_1^2 + x_2^2) \leq 10$$

$$x_1 - 1.25\,x_2 \leq 0$$

$$x_1, x_2 \geq 0$$

11.4 Draw the feasible region for each of the following nonlinear optimization problems. Then determine the optimum solution by superimposing several lines or curves that represent different values of the objective function over the feasible region. Determine both the optimum value of the objective function and the corresponding optimum policy.

(a) Minimize

$$y = (x_1 - 2\,x_2)^2$$

subject to the following constraints:

$$2\,x_1 - x_2 \geq 3$$

$$x_1, x_2 \geq 0$$

(b) Minimize

$$y = x_1 - 2\,x_2$$

subject to the following constraints:

$$(2\,x_1 - x_2)^2 \geq 3$$

$$x_1, x_2 \geq 0$$

(c) Maximize

$$y = (x_1 - 2\,x_2)^2$$

subject to the following constraints:

$$2\,x_1 - x_2 \leq 3$$

$$x_1, x_2 \geq 0$$

11.2 SOLVING OPTIMIZATION PROBLEMS IN EXCEL

Excel includes mathematical procedures for solving optimization problems within its **Solver** feature. However, **Solver** is an Excel add-in, like the **Analysis Toolpak** used to generate histograms (see Chap. 4). To install the **Solver**, choose **Add-Ins** from the **Tools** menu. Then select **Solver Add-In** from the resulting **Add-Ins** dialog box. (Once the **Solver** feature has been installed, it will remain installed unless it is removed by reversing the above procedure.)

To solve an optimization problem in Excel, proceed as follows:

1. Enter a value for each independent variable within a block of adjacent cells. These values will be used as an initial guess. (Initial values are not required when solving linear problems.)

2. Enter an equation for the objective function, expressed as an Excel formula. Within this formula, express the independent variables in terms of their cell addresses.

3. Enter an equation for each constraint, expressed as an Excel formula. The independent variables should again be expressed in terms of their cell addresses.

4. Select **Solver** from the **Tools** menu.

5. When the **Solver Parameters** dialog box appears, enter the following information:

   (a) Enter the address of the cell containing the objective function in the **Set Target Cell** location.

   (b) Select either **Max** or **Min** in the area labeled **Equal to**, beneath the **Target Cell** location.

   (c) Enter the range of cell addresses containing the independent variables in the area labeled **By Changing Cells**.

   (d) Enter the cell address containing each constraint, the type of constraint, and the value of the right-hand side (details given below). To add a constraint, click on the **Add...** button and provide the following information:

      (i) Enter the cell address of the constraint in the **Cell Reference** location.

      (ii) Specify the type of constraint (i.e., $\leq$, $\geq$, or =) from the pull-down menu.

      (iii) Enter the value of the right-hand side in the **Constraint** location.

      Note that you can change or delete any constraint after it has been added if you wish.

   (e) If the objective function and the constraints are linear, click on the **Options...** button and select **Assume Linear Model** from the **Solver Options** dialog box. Then select **OK**.

(f)   When all of the required information has been added correctly, select Solve. This will initiate the actual solution procedure.

6. Once a solution has been obtained, the optimal values of the independent variables, the corresponding value of the objective function, and the value of each constraint will appear within their respective cells. A dialog box labeled Solver Results will also appear, with various options for displaying and saving the optimum solution. Respond by selecting Keep Solver Solution and by requesting an Answer report. The Answer report will then be generated and placed in a separate worksheet.

You may wish to select the feature labeled Show Iteration Results from the Solver Options dialog box. This feature will cause the solution procedure to pause after each iteration (i.e., each repeated step), thus providing a step-by-step history of the computation. This feature is particularly useful if you are taking a course or have had a course in mathematical optimization procedures.

You may also wish to select the Automatic Scaling feature in the Solver Options dialog box. This feature is very useful if the magnitude of the independent variables differs substantially from the magnitude of the objective function or the right-hand side of the constraints.

Example 11.3 Solving the Production Scheduling Problem in Excel

Solve the production scheduling problem given in Example 11.1 using Excel's Solver feature.

To recap briefly, we wish to solve the following problem:
Maximize

$$y = 60\,x_1 + 44\,x_2 \tag{11.6}$$

subject to the following constraints:

$$x_1 + x_2 \geq 1000 \tag{11.7}$$

$$5x_1 + 3x_2 \leq 8000 \tag{11.8}$$

$$x_1, x_2 \geq 0 \tag{11.9}$$

This problem is entirely linear. Hence, we will solve the problem using Solver's optional Assume Linear Model feature.

We begin by entering the model into an Excel worksheet, as shown in Fig. 11.5. Within this worksheet, column A contains labels for the items provided in column B. Within column B, notice the assumed values for the independent variables in cells B3 and B4, the corresponding value of the objective function in cell B6, and the values of the constraints represented by Equations (11.7) and (11.8) in cells B8 and B9, respectively.

	B6	▼	■	=60*B3+44*B4							
	A			B	C	D	E	F	G	H	I
1	Maximum Profit Production Schedule										
2											
3	Units of A mfg. per month (x1):			1							
4	Units of B mfg. per month (x2):			1							
5											
6	Profit (y):			104							
7											
8	Minimum Production Requirement (g1):			2							
9	Labor Availability (g2):			8							
10											
11											
12											
13											
14											
15											
16											
17											

Ready

Figure 11.5

The values shown in cells B6, B8, and B9 result from cell formulas corresponding to Equations (11.6) through (11.8). These values result from the initial values of the independent variables provided in cells B3 and B4. Figure 11.6 shows the cell formulas.

	B6	▼	■ =60*B3+44*B4		
	A			B	C
1	Maximum Profit Production Schedule				
2					
3	Units of A mfg. per month (x1):			1	
4	Units of B mfg. per month (x2):			1	
5					
6	Profit (y):			=60*B3+44*B4	
7					
8	Minimum Production Requirement (g1):			=B3+B4	
9	Labor Availability (g2):			=5*B3+3*B4	
10					
11					
12					
13					
14					
15					
16					
17					

Ready

Figure 11.6

Once the problem specification has been entered into the worksheet, the Solver is invoked from the Tools menu. Figure 11.7 shows the resulting dialog box. Note that the address of the objective function (B6) is entered at the top, in the area labeled Set Target

Cell. The type of problem is entered beneath the cell address. For this problem, we select Max. The range of cell addresses containing the independent variables (B3:B4) is then entered in the area labeled By Changing Cells.

The constraints are then added, one at at time, by pressing the Add button and providing the requested information. The cell address, the type of constraint (≤, ≥, or =), and the value of the right-hand side are added for each constraint. (The cell addresses will automatically be rearranged in ascending order once all of the constraints have been entered.) Four individual constraints, corresponding to Equations (11.7) through (11.9) are specified for this problem. Note that Equation (11.9) involves two separate constraints.

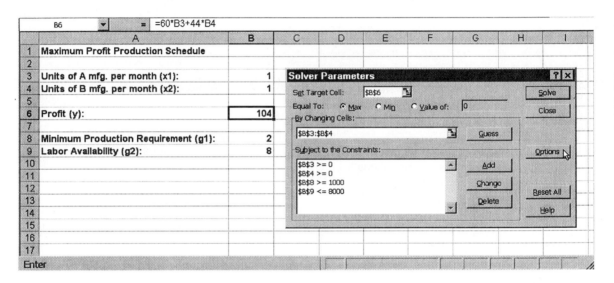

Figure 11.7

Since this is a linear optimization problem, we select the Options button after entering all of the information requested by the Solver Parameters dialog box. This results in the Solver Options dialog box shown in Fig. 11.8. Most of the items in this dialog box should not be changed since they refer to options that affect the mathematical solution techniques. For this problem, however, we select Assume Linear Model, thus invoking the more efficient solution procedure that is designed for purely linear problems. We then select OK to return to the Solver Parameters dialog box.

From the Solver Parameters dialog box, we select Solve to initiate the actual computation. The optimum values then appear in the various cells within the worksheet, as shown in Fig. 11.9. Thus, we see that the maximum profit is $117,333, obtained by manufacturing 0 units of A and 2667 units of B per month (the production rate of B is rounded up to the nearest integer). Moreover, we see that the left-hand side of Equation (11.7) is 2667 (rounded up), and the left-hand side of Equation (11.8) is 1000. We therefore conclude that the first constraint is inactive (not binding), but the second constraint is active (binding).

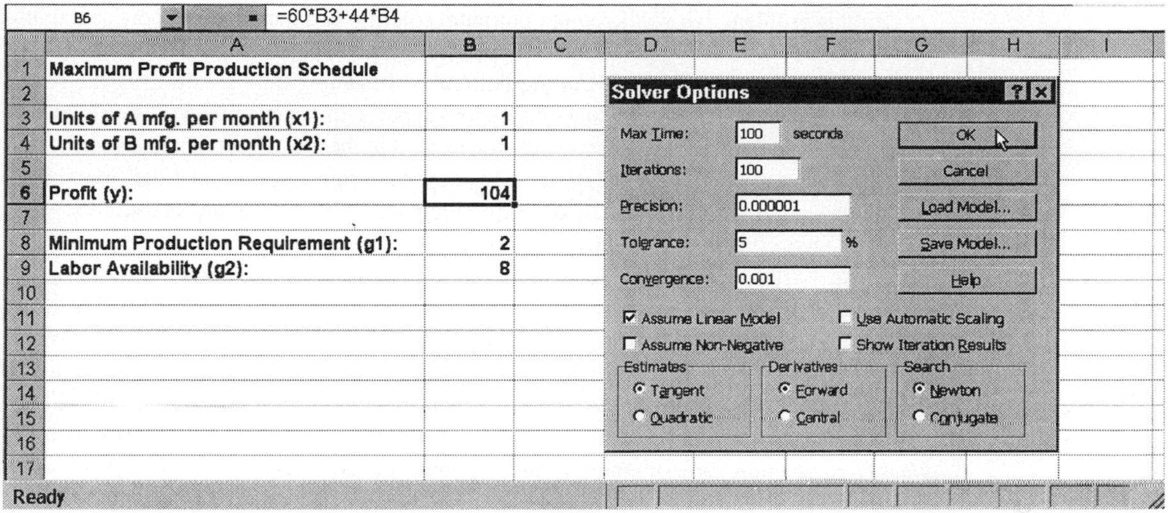

Figure 11.8

Superimposed over the worksheet containing the optimum solution is the Solver Results dialog box, as shown in Fig. 11.9. This dialog box allows us to either keep the optimum solution within the worksheet or restore the original values. It also allows us to generate additional reports in separate worksheets if we wish.

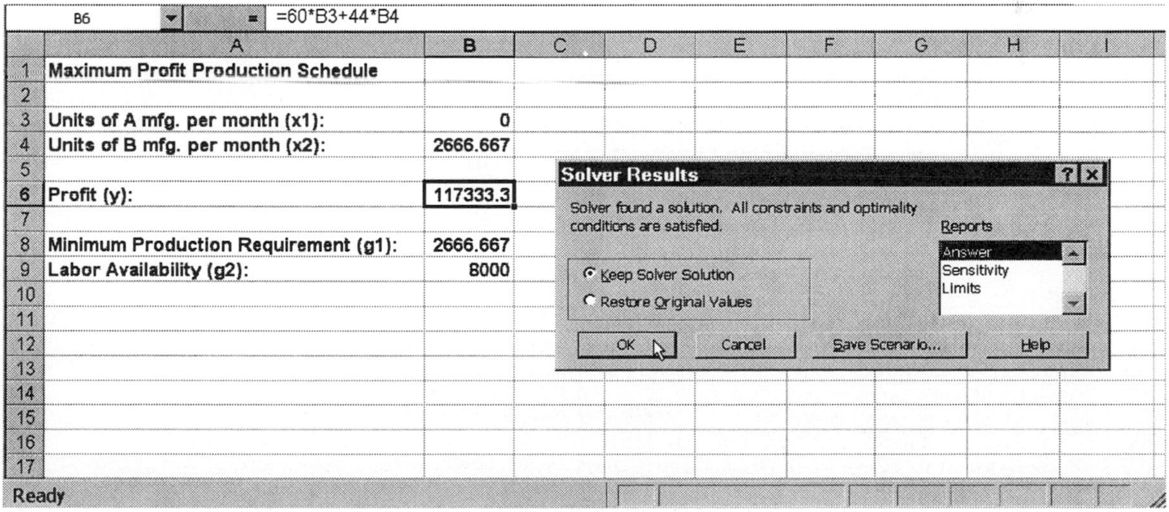

Figure 11.9

In this problem, we will keep the optimum solution and generate a report containing more detailed information about the final solution. The appropriate responses are shown in Fig. 11.9. We then select OK at the bottom of the dialog box, resulting in the report shown in Fig. 11.10. Notice that this report appears in a separate worksheet labeled Answer Report 1. (The original worksheet appeared in the worksheet labeled Sheet 1.)

The report shown in Fig. 11.10 contains both the original values and the final values for the independent variables, the objective function, and each of the constraints. The status of each constraint (binding or nonbinding) is also indicated. In addition, a value called *slack* is shown for each constraint. For an inequality-type constraint, this is the difference in magnitude between the limiting value expressed by the right-hand side of the constraint equation and the actual constraint value. Slack can be thought of as excess capacity, for a ≤ type constraint, or as excess production, for a ≥ type constraint. (Equality constraints, on the other hand, are always binding. Hence, the slack associated with an equality constraint is always zero.) The slack values reported in Fig. 11.10 indicate a weekly production rate of 1667 (rounded) units in excess of the minimum required production rate, with labor working to full capacity (zero slack).

Note that the report shown in Fig. 11.10 is nicely formatted, making it suitable for printing and inclusion within a more comprehensive written report. The attractive formatting is provided automatically when the report is generated.

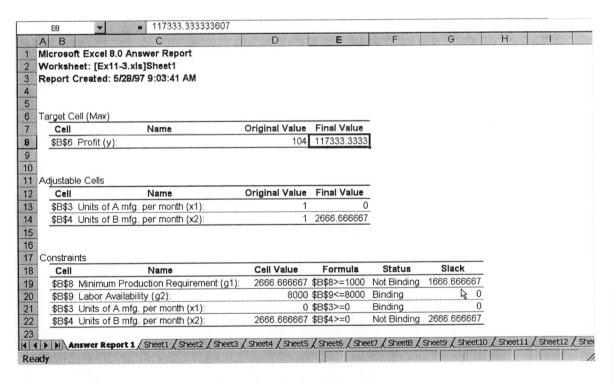

Figure 11.10

Remember that Solver includes different procedures for finding the optimum of linear and of nonlinear problems. Since the solution procedure used to solve linear problems is more efficient than the solution procedure used for nonlinear problems, it should always be used when solving linear optimization problems. However, this method *cannot* be used to solve nonlinear problems.

Moreover, if a nonlinear optimization problem has multiple optima, the solution that is obtained may be dependent upon the initial guess. Thus, an initial guess should be specified, and it should be as close as possible to the desired optimum.

Example 11.4 Solving a Nonlinear Minimization Problem in Excel

Solve the following nonlinear optimization problem using Excel:

Minimize

$$y = x_1 + x_2$$

subject to

$$2 x_1^2 + 3 x_2^2 \geq 12$$

$$x_1, x_2 \geq 0$$

The solution to this problem is shown graphically in Fig. 11.11. The feasible region lies above the elipse represented by the first constraint. This particular problem has two local minima—one at each corner. The values associated with each local minimum are shown below.

First Local Minimum	Second Local Minimum
$y = 2$	$y = 2.45$
$x_1 = 0$	$x_1 = 2.45$
$x_2 = 2$	$x_2 = 0$

The first point, $y = 2$ at $x_1 = 0$, $x_2 = 2$, represents the global (absolute) minimum.

Now let us see how this problem can be solved in Excel. The procedure is very similar to that used in Example 11.3, except that we do *not* select the Assume Linear Model option. We will see that the optimum solution will depend upon the initial values selected for x_1 and x_2.

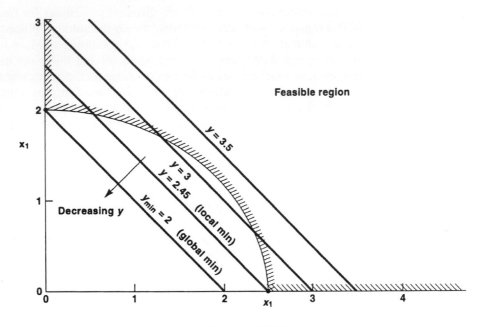

Figure 11.11

Figure 11.12 shows the worksheet containing the problem specification, with initial values of $x_1 = x_2 = 2$. Note that the objective function shown in cell B6 and the nonlinear constraint shown in cell B8 are generated from formulas. The cell formulas are shown in Fig. 11.13.

	B6	▼	=	=B3+B4				
	A	B	C	D	E	F	G	H	I
1	**Nonlinear Optimization Problem**								
2									
3	x1 =	2							
4	x2 =	2							
5									
6	y =	4							
7									
8	g1 =	20							
9									
10									
Ready									

Figure 11.12

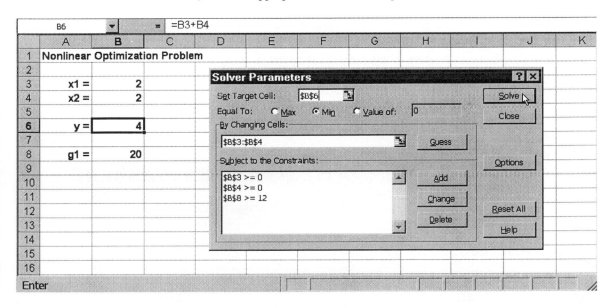

Figure 11.13

Figure 11.14 shows the Solver dialog box (obtained by selecting Tools/Solver from the menu bar), with the appropriate entries for this problem.

Figure 11.14

Once all of the information has been entered correctly, we press the Solve button, resulting in the solution shown in Fig. 11.15. Within this figure, we see that the desired minimum solution is $y = 2$ at $x_1 = 0$, $x_2 = 2$. This is one of the two local minima (in fact, the better of the two solutions) that we found at the beginning of this example when we solved the problem graphically.

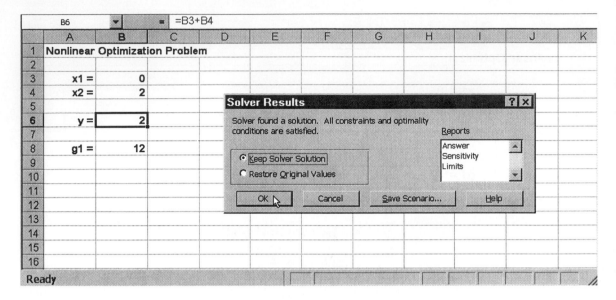

Figure 11.15

Now let us see what happens if we repeat the entire solution procedure from a starting point of $x_1 = 2$ and $x_2 = 0$. From our earlier graphical analysis, we know that this starting point is close to the other local optimum, at $x_1 = 2.45$, $x_2 = 0$. (In more realistic, multidimensional problems, however, we might not have this same insight as to where the minimum solution is likely to be found.)

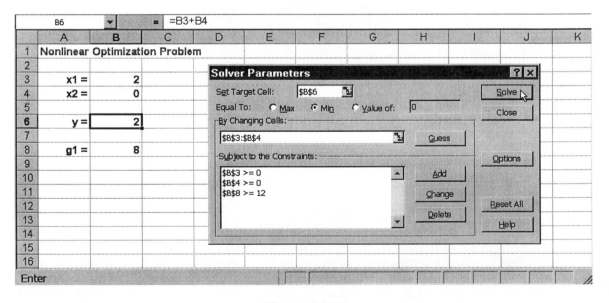

Figure 11.16

Figure 11.16 shows the worksheet containing the problem setup. The resulting solution is shown in Fig. 11.17 and Fig. 11.18 (the latter contains results rounded to two decimals). The solution (rounded) is $y = 2.45$ at $x_1 = 2.45$, $x_2 = 0$, as expected.

Unfortunately, there is no indication that this solution represents a local minimum, and is not the global minimum. This can be a serious deficiency when working with higher dimensionality problems. It is not a shortcoming of Excel but is a characteristic of the complex nature of nonlinear optimization problems.

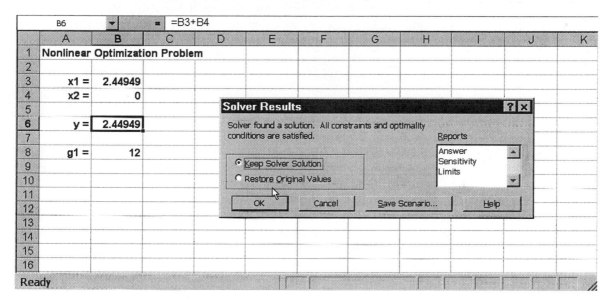

Figure 11.17

Figure 11.18

Once an optimum solution has been found, the Solver Results dialog box allows you to create three different reports, referred to as Answer, Sensitivity, and Limits, respectively. We have already discussed the Answer report and seen an example of an Answer report in Fig. 11.10. The Sensitivity report provides information concerning the sensitivity of the objective function to changes in the independent variables or changes in the constraints. For linear problems, the Sensitivity report also provides information concerning the relationships between changes in the independent variables and changes in the constraints.

The Limits report indicates the extent to which the independent variables can be changed before the constraints are violated. The accompanying changes in the objective function are also given.

Problems

11.5   Solve each of the linear optimization problems given in Prob. 11.1 with Excel's Solver feature, using the Assume Linear Model option. Compare each solution with the corresponding graphical solution obtained in Prob. 11.1.

11.6   Solve each of the linear optimization problems given in Prob. 11.2 with Excel's Solver feature, using the Assume Linear Model option. Compare each solution with the corresponding graphical solution obtained in Prob. 11.2.

11.7   Solve each of the linear optimization problems given in Prob. 11.2 with Excel's Solver feature, *without* the Assume Linear Model option. Compare your solutions with those obtained in Prob. 11.6.

11.8   Solve each of the nonlinear optimization problems given in Prob. 11.3 using Excel's Solver feature. Compare each solution with the corresponding graphical solution obtained in Prob. 11.3.

11.9   Solve each of the nonlinear optimization problems given in Prob. 11.4 using Excel's Solver feature. Compare each solution with the corresponding graphical solution obtained in Prob. 11.4.

11.10 Solve each of the following linear optimization problems using Excel's Solver feature.

(*a*) Maximize

$$y = 4 x_1 + 2 x_2 + 3 x_3$$

subject to the following constraints:

$$3 x_1 + 4 x_2 + 6 x_3 \leq 24$$

$$10\,x_1 + 5\,x_2 - 6\,x_3 \le 30$$
$$x_2 \le 4$$
$$x_1, x_2, x_3 \ge 0$$

(b)   Minimize

$$y = x_1 + 2\,x_2 + 2\,x_3 - x_4$$

subject to the following constraints:

$$2\,x_1 - x_2 + 3\,x_3 + x_4 \le 10$$
$$x_1 + x_2 - x_3 + 2\,x_4 \ge 1$$
$$-x_1 - x_3 + x_4 \le 12$$
$$3\,x_1 + 2\,x_2 + x_4 = 5$$
$$x_1, x_2, x_3, x_4 \ge 0$$

(c)   Maximize

$$y = 4\,x_1 + 5\,x_2 - x_3 - 2\,x_4$$

subject to the following constraints:

$$4\,x_1 - x_2 + 2\,x_3 - x_4 = 6$$
$$x_1 + 2\,x_2 - x_3 - x_4 = 3$$
$$-3\,x_1 + 3\,x_2 - 2\,x_3 + x_4 \le 6$$
$$x_1, x_2, x_3, x_4 \ge 0$$

11.11 Solve each of the following nonlinear optimization problems using Excel's **Solver** feature.

(a)   Maximize

$$y = x \sin x$$

within the interval

$$0 \le x \le \pi$$

(b)   Minimize

$$y = (x_1 - 3)^2 + 9\,(x_2 - 5)^2$$

within the region

$$x_1, x_2 \ge 0$$

(c) Minimize

$$y = 100 \, (x_2 - x_1^2)^2 + (1 - x_1)^2$$

within the region

$$x_1, x_2 \geq 0$$

(d) Minimize

$$y = 3 \, (x_1 - 1)^2 + (x_2 + 2)^2 + 5 \, (x_3 - 4)^2$$

within the region

$$x_1, x_2, x_3 \geq 0$$

11.12 Steel is often blended with several different alloys, such as nickel and manganese, in order to give it special properties. In addition, the carbon content, which is present in all steel, will affect the properties of the steel.

Suppose a specialty steel company receives an order for 200 tons of steel containing at least 3 percent nickel and 2 percent manganese, with a carbon content not exceeding 3.5 percent. The steel will sell for $25 a ton. Assume the steel can be made from any combination of three different alloys, whose properties are given below.

	Nickel	*Manganese*	*Carbon*	*Cost per Ton*
Alloy 1	5%	8%	3%	$18
Alloy 2	4	2	3	15
Alloy 3	2	2	5	10

Develop a linear optimization model to determine the least expensive blend of alloys that will satisfy the requirements. (*Hint*: Let x_i be the number of tons of alloy i used to make the steel.) Solve the problem using Excel's Solver feature.

11.13 In the previous problem, suppose only 50 tons of alloy 2 were available. How would this affect the optimum blend?

11.14 A manufacturer of office furniture makes conference tables in two styles, Premier and Executive. Each style is available in either walnut or oak. Matching chairs are also manufactured in both styles, in either walnut or oak.

The material and labor requirements for each product are given below. Assume there is no distinction between walnut and oak in these requirements.

	Premier Table	Executive Table	Premier Chair	Executive Chair
Wood (sq ft)	100	140	16	20
Labor (hrs)	9	12	4	5

The selling prices for each product are given below.

	Premier Table	Executive Table	Premier Chair	Executive Chair
Walnut	$1200	$1500	$200	$250
Oak	1000	1200	180	220

Furniture-grade walnut costs $5 per square foot, and oak costs $4. Labor costs $18 per hour. Only 3000 hours of labor arc available per week.

Develop a linear optimization model to determine the most profitable weekly production schedule. (*Hint*: There are eight different products: two walnut tables, two oak tables, two walnut chairs, and two oak chairs. Let x_i be the number of units of the ith product manufactured per week, where $i = 1, 2, \ldots, 8$.) Solve the problem using Excel's Solver feature.

11.15 Solve Prob. 11.14 with the additional restriction that at least four matching chairs must be manufactured for each table (i.e., at least four Premier walnut chairs for each Premier walnut table, etc.).

11.16 Solve Prob. 11.15 with the additional restriction that walnut cannot be obtained in quantities exceeding 10,000 sq ft per week.

11.17 A manufacturer of television sets has three factories, located in Philadelphia, St. Louis, and Phoenix. The company has four distribution centers, located in Atlanta, Chicago, Denver, and Seattle. All of the products are shipped from a factory to a distribution center, from which they are distributed to various customers across the country.

The plant capacities and distribution requirements (demands), in units per month, and shipping costs, in dollars per unit, are given below.

Philadelphia	St. Louis	Phoenix
5,000	8,000	12,000

Atlanta	Chicago	Denver	Seattle
6,000	8,000	5,000	6,000

From: →	Philadelphia	St. Louis	Phoenix
To: Atlanta	$20	$25	$60
Chicago	25	12	50
Denver	40	20	15
Seattle	65	45	35

Develop a linear optimization model to determine the least expensive way to supply the distribution centers with products from the factories. (*Hint*: Let x_{ij} represent the number of units shipped from factory i to distribution center j per month.) Solve the problem using Excel's **Solver** feature.

11.18 An oil refinery produces a stream of hydrocarbon base fuel from each of four different processing units. The processing units have capacities of C_1, C_2, C_3, and C_4 barrels per day, respectively. A portion of each base fuel is fed to a central blending station where the base fuels are mixed into three grades of gasoline. The remainder of each base fuel is sold "as is" at the refinery. Storage facilities are available at the blending station for temporary storage of the blended gasolines, if required.

Let N_1, N_2, N_3, and N_4 be the octane ratings of the respective base fuels, and let S_1, S_2, S_3, and S_4 be the profit derived from their direct sale. Let O_1, O_2, and O_3 be the octane numbers of the three grades of gasoline that must be blended on any given day to satisfy customer demands D_1, D_2, and D_3. Let R_1, R_2, and R_3 be the minimum octane requirements of the three grades of blended gasoline, and let P_1, P_2, and P_3 be the profit per gallon that is realized from the sale of each grade of blended gasoline. Finally, let x_{ij} be the quantity of the ith base fuel used to blend the jth gasoline.

The octane number of each gasoline can be expressed as the weighted average of the octane numbers of the constituent base fuels. The weighting factors are the fractions of the base fuels in each gasoline. Thus,

$$O_1 = \frac{x_{11}N_1 + x_{21}N_2 + x_{31}N_3 + x_{41}N_4}{x_{11} + x_{21} + x_{31} + x_{41}}$$

$$O_2 = \frac{x_{12}N_1 + x_{22}N_2 + x_{32}N_3 + x_{42}N_4}{x_{12} + x_{22} + x_{32} + x_{42}}$$

$$O_3 = \frac{x_{13}N_1 + x_{23}N_2 + x_{33}N_3 + x_{43}N_4}{x_{13} + x_{23} + x_{33} + x_{43}}$$

Develop a linear optimization model to determine how much of each base fuel should be blended into gasoline and how much should be sold "as is" in order to maximize profit. Solve the model using the following numerical data:

$C_1 = 13{,}000$ bbl/day	$N_1 = 82$ octane	$S_1 = \$0.90$/bbl
$C_2 = 7{,}000$	$N_2 = 95$	$S_2 = 1.05$
$C_3 = 25{,}000$	$N_3 = 102$	$S_3 = 1.25$
$C_4 = 15{,}000$	$N_4 = 107$	$S_4 = 1.60$
$D_1 = 13{,}000$ bbl/day	$R_1 = 87$ octane	$P_1 = 3.5$¢/gal
$D_2 = 25{,}000$	$R_2 = 89$	$P_2 = 4.5$
$D_3 = 18{,}000$	$R_3 = 93$	$P_3 = 6.0$

Note: 42 gal $= 1$ bbl

11.19 How would the profit and the optimum policy be affected if the selling prices of the gasolines were changed to 3¢/bbl, 4.5¢/bbl (as before), and 5.5¢/bbl, respectively? What would be the consequences of decreasing D_2 by 7,000 bbl/day while simultaneously increasing D_3 by 2,000 bbl/day?

11.20 The cost associated with a set of electrical transmission lines is given by

$$C_1 = 10{,}000\, n\, (1 + 2d^2)$$

where n is the number of lines in the set and d is the diameter of one line, in inches. In addition, the cost of transferring electrical power over the lines over their entire lifetime is given by

$$C_2 = 150{,}000/(nd^2)$$

To prevent major power disruptions in the event of a line breakage, it is required that $n \geq 10$.

Develop a nonlinear optimization model to determine the lowest overall cost $C = C_1 + C_2$. Assume that n is a continuous variable. Solve using Excel's **Solver** feature.

11.21 A pipeline is to transfer crude oil from a tanker docking area to a large oil refinery. The power required to pump the oil may be determined as

$$P = 0.4 \times 10^{-12} w^3/(\rho^2 d^5)$$

where w is the oil flow, in lb_m/hr, ρ is the density of the oil, in lb_m/ft^3, and d is the pipe diameter, in ft. The cost incurred, in dollars, is given by

$$C_p = 10{,}000 d^2 + 170P$$

where the first term represents the cost of the pipeline and the second term represents the cost of the pump and the present value of the pumping power over the life of the refinery.

Suppose the oil flow is maintained at 10^7 lb_m/hr, and the oil density is 50 lb_m/ft^3. In order to keep sludge from settling in the pipeline, the oil velocity, v, must be at least 9 ft/sec. (*Note:* $w = 3600\pi d^2 \rho v/4$.) Develop a nonlinear optimization model to find the pipe diameter that will minimize the cost. Solve using Excel's Solver feature.

11.22 The oil carried by the pipeline described in Prob. 11.21 will be transported at an elevated temperature in an insulated pipeline. The heat loss is determined as

$$Q = \frac{500}{2.5 \times 10^{-4} + 0.0025t}$$

where Q is the heat loss in BTU/hr and t is the thickness of the insulation, in ft.

The cost of the insulation is

$$C_i = 5 \times 10^5 td$$

where d is the diameter of the pipe, in ft. The present value of the energy loss over the life of the refinery can be assumed to be

$$C_e = 0.5Q$$

Assuming the pipeline has not yet been built, develop a nonlinear optimization model to determine the pipe diameter and the thickness of insulation that will minimize the total cost, which is given by

$$C = C_p + C_i + C_e,$$

Solve using Excel's Solver feature, with the restrictions that $d \geq 2$ ft and $v \geq 9$ ft/sec.

ADDITIONAL READING

Fox, R. L. *Optimization Methods for Engineering Design.* Reading, MA: Addison-Wesley, 1971.

Hillier, F. S. and G. J. Lieberman. *Introduction to Mathematical Programming.* 2d ed. New York: McGraw-Hill, 1995.

Nicholson, T. A. J. *Optimization in Industry, Vol II Industrial Applications.* Chicago: Aldine Atherton, Inc., 1971.

Rao, S. *Optimization Theory and Application.* New Delhi: Wiley Eastern Ltd., 1979.

Wagner, H. *Principles of Operations Research.* 2d ed. Englewood Cliffs, NJ: Prentice Hall, 1975.

Winston, W. L. *Operations Research, Applications and Algorithms.* 3d ed. Boston: PWS-Kent Publishing Company (Wadsworth, Inc.), 1994.

Wu, N. and R. Coppins. *Linear Programming and Extensions.* New York: McGraw-Hill, 1981.

CHAPTER 12

SORTING AND RETRIEVING DATA

Engineers often store information in *lists*. Within the context of a spreadsheet, a list is a block of contiguous rows and columns. The list may include both *numerical data* and *alphanumeric data* (i.e., words, names, and so on.). Thus, a professor may maintain a class roster in the form of a list containing student names in one column, semester averages (expressed as percentages) in the second column, and corresponding letter grades in the third column. The columns may include *headings* in the first row. Each additional row will contain a complete set of data for one student; that is, a student name, semester average, and letter grade. The information within each of these rows is often referred to as a *record*.

A list can be created and edited in Excel using the standard worksheet editing procedures described in Chap. 2. Once a list has been created, we can *sort* the list; that is, rearrange the rows so that the information in a designated column increases or decreases. Or we can *filter* the data; that is, determine the rows for which the information in a designated column satisfies some particular criterion. For example, most professors create class rosters with their students listed alphabetically by their last names. At the end of the semester, a professor will usually rearrange (*sort*) the student records by exam scores, from highest to lowest. In addition, the professor may want to retrieve (*filter*) the names of those students whose exam scores exceed a certain value or fall within a certain range. Such procedures are generally referred to as *database operations*.

Excel includes a number of features that allow database operations to be carried out quickly and easily. In this chapter, we will consider the most common of these features, which allow the data within a list to be sorted and filtered.

12.1 CREATING A LIST IN EXCEL

To create a list within an Excel worksheet, identify a block of cells that can be used to store the required rows and columns. Often, it is most convenient to place the list in the upper left portion of the worksheet so that there is room to expand the list into the worksheet (i.e., below and to the right). Each column should contain information of the same type. Each row should should contain one individual record; thus, each row will include one value from each of the columns.

The top row is often used for column headings; in fact, two or more rows can be used for this purpose if the headings are lengthy. Excel is usually able to distinguish automatically between the column headings and the actual information within the columns. This distinction may be based upon differences in the data types (e.g., alphanumeric headings and numerical data) or appearance (e.g., selection of a different font; larger type; or the use of boldface, italic, or underlining in the headings). If the headings are recognized, Excel will automatically omit the column headings from its various database operations.

Example 12.1 Creating a List in Excel

Consider the table shown below, containing information on several states within the United States. Each row (each record) contains the name, capital, population (based upon the U.S. 1990 census), and size of a state.

State	Capital	Population	Size (square miles)
Alaska	Juneau	550,043	615,230
California	Sacramento	29,760,021	158,869
Colorado	Denver	3,294,394	104,100
Florida	Tallahassee	12,937,926	59,988
Missouri	Jefferson City	5,117,073	69,709
New York	Albany	17,990,455	53,989
North Carolina	Raleigh	6,628,637	52,672
North Dakota	Bismarck	638,800	70,704
Pennsylvania	Harrisburg	11,881,643	45,759
Rhode Island	Providence	1,003,464	1,231
Texas	Austin	16,986,510	267,277
Virginia	Richmond	6,187,358	42,326
Washington	Olympia	4,866,692	70,637

Enter this information as a list within an Excel worksheet. Determine the population density (population per square mile) for each state, based upon the given information. Include a heading for each column.

Figure 12.1 shows the Excel worksheet containing this information. Rows 1 and 2 contain various column headings. The remaining rows each contain the information for one state. Note that the states are listed alphabetically, as in the given table.

Also, note that the population densities are determined by a formula. For example, the population density for Alaska, shown in cell E3, is determined by the formula =C3/D3. Thus, the population density, in persons per square mile, is determined by dividing each state's population by its area.

Finally, note that the list includes column headings. Excel will be able to distinguish automatically between the headings and the information within columns A and B based upon format differences (the headings are centered, underlined, and shown in boldface). The headings in columns C, D, and E will be recognized based upon differences in data type (alphanumeric headings versus numerical data).

	E3		=	=C3/D3			
	A	B	C	D	E	F	G
1				Area	Population		
2	State	Capital	Population	(sq. miles)	Density		
3	Alaska	Juneau	550,043	615,230	0.9		
4	California	Sacramento	29,760,021	158,869	187.3		
5	Colorado	Denver	3,294,394	104,100	31.6		
6	Florida	Tallahassee	12,937,926	59,988	215.7		
7	Missouri	Jefferson City	5,117,073	69,709	73.4		
8	New York	Albany	17,990,455	53,989	333.2		
9	North Carolina	Raleigh	6,628,637	52,672	125.8		
10	North Dakota	Bismarck	638,800	70,704	9.0		
11	Pennsylvania	Harrisburg	11,881,643	45,759	259.7		
12	Rhode Island	Providence	1,003,464	1,231	815.2		
13	Texas	Austin	16,986,510	267,277	63.6		
14	Virginia	Richmond	6,187,358	42,326	146.2		
15	Washington	Olympia	4,866,692	70,637	68.9		
16							
17							
Ready							

Figure 12.1

Once a list has been created, new records can be added by inserting new rows between existing rows or by adding new rows at the bottom of the list. Existing records can be modified simply be retyping the appropriate material, and they can be deleted by removing the appropriate rows.

Another way to add, delete, or modify a record, however, is to use a *form*. This is basically a dialog box that requests all of the information required to add, change, or delete a record, based upon the structure of the list. Forms can be accessed by choosing Form from the Data menu and then supplying the information requested by the resulting dialog box, as shown in Fig. 12.2.

Figure 12.2 shows a *form* dialog box superimposed over the worksheet containing the list. The left portion of the dialog box shows four data entry areas, corresponding to columns A through D of the list. A *calculated* value, based upon the formula used in column E, is also shown. This value is displayed differently, however, because calculated values cannot be altered directly within a *form* dialog box.

The right portion of the dialog box contains a number of buttons that are used to create a new record (New), delete the existing record (Delete), move backward or forward (Find Prev, Find Next), or modify the contents of the current record (Criteria). The Restore button is used to restore the original data within a record after one or more changes have been entered into the data entry areas (using the Criteria button).

The use of forms does not provide any new capabilities within Excel. They are useful, however, for persons who may lack experience or confidence in working directly with the rows and columns in a list.

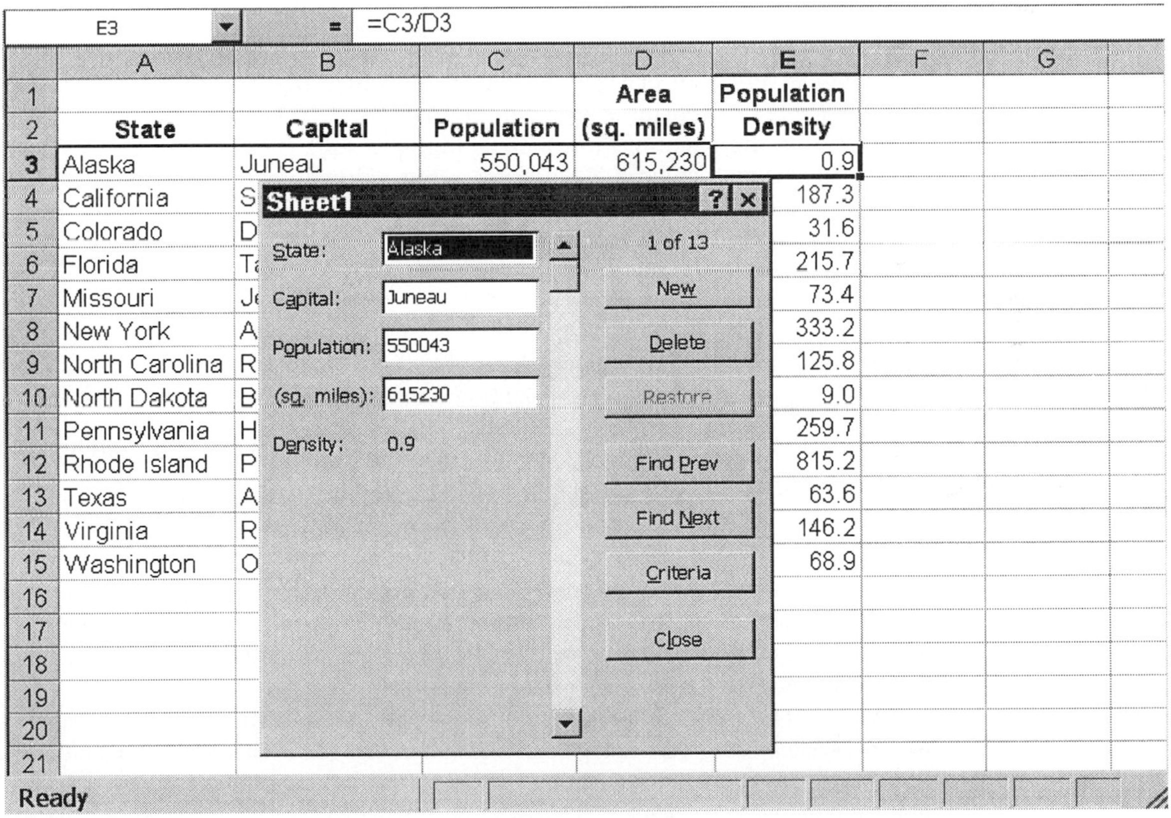

Figure 12.2

12.2 SORTING DATA IN EXCEL

Once a dataset has been correctly entered into a Worksheet as a list, the records can easily be sorted into ascending or descending order, based upon the contents of any column. To do so, simply select any cell within the desired column (including the column heading) and then click on either the *Ascending Sort Button* (lowest to highest, or A to Z) or the *Descending Sort Button* (highest to lowest, or Z to A) in the Standard Toolbar (see Fig. 12.3).

Figure 12.3

Alternatively, you can sort a chart by selecting any cell within the desired column and then choosing Sort from the Data menu. Then provide the required values within the resulting Sort menu. Multiple sorts (i.e., sorting records with respect to a second column if the primary sort column contains identical entries) can be carried out in this manner. You can also rearrange the *columns* based upon the contents of a *row* as a sort option.

Example 12.2 Sorting a List in Excel

Sort the list of states created in Example 12.1 by population density, from highest to lowest.

Figure 12.4 shows the results of the sorting operation. These results were obtained by first selecting a cell within column E (in this case, cell E3) and then clicking on the Descending Sort button. The results indicate that Rhode Island has the highest population density (815.2 persons per square mile), followed by New York (333.2), and so on, down to Alaska, which has the lowest population density (0.9).

We could have obtained these same results by selecting Sort from the Data menu and then responding to the Sort dialog box, as shown in Fig. 12.5. This method is a bit more complicated, but it offers the possibility of several options that are not available using the Sort buttons.

The original list shown in Fig. 12.1 can be restored by selecting any cell within column A and then rearranging the list in ascending order.

E3   =C3/D3

	A	B	C	D	E	F	G
1				**Area**	**Population**		
2	**State**	**Capital**	**Population**	**(sq. miles)**	**Density**		
3	Rhode Island	Providence	1,003,464	1,231	815.2		
4	New York	Albany	17,990,455	53,989	333.2		
5	Pennsylvania	Harrisburg	11,881,643	45,759	259.7		
6	Florida	Tallahassee	12,937,926	59,988	215.7		
7	California	Sacramento	29,760,021	158,869	187.3		
8	Virginia	Richmond	6,187,358	42,326	146.2		
9	North Carolina	Raleigh	6,628,637	52,672	125.8		
10	Missouri	Jefferson City	5,117,073	69,709	73.4		
11	Washington	Olympia	4,866,692	70,637	68.9		
12	Texas	Austin	16,986,510	267,277	63.6		
13	Colorado	Denver	3,294,394	104,100	31.6		
14	North Dakota	Bismarck	638,800	70,704	9.0		
15	Alaska	Juneau	550,043	615,230	0.9		
16							
17							

Ready

Figure 12.4

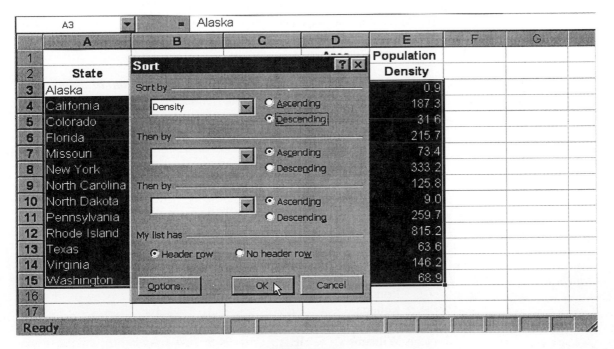

Figure 12.5

A sort need not apply to an entire list. You can sort an individual column, a block of adjacent columns, or a block of adjacent rows. To do so, simply select the block of cells that will be sorted, starting with the mouse pointer in the column that the sort will be based upon. Then use either of the sort procedures described above.

Exercises

12.1 Enter the data given in Example 12.1 into an Excel worksheet. Then carry out the following sorting operations:

  (*a*) Sort the records in reverse order, beginning with Washington and ending with Alaska. Then restore the original list.

  (*b*) Sort the records so that the state capitals are arranged alphabetically.

  (*c*) Sort the records by population, from smallest to largest.

  (*d*) Sort the records by size (area), from largest to smallest.

  (*e*) Sort the column containing the state capitals alphabetically, leaving all other columns in their original order. Then sort the capitals in reverse order.   Then restore the original list of capitals (so that they correspond to the correct states) by pressing the "undo" arrow (the counterclockwise arrow in the Standard toolbar) twice.

  (*f*) Select the last six records (North Dakota through Washington). Sort these records by population, from largest to smallest. Then restore their original order.

12.2 The final scores for Professor Bohring's *Introduction to Engineering* course are shown below (all names are fictitious):

Student	Departmental Major	Final Score	Final Grade
Barnes	EE	87.2	B
Davidson	ChE	93.5	A
Edwards	ME	74.6	C
Graham	ME	86.2	B
Harris	ChE	63.9	D
Jones	IE	79.8	B
Martin	EE	99.2	A
O'Donnell	CE	80.0	B
Prince	ChE	69.2	C
Roberts	EE	48.3	F
Thomas	ME	77.5	C
Williams	IE	94.5	A
Young	CE	73.2	C

Enter the data into an Excel worksheet. Then carry out the following sorting operations:

(*a*)   Sort the student records by final score, from highest to lowest.

(*b*)   Sort the student records by departmental major. List the student names alphabetically within each departmental category. To do so, select Sort from the Data menu to access the Sort dialog box. Specify Sort by Major/Ascending, Then by Student/Ascending.

(*c*)   Sort the student records by departmental major. Now list the students by final scores, from highest to lowest, within each departmental category. Use the Sort dialog box, as in part (*b*). Consider each Ascending/Descending choice carefully.

(*d*)   Rearrange the student records so that the names are listed alphabetically, thus restoring the list to its original order.

12.3 RETRIEVING (FILTERING) DATA IN EXCEL

One of the most common database activities involves the retrieval of information that satisfies a certain condition. For example, a professor may wish to retrieve the names of those students whose exam scores exceed 90 percent. Similarly, the professor may wish to determine which students are electrical engineering (EE) majors or, combining criteria, which EE majors have exam scores exceeding 90 percent.

Operations of this type can easily be carried out in Excel by *filtering* a list of data. Filtering shows only those records that satisfy the stated criteria. To filter the data, select any cell within the list and then choose Filter from the Data menu. You may then use either the AutoFilter or the Advanced Filter features. We will restrict our attention to AutoFilter since it can be used for most common data-retrieval operations.

When you choose AutoFilter, a downward-pointing arrow will appear at or near the top of each column. Clicking on one of these arrows will permit you to choose any of several different retrieval criteria for the data within the column. The criteria include selection of the top 10 records, the selection of those records that satisfy various equality conditions based upon the contents of the column, or the selection of all records (which is a useful way to restore the original list). More complicated "customized" selections are also possible. The details are illustrated in the following example.

Example 12.3 Filtering a List in Excel

In this example, we will perform several filtering operations on the data within the list of states, originally created in Example 12.1. Using Excel's AutoFilter feature, determine:

1.   The 10 states having the highest population density.

2.   Which state has its capital in Richmond.

3.   Which states have areas exceeding 100,00 square miles.

4.   Which states have populations between 10 and 20 million people.

To answer part 1, we first select an arbitrary cell within the list (in this case, cell E3) and then choose Filter/AutoFilter from the Data menu.   This results in a series of downward-pointing arrows within the column heading area, as shown in Fig. 12.6.

	E3 ▼	=	=C3/D3			
	A	B	C	D	E	F	G
1					Area	Population	
2	State ▼	Capital ▼	Populatio ▼	(sq. mile ▼	Density ▼		
3	Alaska	Juneau	550,043	615,230	0.9		
4	California	Sacramento	29,760,021	158,869	187.3		
5	Colorado	Denver	3,294,394	104,100	31.6		
6	Florida	Tallahassee	12,937,926	59,988	215.7		
7	Missouri	Jefferson City	5,117,073	69,709	73.4		
8	New York	Albany	17,990,455	53,989	333.2		
9	North Carolina	Raleigh	6,628,637	52,672	125.8		
10	North Dakota	Bismarck	638,800	70,704	9.0		
11	Pennsylvania	Harrisburg	11,881,643	45,759	259.7		
12	Rhode Island	Providence	1,003,464	1,231	815.2		
13	Texas	Austin	16,986,510	267,277	63.6		
14	Virginia	Richmond	6,187,358	42,326	146.2		
15	Washington	Olympia	4,866,692	70,637	68.9		
16							
17							
Ready							

Figure 12.6

We then click on the downward-pointing arrow in cell E2 and select Top 10 from the dropdown menu, as shown in Fig. 12.7. This results in another dialog box, from which we can select various groupings, such as top 5, bottom 10, bottom 3, and so on. In our case, we simply accept the Top 10 default selection.

The results, showing the states with the top 10 population densities, are shown in Fig. 12.8. Notice that Fig. 12.8 contains only 10 records, corresponding to the 10 states having the highest population densities. The records are *not* automatically sorted by population density. We could, of course, sort them using the techniques discussed in Sec. 12.2 if we wished.

When finished, we click on the arrow in cell E2 again and select (All), thereby restoring the original list in preparation for the next data-retrieval operation.

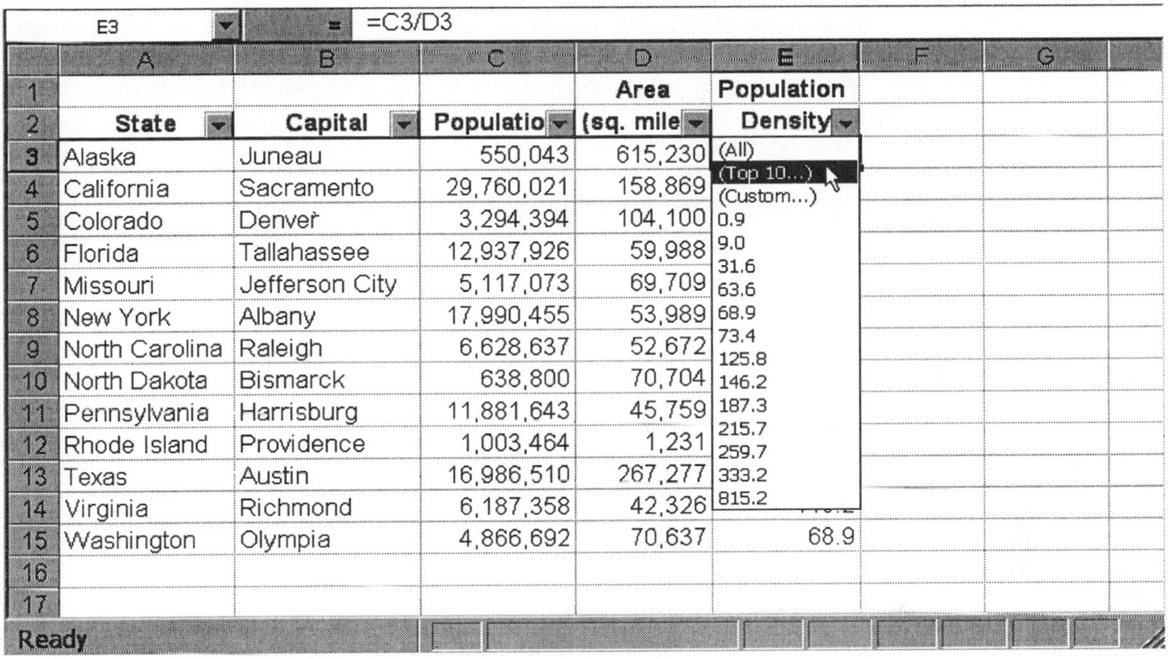

E3   =C3/D3

	A	B	C	D	E	F	G
1				Area	Population		
2	State	Capital	Populatio	(sq. mile	Density		
3	Alaska	Juneau	550,043	615,230	(All)		
4	California	Sacramento	29,760,021	158,869	(Top 10...)		
5	Colorado	Denver	3,294,394	104,100	(Custom...) 0.9		
6	Florida	Tallahassee	12,937,926	59,988	9.0		
7	Missouri	Jefferson City	5,117,073	69,709	31.6 / 63.6		
8	New York	Albany	17,990,455	53,989	68.9		
9	North Carolina	Raleigh	6,628,637	52,672	73.4 / 125.8		
10	North Dakota	Bismarck	638,800	70,704	146.2		
11	Pennsylvania	Harrisburg	11,881,643	45,759	187.3		
12	Rhode Island	Providence	1,003,464	1,231	215.7 / 259.7		
13	Texas	Austin	16,986,510	267,277	333.2		
14	Virginia	Richmond	6,187,358	42,326	815.2		
15	Washington	Olympia	4,866,692	70,637	68.9		
16							
17							

Ready

Figure 12.7

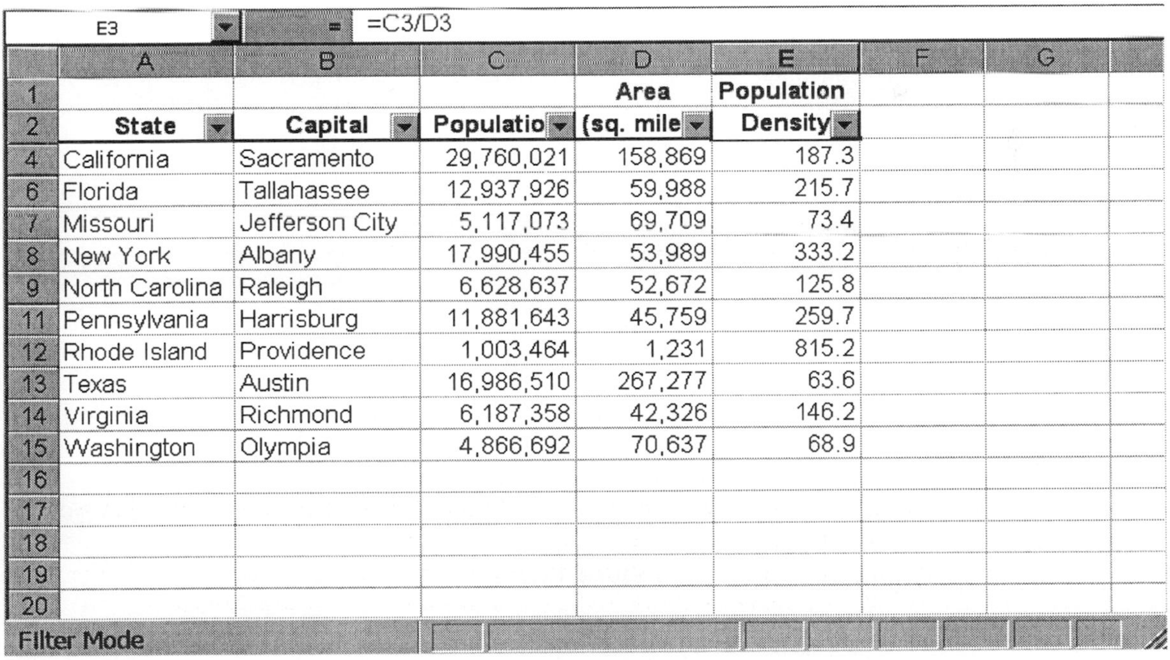

E3   =C3/D3

	A	B	C	D	E	F	G
1				Area	Population		
2	State	Capital	Populatio	(sq. mile	Density		
4	California	Sacramento	29,760,021	158,869	187.3		
6	Florida	Tallahassee	12,937,926	59,988	215.7		
7	Missouri	Jefferson City	5,117,073	69,709	73.4		
8	New York	Albany	17,990,455	53,989	333.2		
9	North Carolina	Raleigh	6,628,637	52,672	125.8		
11	Pennsylvania	Harrisburg	11,881,643	45,759	259.7		
12	Rhode Island	Providence	1,003,464	1,231	815.2		
13	Texas	Austin	16,986,510	267,277	63.6		
14	Virginia	Richmond	6,187,358	42,326	146.2		
15	Washington	Olympia	4,866,692	70,637	68.9		
16							
17							
18							
19							
20							

Filter Mode

Figure 12.8

To determine which state has its capital in Richmond, we click on the downward-pointing arrow in cell B2 and select Richmond from the resulting dropdown menu (see Fig. 12.9). The results are shown in Fig. 12.10. Thus, Virginia is the state whose capital is Richmond.

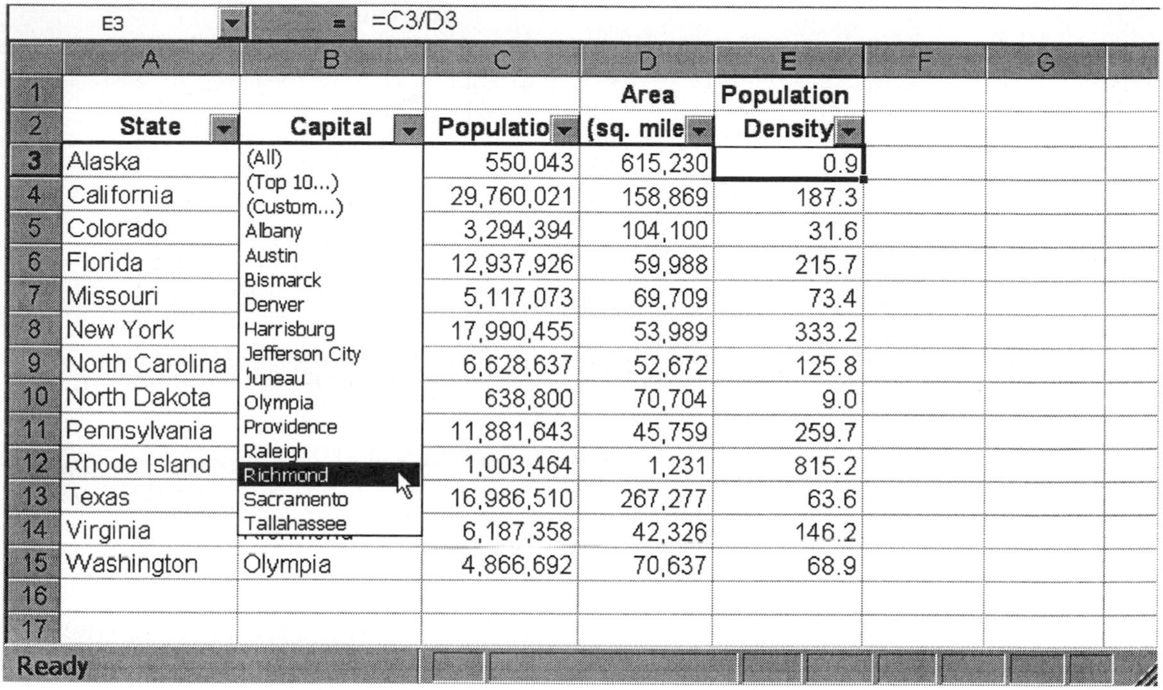

| | E3 | | = | =C3/D3 | | |

	A	B	C	D	E	F	G
1				Area	Population		
2	State	Capital	Populatio	(sq. mile	Density		
3	Alaska	(All)	550,043	615,230	0.9		
4	California	(Top 10...) / (Custom...)	29,760,021	158,869	187.3		
5	Colorado	Albany	3,294,394	104,100	31.6		
6	Florida	Austin / Bismarck	12,937,926	59,988	215.7		
7	Missouri	Denver	5,117,073	69,709	73.4		
8	New York	Harrisburg	17,990,455	53,989	333.2		
9	North Carolina	Jefferson City / Juneau	6,628,637	52,672	125.8		
10	North Dakota	Olympia	638,800	70,704	9.0		
11	Pennsylvania	Providence	11,881,643	45,759	259.7		
12	Rhode Island	Raleigh / Richmond	1,003,464	1,231	815.2		
13	Texas	Sacramento	16,986,510	267,277	63.6		
14	Virginia	Tallahassee	6,187,358	42,326	146.2		
15	Washington	Olympia	4,866,692	70,637	68.9		
16							
17							

Ready

Figure 12.9

| | E3 | | = | =C3/D3 | | |

	A	B	C	D	E	F	G
1				Area	Population		
2	State	Capital	Populatio	(sq. mile	Density		
14	Virginia	Richmond	6,187,358	42,326	146.2		
16							
17							
18							

Filter Mode

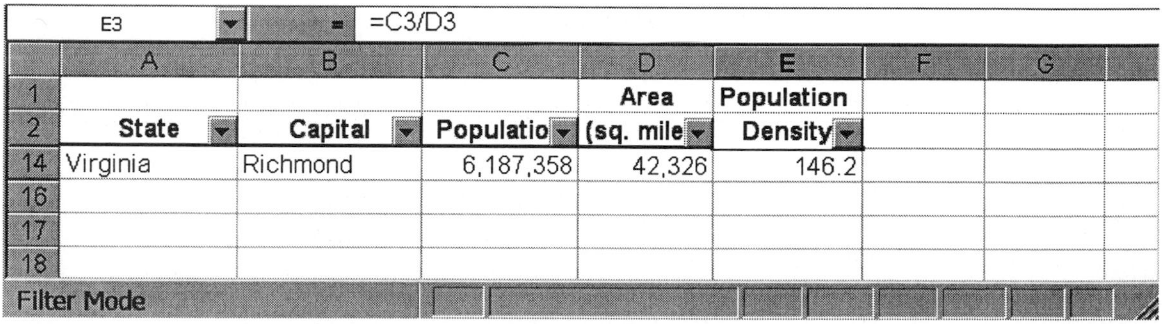

Figure 12.10

Now let us consider which states have areas exceeding 100,000 square miles. To do so, we restore the original list, click on the downward-pointing arrow in cell D2, and select Custom, resulting in the dialog box shown in Fig. 12.11.

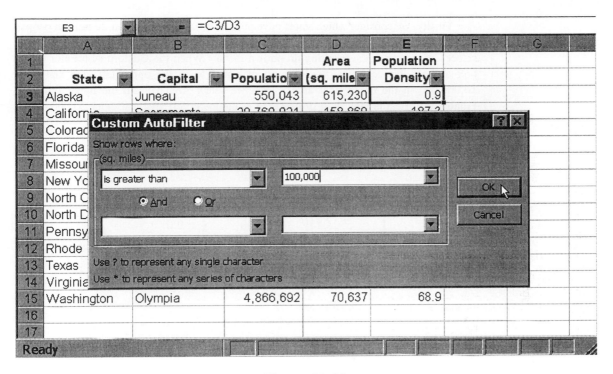

Figure 12.11

Within this dialog box, we select is greater than from the selections available in the upper left data entry area. (Click on the downward-pointing arrow to see the selections.) We then *type* the value 100,000 in the upper right data entry area and click on OK.

Figure 12.12 shows the results. Thus, Alaska, California, Colorado, and Texas have areas that exceed 100,000 square miles. Note that the states are not ranked by size; rather, they retain their original alphabetical sort order.

	E3		=	=C3/D3			
	A	B	C	D	E	F	G
1				Area	Population		
2	State	Capital	Populatio	(sq. mile	Density		
3	Alaska	Juneau	550,043	615,230	0.9		
4	California	Sacramento	29,760,021	158,869	187.3		
5	Colorado	Denver	3,294,394	104,100	31.6		
13	Texas	Austin	16,986,510	267,277	63.6		
16							
17							

Filter Mode

Figure 12.12

Finally, we consider which states have populations between 10 and 20 million people. We again restore the original list, click on the downward-pointing arrow in cell C2, and select Custom from the resulting dropdown menu. When the Custom AutoFilter dialog box appears, we select is greater than or equal to in the upper left area and enter 10,000,000 in the upper right area. Then we select And beneath the upper left area. Finally, we select is less than or equal to in the lower left area and enter 20,000,000 in the lower right area. The completed dialog box is shown in Fig. 12.13.

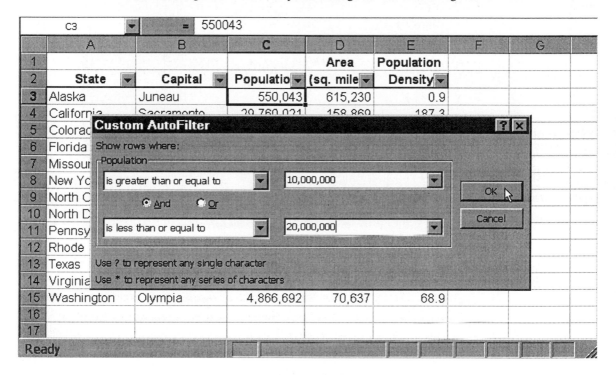

Figure 12.13

Clicking OK provides the desired results, as shown in Fig. 12.14. The figure shows us that Florida, New York, Pennsylvania, and Texas have populations between 10 and 20 million people. The records are shown in their original alphabetical order; they are not ranked by population.

Exercises

12.3 Carry out the following filtering operations on the list containing the data given in Example 12.1.

(a) Determine the five states having the smallest area.

(b) Determine which state has its capital in Jefferson City.

C3		=	550043			
A	B	C	D	E	F	G
			Area	Population		
State ▾	**Capital** ▾	**Populatio** ▾	**(sq. mile** ▾	**Density** ▾		
6 Florida	Tallahassee	12,937,926	59,988	215.7		
8 New York	Albany	17,990,455	53,989	333.2		
11 Pennsylvania	Harrisburg	11,881,643	45,759	259.7		
13 Texas	Austin	16,986,510	267,277	63.6		
16						
17						
Filter Mode						

Figure 12.14

(c) Determine the total population of the three states having the largest area. (*Hint*: use AutoSum after the filtering operation.)

(d) Determine the total area of the four states having the largest population.

(e) Determine the average population density of the five states having the smallest population.

(f) Determine which states have populations exceeding eight million people.

(g) Determine which states have areas less than 60,000 square miles.

(h) Determine which states have populations exceeding 16 million people or areas less than 70,000 square miles.

(i) Determine which states have population densities between 65 and 200 persons per square mile.

(j) Determine which states have state capitals whose names begin with either A or B.

(k) Determine which states have populations between three and nine million people and areas between 60,000 and 80,000 square miles.

12.4 Enter Professor Bohring's class roster, given in Exercise 12.2, into an Excel worksheet. Then carry out the following filtering operations:

(a) Determine which students are industrial engineering (IE) majors.

(b) Determine which students are majoring in either chemical engineering (ChE) or mechanical engineering (ME).

(c) Determine the average scores of the civil engineering (CE) students.

(d) Determine which students have final grades of C or better.

(e)   Determine which students have final scores less than 70.

(f)   Determine which electrical engineering (EE) students have final scores less than 70.

(g)   Determine which students have final scores between 70 and 90.

(h)   Determine which students are either industrial engineering majors or have final scores of at least 90.

ADDITIONAL READING

Dodge, M., C. Kinata, and C. Stinson. *Running Microsoft Excel 97*. Microsoft Press, 1997.

ANSWERS TO SELECTED PROBLEMS

Chapter 3

3.16 The data should be represented by a power equation: $V = at^b$

3.17 Type A bacteria.

Chapter 4

4.2 Mean = 70.2 in, median = 70.1 in, mode = 71.1 in, min = 66.5 in, max = 73.8 in, variance = 4.21 in^2, standard deviation = 2.05 in.

4.3 Same numerical results as Prob. 4.2.

4.5 (*a*) Mean = 70.1, median = 72.5, min = 41, max = 96. The mean is pulled down by a few very low scores, in the 40s.

(*b*)

0 - 10:	0	51 - 60:	2
11 - 20:	0	61 - 70:	6
21 - 30:	0	71 - 80:	11
31 - 40:	0	81 - 90	3
41 - 50:	5	91 - 100	3

(*c*) 11 students have exam scores ranging from 71 to 80; 3 are above 90; 5 are 50 or less.

4.9

Bin	(interpretation)	Frequency	Cum. Distr.
64 in	≤ 64 in	0	0 %
65	> 64, ≤ 65	0	0
66	> 65, ≤ 66	0	0
67	> 66, ≤ 67	3	15
68	> 67, ≤ 68	0	15
69	> 68, ≤ 69	3	30
70	> 69, ≤ 70	4	50
71	> 70, ≤ 71	2	60
72	> 71, ≤ 72	5	85
73	> 72, ≤ 73	1	90
74	> 73, ≤ 74	2	100
More	> 74	0	100

4.11 (*a*)   50 %   (*b*)   15 %

(*c*)   Equivalent male student = 64" / 9 = 71.1"
Probability of exceeding 71.1" = 30 %

(*d*)   Equivalent male student = 65" / 9 = 72.2"
Probability of not exceeding 72.2" = 86.5 %
Probability that male student exceeds 69" = 68 %
Probability that both events occur = (0.865)(0.68) × 100 = 58.8 %

4.12 (*a*)   Mean = 23.2 mpg, median = 23.1 mpg, mode = 23.9 mpg, min = 19.2 mpg, max = 27.1 mpg, standard deviation = 1.85 mpg.

4.13 (*a*)   The mean is pulled up by a few cars with relatively high mpg (compared to the median).

(*c*)   Probability that the fuel efficiency not exceed 20 mpg = 4.2 % (based upon 2-mpg intervals).

Probability that the fuel efficiency not exceed 22 mpg = 29.2 %.

Probability that the fuel efficiency *will* exceed 25 mpg = (100 − 85.4) = 14.6 %.

4.14 (*a*)   Mean = 988.6 ohms, median = 989.5 ohms, mode = 1006 ohms, min = 935 ohms, max = 1045 ohms, standard dev. = 23.8 ohms.

(*d*)   Probability ≤ 980 ohms = 40 %. Probability > 1020 ohms = 8 %.

Chapter 5

5.3 $y = -2.605\,x + 51.605$; $r^2 = 0.9847$
where y represents concentration and t represents time.

5.4 $y = -1.415\,x + 37.859$; $r^2 = 0.5695$
The earlier data set (in Prob. 5.3) results in a better fit, as can be seen visually and from the higher r^2 value associated with the earlier data set.

5.5 $y = -0.0071\,x + 45.57$; $r^2 = 0.6423$
where y represents mileage and x represents weight. A better fit might be obtained by passing a curve (e.g., a polynomial) through the data, though the data show a lot of scatter.

5.6 (*a*) $SSE = 0.045825$ (*b*) $SSE = 0.087062$

(*c*) $SST = 104.97$ (*d*) $SSE = 0.125967$

The calculated *SST* value in part (c) is based upon the values of $(y_i - \bar{y})^2$, but the exponential curve fit is based upon the use of ln y_i rather than y_i. Hence, the value calculated for *SSE* in part (d) is not meaningful.

5.8   Exponential fit: $y = 32.951\, e^{0.0002x}$; $r^2 = 0.4975$

Power fit: $y = 27.508\, x^{0.0841}$; $r^2 = 0.9868$

Logarithmic fit: $y = 2.9941\, \ln(x)$; $r^2 = 1$

Polynomial fit: $y = 3 \times 10^{-14}\, x^5 - 2 \times 10^{-10}\, x^4 + 4 \times 10^{-7}\, x^3 -$
$0.003\, x^2 + 0.1401\, x + 28.47$; $r^2 = 0.8988$

In each of these equations, y represents temperature and x represents distance. The log function is the best, though the power function also provides a reasonably good fit. The remaining functions are poor.

5.11   (a) $y = 0.0018\, x^{1.599}$; $r^2 = 1$

(b)   Straight line: $y = 0.0282\, x - 0.2284$; $r^2 = 0.9706$

Exponential: $y = 0.0196\, e^{0.0596x}$; $r^2 = 0.7632$

Log: $y = 0.5331\, \ln(x)$; $r^2 = 0.6311$

(c)   3rd-degree polynomial:

$y = -6 \times 10^{-7}\, x^3 + 0.003\, x^2 + 0.0064\, x - 0.0123$; $r^2 = 1$

5th-degree polynomial:

$y = -5 \times 10^{-10}\, x^5 + 1 \times 10^{-7}\, x^4 - 1 \times 10^{-5}\, x^3 +$
$0.0007\, x^2 + 0.0009\, x + 0.0017$; $r^2 = 1$

In each of these equations, y represents reaction rate and x represents concentration.

5.12   Quadratic: $y = 0.0015\, x^2 - 0.7798\, x + 101.7$; $r^2 = 0.9939$

Power function: $y = 1 \times 10^{-43}\, x^{17.518}$; $r^2 = 0.9995$

In each of these equations, y represents reaction rate and x represents temperature. The power function appears to be a better fit, though the equation involves the product of a very small quantity and a very large quantity. The use of this equation can lead to large errors.

5.13   (a) $y = -108573\, x^5 + 14476\, x^4 + 1 \times 10^6\, x^3 - 3 \times 10^6\, x^2 +$
$590199\, x + 2 \times 10^6$; $r^2 = 11.769$

(b) $y = 1 \times 10^{-8}\, x^5 - 9 \times 10^{-6}\, x^4 - 0.0966\, x^3 - 17.191\, x^2 +$
$702348\, x - 7 \times 10^8$; $r^2 = 0.9$

In each of these equations, y represents ozone layer thickness and x represents time. Clearly, (b) is better than (a), though neither is as accurate as the original fit given in the example (based upon the r^2 values). Both of these results are affected by numerical errors in fitting the polynomial to the data.

5.14 The data can be fit reasonably well by the following cubic equation:

$$y = 0.0009\,x^3 - 0.0111\,x^2 - 0.0673\,x + 1.0269; \qquad r^2 = 0.9971$$

where y represents ozone layer thickness and x represents time.

5.15 The trendline equation shown in the graph is the same as the equation in Equation 5.8. However, the tabulated results provide more accurate coefficients, resulting in

$$y = 1.32353 \times 10^{-8}\,x^5 - 5.12383 \times 10^{-6}\,x^4 + 7.61652 \times 10^{-4}\,x^3 -$$
$$5.25453 \times 10^{-2}\,x^2 + 1.77701\,x - 20.95804; \qquad r^2 = 0.999754$$

where y represents time and x represents top speed.

5.17 Because of the scatter in the data, a meaningful curvefit cannot be obtained.

5.19 $y = 3 \times 10^{-8}\,x^6 - 3 \times 10^{-6}\,x^5 + 1 \times 10^{-5}\,x^4 + 0.0052\,x^3 -$
$$0.074\,x^2 + 0.4553\,x; \qquad r^2 = 0.9994$$

where y represents power and x represents wind velocity. Using this equation, we obtain

(*a*)   51.2 watts   (*b*)   40.5 mph

5.20 $y = 0.0001\,x^6 - 0.0039\,x^5 + 0.0455\,x^4 - 0.2845\,x^3 + 1.2871\,x^2 -$
$$4.7843\,x + 4.7727; \qquad r^2 = 0.9986$$

where y represents voltage and x represents time. Using this equation, we obtain

(*a*)   4.7 volts   (*b*)   −251.6b volts

(*c*)   3.56 sec   (*d*)   5.31 sec

The answer to part (*b*) is not meaningful because of the extrapolation well beyond the range of the data. The answers to parts (*c*) and (*d*) were obtained using Excel's **Goal Seek** feature, as explained in Chap. 7.

5.21 (*a*) $y = -5 \times 10^{-7}\,x^6 + 5 \times 10^{-5}\,x^5 - 0.0019\,x^4 + 0.0347\,x^3 -$
$$0.3256\,x^2 + 1.2584\,x; \qquad r^2 = 0.9952$$

(b)   For $0 \leq x \leq 10$, $y = -10^{-5} x^6 + 0.0005 x^5 - 0.0019 x^4 + 0.0347 x^3 - 0.3256 x^2 + 1.2584 x$; $r^2 = 1$

For $10 \leq x \leq 30$, $y = 4.8123 e^{-0.1987 x}$; $r^2 = 0.9997$

In the above equations, y represents current and x represents time.

(c)   Using the equation obtained in part (a), $y = 0.552$ ma when $x = 0.5$ sec, and $y = -4.91$ ma when $x = 22.8$ sec.

Using the equations obtained in part (b), $y = 0.627$ ma when $x = 0.5$ sec (from the first equation), and $y = 0.0519$ ma when $x = 22.8$ sec (from the second equation).

(d)   Using the equation obtained in part (a), $x = 1.049$ sec and 6.91 sec (via Goal Seek).

Using the first equation obtained in part (b), $x = 0.919$ sec (via Goal Seek). The second value could not be obtained.

5.24   (a) $y_A = 4.9919 e^{-0.1002 x}$; $r^2 = 0.985$

$y_B = 10^{-7} x^5 - 2 \times 10^{-5} x^4 + 0.0011 x^3 - 0.0357 x^2 + 0.4765 x$; $r^2 = 0.9986$

$y_C = -10^{-4} x^3 + 0.0051 x^2 + 0.0585 x$; $r^2 = 0.9971$

where y_A, y_B and y_C represent the concentrations of A, B, and C, respectively, and x represents time.

(b)   Using the above equations, we obtain $y_A = 0.498$, $y_B = 0.505$ and $y_C = 2.774$ moles per liter when $x = 23$ sec. Note that the value of y_B is considerably in error. This sometimes happens when using high-degree polynomials with rounded coefficients (as provided by Excel) and relatively large x-values. Unfortunately, this is not seen in the trendlines.

5.25   (b) $y = 3 \times 10^7 x$

where y represents stress and x represents strain.

(c)   For this material, the modulus of elasticity is 3×10^7 (the slope of the straight line obtained in part (b) above).

5.26   (b) $y = -10^{10} x^6 + 10^{10} x^5 - 4 \times 10^9 x^4 + 6 \times 10^8 x^3 - 5 \times 10^7 x^2 + 2 \times 10^6 x + 1753.2$; $r^2 = 0.9803$

(c)   Using the above equation, we obtain $y = -38569$ psi when $x = 0.115$. This value is considerably in error because of the rounded coefficients provided by Excel.

The elongation corresponding to a strain of 0.115 is determined as stain $\times$ original length $= 0.115 \times 4 = 0.460$ in.

Chapter 6

6.1 (*a*) 0.11452 Kcal / (kg)(°C) (*c*) 0.15204 Kcal / (kg)(°C)
    (*b*) 0.14356 Kcal / (kg)(°C) (*d*) 0.17880 Kcal / (kg)(°C)

6.2 (*a*) 35.7 °F (*c*) − 8.0 °F
    (*b*) 14.8 °F (*d*) −31.3 °F

6.3 (*a*) 0.978574 m / sec^2 (*c*) 0.978489 m / sec^2
    (*b*) -8.5×10^{-5} m / sec^2 (*d*) 0.980143 m / sec^2

6.4 (*a*) $2290.40 (*c*) 10.8 years
    (*b*) $3423.60 (*d*) $68,765.60

6.5 (*a*) 0.742 ma (*c*) 7.5 sec
    (*b*) 0.388 ma (*d*) 19.6 sec, 30.5 sec

6.6 (*a*) 1.382813 (*c*) 4.242188
    (*b*) 1.298948

6.7 2.59195 The result is inaccurate since it was obtained by extrapolating behind the base row.

6.8 3.96875 This result is based upon fewer data points (hence, a lower-degree interpolating polynomial). Therefore, it is less preferable.

6.9 1.531248 using the second row as the base row; 1.531247 using the fifth row. Both results are very close to the result obtained in Example 6.5, though they differ markedly from the (erroneous) result obtained in Prob. 6.7.

6.10 (*a*) 1.298948 (*c*) 652.9931
    (*b*) 56.74656 (*d*) 4.242188
The answers to parts (*a*) and (*d*) are virtually identical to those obtained in Prob. 6.6.

6.11 0.114529 Kcal / (kg)(°C), which is very close to the value obtained in Prob. 6.1(*a*).

6.12 0.151909 Kcal / (kg)(°C), which is somewhat lower than the value obtained in Prob. 6.1(*c*).

6.13 (*a*) 0.80605 ma   (*c*) 0.823573 ma
(*b*) 0.81879 ma

All of these values are higher than the value obtained in Prob. 6.5(*a*).

6.14 (*a*) 27   (*c*) 4.913
(*b*) 1.728   (*d*) 125

If an extended forward difference table is used (as in Example 6.6), it does not matter which row is selected as the base row.

6.15 0.382813 ma, which is close to the value obtained using linear interpolation (0.388 ma).

6.16 (*a*) 120.259   (*c*) 173.075
(*b*) 21.97656

6.17 275.115   The result is inaccurate, since it was obtained by extrapolating ahead of the base row.

6.18 (*a*) 120.475   (*c*) 172.483
(*b*) 21.859

These results are based upon fewer data points (hence, a lower-degree interpolating polynomial). Therefore, they are less preferable.

6.19 279.773 using the sixth row as the base row; 276.249 using the second row. The first results is very close to that obtained in Example 6.8 (279.771). The second result is similar to the less accurate result obtained in Prob. 6.17 (275.115)

6.20 (*a*) 120.202   (*c*) 2.815
(*b*) 21.979   (*d*) −21.738

The first two values are very similar to those obtained in Prob. 6.16.

6.21 0.152032 Kcal / (kg)(°C). These values are very close to those obtained earlier.

6.22 Three data points: 0.380469 ma
Four data points: 0.381543 ma
Five data points: 0.382819 ma

These values are similar to the value 0.388 ma, obtained in Prob. 6.5 (*b*). The additional data points generally contribute to higher accuracy. (Remember, however, that high-degree polynomials can sometimes produce wild oscillations.)

6.23   Three data points: 0.382813 ma

Four data points: 0.382813 ma

Five data points: 0.384021 ma

The results should be more accurate because the given point is now closer to the base row.

6.24   (*a*)   27   (*c*)   4.913

(*b*)   1.728   (*d*)   125

For this data set, it does not matter which row is selected as the base row. The results are identical with those obtained in Prob. 6.14 using forward differences.

6.25   (*a*)   29.25   (using linear interpolation)

(*b*)   27.0   (using Lagrangian interpolation)

The second answer (Lagrangian interpolation) is identical with the result obtained in Prob. 6.24(*a*).

6.26   (*a*)   0.0796 ma   (*c*)   0.0721 ma

(*b*)   0.0508 ma

6.27   (*a*)   577.52 psia   (using Linear interpolation)

(*b*)   566.06 psia   (using Lagrangian interpolation)

The second answer is very close to the known value.

Chapter 7

7.1   (*a*)   One real root.

(*b*)   No real roots, two complex roots.

(*c*)   One real root, two complex roots.

(*d*)   Two real roots, including one at the origin.

7.2   The result is a cubic polynomial; hence, there will be at least one, and possibly three, real roots.

7.3   Many real roots (sinusoidal function).

7.4 $x = -1.5$ (approximately)

7.5 $x = 2.4$ (approximately)

7.6   (*a*) $V = 24.5$ liters / mole   Ideal gas law: $V = 24.6$ liters / mole
     (*b*) $V = 3.21$ liters / mole   Ideal gas law: $V = 3.28$ liters / mole
     (*c*) $V = 246.1$ liters / mole   Ideal gas law: $V = 246.2$ liters / mole

7.7   (*a*) $V = 23.1$ liters / mole   Ideal gas law: $V = 24.6$ liters / mole
     (*b*) $V = 0.25$ liters / mole   Ideal gas law: $V = 3.28$ liters / mole
     (*c*) $V = 244.7$ liters / mole   Ideal gas law: $V = 246.2$ liters / mole

7.8   (*a*)   1.08, 3.64, 6.58   (Do not mistake discontinuities for roots.)
     (*b*)   −1.08, −3.64

7.9   −1.49381

7.10   2.411917

7.11 $V = 0.250822$ liters/mole

7.12   1.076874, −1.076874

7.13   55.56°

7.14   (*a*)   \$313.37
      (*b*)   1.3068% (15.68% per year)
      (*c*)   32 months
      (*d*)   Parts (*b*) and (*c*) require Goal Seek (or Solver).

7.15   3.137833   (Do not mistake a discontinuity for a root.)

7.16   (*a*)   3.40535 in   (*b*)   2.55401 in

7.17   (*a*)   82.2°   (1.434 radians)   (*b*)   126.3°   (2.205 radians)

7.18 $f = 0.037393, 0.029441, 0.025883$.   Note that *f* decreases as *Re* increases.

7.19   (*b*) $t = 3.54$ sec   (*c*) $t = 9.82$ sec   (*d*) $\omega = 0.686$ sec^{-1}

7.20   (*b*) $t = 0.049$ sec   (*c*) $t = 0.149$ sec   (*d*) $R = 2963$ ohms

7.21   The value of Equation (7.7) is 6.54×10^{-7} at the root. This suggests very rapid convergence.

7.22   (*a*) $x = -1.49268$ (*b*) $x = -1.49380$

The Newton-Raphson method requires less computation. Both results are close to those obtained in the earlier problems. In particular, the solution obtained with the Newton-Raphson method is essentially the same as that obtained in Prob. 7.9.

7.23   (*a*)   2.411622   (*b*)   2.412011

The Newton-Raphson method again requires less computation. Both results are close to those obtained in the earlier problems.

7.24   (*a*)   0.250734 liters / mole   (*b*)   0.250822 liters / mole

The results are close to, but not identical with, those obtained earlier.

7.25   (*a*)   1.077052, -1.08 (*b*)   1.076874, -1.076874

The solutions obtained with the Newton-Raphson method are identical to those obtained in Prob. 7.12. The results obtained using the method of bisection are less accurate.

Chapter 8

8.1   (*a*) $AX = B$, where

$$A = \begin{bmatrix} 1 & -2 & 3 \\ 3 & 1 & -2 \\ 2 & 3 & 1 \end{bmatrix}, \qquad X = \begin{bmatrix} x_1 \\ x_2 \\ x_3 \end{bmatrix}, \qquad B = \begin{bmatrix} 17 \\ 0 \\ 7 \end{bmatrix}$$

(*b*) $AX = B$, where

$$A = \begin{bmatrix} 0.1 & -0.5 & 0 & 1 \\ 0.5 & -2.5 & 1 & -0.4 \\ 1 & 0.2 & -0.1 & 0.4 \\ 0.2 & 0.4 & -0.2 & 0 \end{bmatrix}, \qquad X = \begin{bmatrix} x_1 \\ x_2 \\ x_3 \\ x_4 \end{bmatrix}, \qquad B = \begin{bmatrix} 2.7 \\ -4.7 \\ 3.6 \\ 1.2 \end{bmatrix}$$

(*c*) $AX = B$, where

$$A = \begin{bmatrix} 11 & 3 & 0 & 1 & 2 \\ 0 & 4 & 2 & 0 & 1 \\ 3 & 2 & 7 & 1 & 0 \\ 4 & 0 & 4 & 10 & 1 \\ 2 & 5 & 1 & 3 & 13 \end{bmatrix}, \qquad X = \begin{bmatrix} x_1 \\ x_2 \\ x_3 \\ x_4 \\ x_5 \end{bmatrix}, \qquad B = \begin{bmatrix} 51 \\ 15 \\ 15 \\ 20 \\ 92 \end{bmatrix}$$

8.2 (a) $C = \begin{bmatrix} 29 & 32 & 35 & 38 \\ 65 & 72 & 79 & 86 \\ 101 & 112 & 123 & 134 \end{bmatrix}$ (c) $C = \begin{bmatrix} 4 & 5 & 6 \\ 8 & 10 & 12 \\ 12 & 15 & 18 \end{bmatrix}$

(b) $C = \begin{bmatrix} 32 \end{bmatrix}$ (d) $C = \begin{bmatrix} 5 \\ 17 \\ 29 \end{bmatrix}$

8.3 $\begin{aligned} 4x_1 - 2x_2 + x_3 &= 6 \\ x_1 + 3x_2 - 7x_3 &= 34 \\ -2x_1 \qquad - 5x_3 &= 14 \end{aligned}$

8.4 (a) $A + B = \begin{bmatrix} 11 & -5 & 4 \\ -1 & 13 & 11 \\ 12 & -6 & 2 \end{bmatrix}$ (d) $2A = \begin{bmatrix} 12 & 0 & 6 \\ -2 & 8 & 18 \\ 16 & -10 & 4 \end{bmatrix}$

(b) $B + A = \begin{bmatrix} 11 & -5 & 4 \\ -1 & 13 & 11 \\ 12 & -6 & 2 \end{bmatrix}$ (e) $AB = \begin{bmatrix} 42 & -33 & -42 \\ 31 & 32 & 15 \\ 48 & -87 & -66 \end{bmatrix}$

(c) $A - B = \begin{bmatrix} 1 & 5 & 10 \\ -1 & -5 & 7 \\ 4 & -4 & 2 \end{bmatrix}$ (f) $BA = \begin{bmatrix} -21 & 15 & -44 \\ 7 & 26 & 853 \\ 25 & -4 & \end{bmatrix}$

8.5 (a) $IA = \begin{bmatrix} 6 & 0 & -3 \\ -1 & 4 & 9 \\ 8 & -5 & 2 \end{bmatrix}$ (b) $AI = \begin{bmatrix} 6 & 0 & -3 \\ -1 & 4 & 9 \\ 8 & -5 & 2 \end{bmatrix}$

8.6 $A^{-1}A = \begin{bmatrix} 1 & 0 \\ 0 & 1 \end{bmatrix} = AA^{-1}$

8.7 $AB = I$ Hence, B is the inverse of A. The results will be imprecise because of numerical roundoff.

8.8 $A^{-1}B = X = \begin{bmatrix} 3 \\ 1 \\ -4 \end{bmatrix}$; hence, $x_1 = 3$, $x_2 = 1$, $x_3 = -4$.

8.9 (a) $A + B = \begin{bmatrix} 11 & -5 & 4 \\ -1 & 13 & 11 \\ 12 & -6 & 2 \end{bmatrix}$ (c) $A - B = \begin{bmatrix} 1 & 5 & 10 \\ -1 & -5 & 7 \\ 4 & -4 & 2 \end{bmatrix}$

(b) $B + A = \begin{bmatrix} 11 & -5 & 4 \\ -1 & 13 & 11 \\ 12 & -6 & 2 \end{bmatrix}$ (d) $2A = \begin{bmatrix} 12 & 0 & 6 \\ -2 & 8 & 18 \\ 16 & -10 & 4 \end{bmatrix}$

8.10 (a) $C = \begin{bmatrix} 29 & 32 \\ 65 & 72 \end{bmatrix}$ (c) $C = \begin{bmatrix} 4 & 5 & 6 \\ 8 & 10 & 12 \\ 12 & 15 & 18 \end{bmatrix}$

(b) $C = 32$ (d) $C = \begin{bmatrix} 5 \\ 17 \\ 29 \end{bmatrix}$

8.11 $A^{-1} = \begin{bmatrix} 0.166667 & 0.166667 \\ 0.222222 & -0.111111 \end{bmatrix}$, $AA^{-1} = AA^{-1} = I$

8.12 (a) $IA = \begin{bmatrix} 6 & 0 & -3 \\ -1 & 4 & 9 \\ 8 & -5 & 2 \end{bmatrix}$ (b) $AI = \begin{bmatrix} 6 & 0 & -3 \\ -1 & 4 & 9 \\ 8 & -5 & 2 \end{bmatrix}$

8.13 $AB = \begin{bmatrix} 1.000002 & 0.054 & 0 \\ -6\times10^{-6} & 0.838002 & -5\times10^{-6} \\ -4\times10^{-6} & -0.036 & 1 \end{bmatrix}$; hence, $B = A^{-1}$

The results are imprecise because of numerical roundoff.

8.14 $x_1 = 3, x_2 = 1, x_3 = -4$

8.15 (a) $x_1 = 3, x_2 = -1, x_3 = 4$
(b) $x_1 = 2, x_2 = 1, x_3 = -2, x_4 = 3$
(c) $x_1 = 2.98, x_2 = 2.22, x_3 = 0.211, x_4 = 0.152, x_5 = 5.72$ (rounded)

8.16 (a) $x_1 = 0.2, x_2 = 1.1$
(b) $x_1 = 3, x_2 = 1, x_3 = -4$
(c) $x_1 = 1, x_2 = 5, x_3 = 7, x_4 = 1$

8.17 $x_1 = 0.199998, x_2 = 1.100001$
The results are close to those obtained in Prob. 8.16.   Numerical error produces some inaccuracy.

8.18 $x_1 = 2.999998, x_2 = 0.999971, x_3 = -3.9999$
$x_1 = 1, x_2 = 5, x_3 = 7, x_4 = 1$

8.19   (a) $x_1 = 1.815748, x_2 = 5.189435$
(b) $x_1 = 1.681515, x_2 = 0.953329$
(c) $x_1 = 1.660263, x_2 = 3/078627, x_3 = 0.997984$

8.20 $i_1 = 0.406429$ amperes $i_6 = 1.369441$ amperes
$i_2 = 0.545575$ $i_7 = 0.409952$
$i_3 = 1.047996$ $i_8 = 1.439454$
$i_4 = 1.77587$ $i_9 = 0.218846$
$i_5 = 0.82915$ $i_{10} = 1.220608$

8.21 $i_1 = 1.369441$ amperes $i_4 = -0.005$ amperes
$i_2 = 0.113077$ $i_5 = -0.04371$
$i_3 = 0.108601$ $i_6 = 0.065175$

8.22 $R_1 = 7538$ lb$_f$ $T_3 = 2409$ lb$_f$
$R_2 = 11,746$ $T_4 = -2409$
$R_3 = 2863$ $T_5 = 1653$
$T_1 = -13,563$ $T_6 = -3306$
$T_2 = -756$ $T_7 = -5398$
Negative internal forces represent compression.   The results are rounded.

8.23 $T_0 = 197°C, T_1 = 49.3°C, T_2 = 43.5°C, T_3 = 23.7°C, Q = 3$ cal/(cm^2)(sec)
The results are rounded.

8.24   Let x_1 = metric tons of alloy A1 manufactured per day
x_2 = metric tons of alloy A2 manufactured per day
x_3 = metric tons of alloy A3 manufactured per day
x_4 = metric tons of alloy A4 manufactured per day

The following equations must be solved:

$$16x_1 + 6x_2 + 3x_3 + 14x_4 = 1200$$
$$7x_1 + 3x_2 + 7x_3 + 9x_4 = 800$$
$$12x_1 + 10x_2 + 11x_3 + 7x_4 = 1000$$
$$3x_1 + 8x_2 + 15x_3 + 22x_4 = 1500$$

The solution is: $x_1 = 28.8$ tons/day $x_3 = 25.3$ tons/day
$x_2 = 6.3$ tons/day $x_4 = 44.7$ tons/day

8.25 $x_b = 62.8$ ft, $w = 92.8$ ft, $v = 80.9$ ft

8.26 $x_0 = 11.2$ in, $\beta = 0.270$ sec^{-1}, $\omega = 0.83$ sec^{-1}
The solution is sensitive to the initial guess.

Chapter 9

9.1 (*a*) 156   (*b*) 157.5   (*c*) 156.24

9.2 The solution must fall between 152.88 and 159.12. Using five equally spaced intervals, we obtain 159.84 (value is too high; hence, too few intervals). Using six equally spaced intervals, we obtain 158.67, which satisfies the criterion. Hence, the solution is six equally spaced intervals.

9.3 (*a*) 2   (*b*) 1.983524   (*c*) 1.997143

9.4 $I = 115692.9\,^\circ$ cm. Hence, $\bar{T} = 115692.9 / (2400 - 0.1) = 48.207383^\circ$.
Simple arithmetic average: $\bar{T} = 40.299^\circ$.

9.5 Fitting a sixth-degree polynomial to the data,

$y = -489{,}770\,x^6 + 813{,}150\,x^5 - 530{,}960\,x^4 + 172{,}082\,x^3 - 28{,}281\,x^2 +$
$\qquad 1978.8\,x + 1.2224$

Integrating this expression from 0 to 0.5 sec, we obtain $I = 10.82727$.
Hence, $L = 10.82727 / 2.15 = 5.03594$ henries.

9.6 (*a*) $I = 10.2358$; $L = 4.761$ henries
     (*b*) $I = 10.67137$; $L = 4.9634$ henries
     (*c*) $I = 10.72223$; $L = 4.987$ henries

9.7 $\overline{RR} = 62.9 / (308 - 253) = 1.143636$ moles / sec
Arithmetic average: $\overline{RR} = 1.2075$ moles / sec

9.8 $I = 2773.05$. Hence, average power $= 2773.05 / 60 = 46.2175$ watts.

9.9 $I = 10{,}980.56$. Hence, average stress $= 10{,}980.56 / 0.24 = 45{,}752.33$ psi.

9.10 (*a*) 156   (*b*) 156   (*c*) 156

9.11 The solution must fall between 152.88 and 159.12. Using only two equally spaced intervals, we obtain 154.32, which satisfies the criterion. When using the trapezoidal rule, six intervals were required.

9.12 (*a*)   2   (*b*)   2.0001   (*c*)   2.000

9.13 (*a*) $I = 10.77629$; $L = 5.012$ henries
        (*b*) $I = 10.82663$; $L = 5.0356$ henries
        (*c*) $I = 10.82727$; $L = 5.03594$ henries

9.14 $I = 2768.57$. Hence, average power = $2768.57 / 60 = 46.1428$ watts.

9.15 (*a*)   2 coulombs   (*b*)   2.626 coulombs   (*c*)   2.118 coulombs

The solutions obtained with the trapezoidal rule and Simpson's rule are each based upon ten equally spaced intervals.

9.16 (*a*)   320 cubic feet (classical calculus)
              320 cubic feet (trapezoidal rule, 10 equally spaced intervals)
              320 cubic feet (Simpson's rule, 10 equally spaced intervals)

        (*b*)   240 cubic feet (classical calculus)
              240 cubic feet (trapezoidal rule, 10 equally spaced intervals)
              240 cubic feet (Simpson's rule, 10 equally spaced intervals)

        (*c*)   191.25 cubic feet (classical calculus)
              191.25 cubic feet (trapezoidal rule, 10 equally spaced intervals)
              191.25 cubic feet (Simpson's rule, 10 equally spaced intervals)

9.17 Using classical calculus, we obtain

for $0 \leq t \leq 8$: $v = 6t^2$, $h = 2t^3$

for $t \geq 8$: $v = 641.6 - 32.2t$, $h = 641.6t - 16.1\,t^2 - 3078.4$

where v represents velocity, h represents height, and t represents time. Using the second equation, set the first derivative of h with respect to time equal to zero, resulting in $h_{max} = 3313.7$ ft at (approximately) 19.9 sec. Then set the second equation for h equal to zero and solve for t_{max}, resulting in $t_{max} = 34.3$ sec (approximately).

To determine the maximum height or the time for the rocket to hit the ground by numerical integration, we must *assume* a value for the final time (i.e., the time when the desired event occurs) and then carry out the integration. (Remember to use an even number of equally spaced intervals with Simpson's rule). This is a tedious trial-and-error process. A better approach is to treat the velocity equation as a *differential equation* and "step ahead" to determine the height as a function of time; for example,

$$h(t + \Delta t) = h(t) + v\,\Delta t$$

To carry out this process accurately, however, is more complicated than it appears. A detailed discussion is beyond the scope of this book.

9.18 (*a*) $Q = 775.3733$ BTU   (*c*) $Q = 774.3733$ BTU
(*b*) $Q = 774.3712$ BTU

The solutions obtained with the trapezoidal rule and Simpson's rule are each based upon ten equally spaced intervals.

9.19 (*a*) $Q = 0.039269$ cu ft / sec   (*c*) $Q = 0.03927$ cu ft / sec
(*b*) $Q = 0.038877$ cu ft / sec

The solutions obtained with the trapezoidal rule and Simpson's rule are each based upon ten equally spaced intervals.

Chapter 10

10.1 (*a*) $3402.97   (*b*) $3425.28   (*c*) $3436.19

10.2 $8292.82

10.3 (*a*)   5.9254%   (*b*)   5.7981%   (*c*)   5.7703%
(*d*)   5.7569%

10.4 (*a*)   14.21 years   (*b*)   55.8 quarters (13.95 years)
(*c*)   166.7 months (13.89 years)
(*d*)   5060.3 days (13.86 years)

10.5 (*a*) $24,569.13   (*b*) $36,490.10   (*c*) $118,849.53

10.6 (*a*) $5372.54   (*b*) $5442.06   (*c*) $5458.19
(*d*) $5466.09

10.7 9.045% per year

10.8 39.5 months (3.29 years)

10.9 10.63 years

10.10 24.68 years

10.11 (*a*) $P = 7129.86 (*b*) $P = 7275.11
Hence, choose (*b*).

10.12 (*a*) $P = 6805.83 (*b*) $P = 6753.36
Hence, choose (*a*).

10.13 (*a*) $P = \$7047.12$ (*b*) $P = \$7140.50$
Hence, choose (*b*).

10.14 (*a*) $P = \$7129.86$ (*b*) $P = \$6799.17$
Hence, choose (*a*).

10.17 $A = \$132,695$ per year

10.18 $A = \$26,155.37$

10.19 10.56% per year

10.20 $A = \$27,101.16$ per year

10.21 26.7 years (realistically, 27 years).

10.22 (*a*) $A = \$643.70$ per month   (*b*)   Interest $= \$151,732$

10.23 (*a*) $\$304.35$ per month   (*b*)   Interest $= \$2608.80$

10.24 (*a*) $\$237,174.56$ (*b*)   33.4 years

10.25 $\$3,794.67$ per year

10.26 $\$3,579.88$ per year. The amount of each yearly payment is less, because it has more time to earn interest.

10.27 62.4 years

10.28 3.46% per year

10.29 The money is compounded monthly. Hence, the effective annual interest rate is $(1 + 0.055/12)^{12} - 1 = 0.056408$ (or 5.6408%). Using this interest rate, the present value of $\$50,000$ per year for 20 years is $\$590,600.79$. Therefore, the student should accept the lump-sum payment of $\$650,000$.

10.30 The first alternative has a present value of $\$11,060,221$, resulting in a profit of $\$1,060,221$. The second alternative has a present value of $\$10,220,538$, resulting in a profit of $\$220,538$. Hence, the company should select the first alternative.

10.31 Now the first alternative has a present value of $\$9,767,238$ and the second alternative has a present value of $\$8,876,470$. Since both of these values are less than the original $\$10$ million investment, the company should not invest in either one.

10.32 *(a)* $1134   *(b)* $1134

10.33 *(a)*   −$70,709   *(b)*   −$70,709

In this situation, the present value of the future revenues is less than the initial cash outlay.

10.34 The present value of investment A is $= \$107,470$, and the present value of investment B is $51, 604$; hence, choose investment A.

If the time value of money is not taken into account (i.e., if $i = 0$), the present value of investment A would be $300,000 and the present value of investment B would be $400,000. Hence, investment B would be more desirable.

10.35 The present value of investment A is $33,276, and the present value of investment B is $31,357. Hence, investment A is more desirable.

10.36 *(a)* $PV_A = \$250,000$; $PV_B = \$320,000$; choose B

(b) $PV_A = \$131,855$; $PV_B = \$150,753$; choose B

(c) $PV_A = \$40,716$; $PV_B = \$25,699$; choose A

10.37 $PV_A = \$23,530$; $PV_B = \$5,005$; Investment A is more desirable.

If the time value of money is not taken into account, $PV_A = \$350,000$ and $PV_B = \$350,000$. Hence, they would be equally desirable.

10.38 Expand current plant: $PV = \$9,047,762$

Build new plant: $PV = \$11,622,602$

Hence, it would be better to expand the current plant.

10.39 Expand current plant: $PV = \$9,047,762$

Build new plant: $PV = \$7,116,632$

Now it would be better to build the new plant.

10.40 Expand now: $PV = -\$10,000,000$

Expand over three years: $PV = -\$10,072,245$

Hence, it would be better to expand now.

10.41 Expand now: $PV = -\$10,000,000$

Expand over three years: $PV = -\$9,703,011$

Hence, it would be better to expand partially now and the remainder in three years.

10.42 $IRR = 13.87\%$

10.43 (a) $NPV(8\%) = \$2,380,000$
$NPV(9\%) = \$1,120,000$
$NPV(10\%) = -\$28,800$

(d) $IRR = 9.97\%$

10.44 $u = 1.138712$. Hence, $i = (u - 1) = 0.138712$ (13.87%).

10.45 (a) $IRR_A = 21.86\%$, $IRR_B = 14.71\%$. Choose A, as in Prob. 10.34.

(b) $IRR_A = 56.25\%$, $IRR_B = 45.37\%$. Choose A, as in Prob. 10.35.

(c) $IRR_A = 8.75\%$, $IRR_B = 8.15\%$. Choose A, as in Prob. 10.37.

10.46 (b) $PV_A = \$149,422$, $PV_B = \$175,446$. Investment A is more attractive.

(c) $IRR_A = 15.24\%$, $IRR_B = 13.48\%$. Investment B is more attractive.

(d)   From the plot of present value versus interest rate, we see that the curves cross each other at approximately $i = 0.085$ (8.5%). Hence, a comparison based upon an interest rate below 8.5% will favor A, whereas a comparison at an interest rate above 8.5% (including the IRR comparison) will favor B.

10.47 $IRR_A = 24.81\%$, $IRR_B = 23.44\%$. Hence, the first proposal (expand the current plant) appears more desirable. This is consistent with the conclusion reached in Prob. 10.40, but it differs with the conclusion reached in Prob. 10.41. (For an explanation, plot the present value versus interest rate for each proposal.)

Chapter 11

11.1 (a) $y_{min} = 44,000$ at $x_1 = 0$, $x_2 = 1000$

(b) $y_{max} = 96,000$ at $x_1 = 1600$, $x_2 = 0$

(c) $y_{max} = 80,000$ at $x_1 = 1600$, $x_2 = 0$

(d) $y_{max} = 117,000$ at $x_1 = 0$, $x_2 = 2670$

(e) $y_{min} = 96,000$ at $x_1 = 1600$, $x_2 = 0$

(f) $y_{max} = 60,000$ at $x_1 = 1000$, $x_2 = 0$.

11.2 (a) $y_{max} = 4.67$ at $x_1 = 3.33$, $x_2 = 1.33$

(b) $y_{min} = 0$ at $x_1 = 0$, $x_2 = 0$

(c) $y_{min} = 4.67$ at $x_1 = 3.33$, $x_2 = 1.33$

(d) $y_{max} =$ unbounded

(e) $y_{min} = 12$ at $x_1 = 6$, $x_2 = 0$

11.3   (a) $y_{min} = 20.5$ at $x_1 = 1.4$, $x_2 = 1.1$
      (b) $y_{min} = 11.6$ at $x_1 = 0$, $x_2 = 1.8$
      (c) $y_{min} = 10$ at $x_1 = 0.5$, $x_2 = 0.5$
      (d) y_{max} = unbounded
      (e) $y_{min} = 10$ at $x_1 = 0.5$, $x_2 = 0.5$
      (f) $y_{max} = 11.2$ at $x_1 = 1.3$, $x_2 = 1.3$

11.4   (a) $y_{min} = 0$ at $x_1 = 2$, $x_2 = 1$
      (b) y_{min} = unbounded
      (c) y_{max} = unbounded

11.5   (a) $y_{min} = 44{,}000$ at $x_1 = 0$, $x_2 = 1000$
      (b) $y_{max} = 96{,}000$ at $x_1 = 1600$, $x_2 = 0$
      (c) $y_{max} = 80{,}000$ at $x_1 = 1600$, $x_2 = 0$
      (d) $y_{max} = 117{,}333.48$ at $x_1 = 0$, $x_2 = 2666.67$
      (e) $y_{min} = 96{,}000$ at $x_1 = 1600$, $x_2 = 0$
      (f) $y_{max} = 60{,}000$ at $x_1 = 1000$, $x_2 = 0$.

11.6   (a) $y_{max} = 4.6667$ at $x_1 = 3.3333$, $x_2 = 1.3333$
      (b) $y_{min} = 0$ at $x_1 = 0$, $x_2 = 0$
      (c) $y_{min} = 4.6667$ at $x_1 = 3.3333$, $x_2 = 1.3333$
      (d) y_{max} = unbounded
      (e) $y_{min} = 12$ at $x_1 = 6$, $x_2 = 0$

11.8   (a) $y_{min} = 20.49806$ at $x_1 = 1.393167$, $x_2 = 1.11453$
      (b) $y_{min} = 11.64898$ at $x_1 = 0$, $x_2 = 1.784125$
      (c) $y_{min} = 10$ at $x_1 = 0.5$, $x_2 = 0.5$
      (d) y_{max} = unbounded
      (e) $y_{min} = 10$ at $x_1 = 0.5$, $x_2 = 0.5$
      (f) $y_{max} = 11.15997$ at $x_1 = 1.261567$, $x_2 = 1.261567$

11.9   (a) $y_{min} = 0$ at $x_1 = 2$, $x_2 = 1$
      (b) y_{min} = unbounded
      (c) y_{max} = unbounded

11.10   (a) $y_{max} = 22.384610$ at $x_1 = 4.153846$, $x_2 = 0$, $x_3 = 1.923077$
      (b) $y_{min} = 5$ at $x_1 = 0$, $x_2 = 2.5$, $x_3 = 0$, $x_4 = 0$
      (c) $y_{max} = 19$ at $x_1 = 4.666667$, $x_2 = 3.666667$, $x_3 = 0$, $x_4 = 9$

11.11 (a) $y_{max} = 1.819706$ at $x = 2.028758$

(b) $y_{min} = 2.97 \times 10^{-11}$ at $x_1 = 3$, $x_2 = 4.999998$

(c) $y_{min} = 9.3 \times 10^{-12}$ at $x_1 = 0.999998$, $x_2 = 0.999996$

This well-known unconstrained optimization problem is particularly difficult to solve. The correct answer is $y_{min} = 0$ at $x_1 = 1$, $x_2 = 1$. Excel handles it very well.

(d) $y_{min} = 4$ at $x_1 = 1$, $x_2 = 0$, $x_3 = 4$

11.12 *Model*:   Maximize $y = 25(x_1 + x_2 + x_3) - (18x_1 + 15x_2 + 10x_3)$

subject to $x_1 + x_2 + x_3 = 200$

$0.05x_1 + 0.04x_2 + 0.02x_3 \geq 0.03(x_1 + x_2 + x_3)$

$0.08x_1 + 0.02x_2 + 0.02x_3 \geq 0.02(x_1 + x_2 + x_3)$

$0.03x_1 + 0.03x_2 + 0.05x_3 \leq 0.035(x_1 + x_2 + x_3)$

$x_1, x_2, x_3 \geq 0$

Solution: $y_{max} = 2250$ at $x_1 = 0$, $x_2 = 150$, $x_3 = 50$

11.13 *Solution*: $y_{max} = 1950$ at $x_1 = 100$, $x_2 = 50$, $x_3 = 50$

11.14 *Model*:   Maximize $y = 538x_1 + 589x_2 + 48x_3 + 60x_4 + 438x_5 + 424x_6 + 44x_7 + 50x_8$

subject to $9(x_1 + x_5) + 12(x_2 + x_6) + 4(x_3 + x_7) + 5(x_4 + x_8) \leq 3000$

all $x_i \geq 0$

where x_1 = number of Premier walnut tables mfg. per week
x_2 = number of Executive walnut tables mfg per week
x_3 = number of Premier walnut chairs mfg. per week
x_4 = number of Executive walnut chairs mfg per week
x_5 = number of Premier oak tables mfg. per week
x_6 = number of Executive oak tables mfg per week
x_7 = number of Premier oak chairs mfg. per week
x_8 = number of Executive oak chairs mfg per week

Solution: $y_{max} = 179{,}154$ at $x_1 = 333$, all other $x_i = 0$

11.15 *Model*:   Add the following constraints to the model in Prob. 11.14:

$x_3 = 4x_1$ $x_4 = 4x_2$ $x_7 = 4x_5$ $x_8 = 4x_6$

Solution: $y_{max} = 87{,}600$ at $x_1 = 120$, $x_3 = 480$, all other $x_i = 0$.

11.16 *Model*:   Add the following constraint to the model in Prob. 11.15:

$100x_1 + 140x_2 + 16x_3 + 20x_4 \leq 10{,}000$

Solution: $y_{max} = 80{,}756$ at $x_1 = 61$, $x_3 = 244$, $x_5 = 59$, $x_7 = 236$, all other $x_i = 0$.

11.17 *Model:* Minimize $y = 20x_{11} + 25x_{12} + 40x_{13} + 65x_{14} + 25x_{21} + 12x_{22} + 20x_{23} + 45x_{24} + 60x_{31} + 50x_{32} + 15x_{33} + 35x_{34}$

subject to $\quad x_{11} + x_{12} + x_{13} + x_{14} \le 5{,}000$
$$x_{21} + x_{22} + x_{23} + x_{24} \le 8{,}000$$
$$x_{31} + x_{32} + x_{33} + x_{34} \le 12{,}000$$
$$x_{11} + x_{21} + x_{31} \ge 6{,}000$$
$$x_{12} + x_{22} + x_{32} \ge 8{,}000$$
$$x_{13} + x_{23} + x_{33} \ge 5{,}000$$
$$x_{14} + x_{24} + x_{34} \ge 6{,}000$$
$$\text{all } x_{ij} \ge 0$$

where x_{11} = units shipped from Philadelphia to Atlanta
x_{12} = units shipped from Philadelphia to Chicago
x_{13} = units shipped from Philadelphia to Denver
x_{14} = units shipped from Philadelphia to Seattle
x_{21} = units shipped from St. Louis to Atlanta
x_{22} = units shipped from St. Louis to Chicago
x_{23} = units shipped from St. Louis to Denver
x_{24} = units shipped from St. Louis to Seattle
x_{31} = units shipped from Phoenix to Atlanta
x_{32} = units shipped from Phoenix to Chicago
x_{33} = units shipped from Phoenix to Denver
x_{34} = units shipped from Phoenix to Seattle

Solution: $y_{min} = 546{,}000$ at $x_{11} = 5{,}000$, $x_{22} = 3{,}000$, $x_{23} = 5{,}000$, $x_{31} = 1{,}000$, $x_{32} = 5{,}000$, $x_{34} = 6{,}000$, all other $x_{ij} = 0$

11.18 *Model:* Maximize $y = P_1(x_{11} + x_{21} + x_{31} + x_{41}) + P_2(x_{12} + x_{22} + x_{32} + x_{42}) + P_3(x_{13} + x_{23} + x_{33} + x_{43}) + (S_1/42)(42C_1 - x_{11} - x_{12} - x_{13}) + (S_2/42)(42C_2 - x_{21} - x_{22} - x_{23}) + (S_3/42)(42C_3 - x_{31} - x_{32} - x_{33}) + (S_4/42)(42C_4 - x_{41} - x_{42} - x_{43})$

subject to $\quad x_{11} + x_{21} + x_{31} + x_{41} \ge 42D_1$
$$x_{12} + x_{22} + x_{32} + x_{42} \ge 42D_2$$
$$x_{13} + x_{23} + x_{33} + x_{43} \ge 42D_3$$
$$x_{11}N_1 + x_{21}N_2 + x_{31}N_3 + x_{41}N_4 \ge R_1(x_{11} + x_{21} + x_{31} + x_{41})$$
$$x_{12}N_1 + x_{22}N_2 + x_{32}N_3 + x_{42}N_4 \ge R_2(x_{12} + x_{22} + x_{32} + x_{42})$$
$$x_{13}N_1 + x_{23}N_2 + x_{33}N_3 + x_{43}N_4 \ge R_3(x_{13} + x_{23} + x_{33} + x_{43})$$

where x_{11} = gallons of base fuel 1 used in gasoline 1
x_{21} = gallons of base fuel 2 used in gasoline 1
x_{31} = gallons of base fuel 3 used in gasoline 1
x_{41} = gallons of base fuel 4 used in gasoline 1
x_{12} = gallons of base fuel 1 used in gasoline 2
x_{22} = gallons of base fuel 2 used in gasoline 2
x_{32} = gallons of base fuel 3 used in gasoline 2
x_{42} = gallons of base fuel 4 used in gasoline 2
x_{13} = gallons of base fuel 1 used in gasoline 3
x_{23} = gallons of base fuel 2 used in gasoline 3
x_{33} = gallons of base fuel 3 used in gasoline 3
x_{43} = gallons of base fuel 4 used in gasoline 3

This model assumes full capacity production.

Solution: y_{max} = 123,356 at x_{11} = 436,800, x_{41} = 109,200, x_{12} = 756,000, x_{42} = 294,000, x_{13} = 423,360, x_{43} = 332,640, all other x_{ij} = 0.

All x_{ij} expressed in terms of gallons per day.

11.19 *Part 1*: y_{max} = 116,844.3 at x_{11} = 436,800, x_{41} = 109,200, x_{12} = 755,904, x_{42} = 294,088, x_{13} = 423,360, x_{43} = 332,640, all other x_{ij} = 0.

All x_{ij} expressed in terms of gallons per day.

Part 2: y_{max} = 120,422 at x_{11} = 436,800, x_{41} = 109,200, x_{12} = 544,320, x_{42} = 211,680, x_{13} = 470,400, x_{43} = 369,600, all other x_{ij} = 0.

All x_{ij} expressed in terms of gallons per day.

This last solution is based upon the *original* gasoline selling prices given in Prob. 11.18.

11.20 C_{min} = 209,544.5 at d = 0.523318 in, n = 10.

11.21 $C_{p\ min}$ = 235,661 at d = 2.80348 ft, resulting in v = 9 ft/sec and P = 924 units.

11.22 C_{min} = 844,283.1 at d = 2.80348 ft and t = 0.167095 ft, resulting in v = 9 ft/sec, P = 924 units, and Q = 748,796.5 BTU/hr.

INDEX